Space Modulation Techniques

Space Modulation Techniques

Raed Mesleh
German Jordanian University, Amman, Jordan

Abdelhamid Alhassi
University of Benghazi, Benghazi, Libya

WILEY

This edition first published 2018

Registered Offices
John Wiley & Sons, Inc., 111 River Street, Hoboken, NJ 07030, USA

Editorial Office
111 River Street, Hoboken, NJ 07030, USA

For details of our global editorial offices, customer services, and more information about Wiley products visit us at www.wiley.com.

Wiley also publishes its books in a variety of electronic formats and by print-on-demand. Some content that appears in standard print versions of this book may not be available in other formats.

Library of Congress Cataloging-in-Publication Data:

Names: Mesleh, Raed, 1978- author. | Alhassi, Abdelhamid, 1986- author.
Title: Space modulation techniques / by Raed Mesleh, Abdelhamid Alhassi.
Description: 1st edition. | Hoboken, NJ : John Wiley & Sons, 2018. |
Identifiers: LCCN 2018000551 (print) | LCCN 2018007146 (ebook) | ISBN 9781119375678 (pdf) | ISBN 9781119375685 (epub) | ISBN 9781119375654 (cloth)
Subjects: LCSH: Amplitude modulation. | Wireless communication systems–Technological innovations.
Classification: LCC TK6553 (ebook) | LCC TK6553 .M474 2018 (print) | DDC 621.382–dc23
LC record available at https://lccn.loc.gov/2018000551

Cover design: Wiley
Cover image: © StationaryTraveller/iStockphoto

Set in 10/12pt Warnock Pro by SPi Global, Chennai, India

Printed in the United States of America

V076074_052418

To my mother and wife for their unending love and support and to my father who could not see this book completed.
Raed Mesleh

To my parents, Houssein and Mareia, and my wife Farah, for their care, love, and support.
Abdelhamid Alhassi

Contents

Preface

The inspiration for this book arose from the desire to enlighten and instill a greater appreciation among wireless engineering society about a very promising technology for future wireless systems. Through this treatise, we aspire to expound the several benefits of space modulation techniques (SMTs) and demonstrate the several opportunities they convey. We believe that this book is also a unique tribute to the many scientists who were involved in the development of SMTs in the past 10 years.

SMT technology has come about from research that began 10 years ago and formed a basis for the work to be applied in what were then termed "beyond 4G" or B4G technologies before any consideration of what will be adopted within 5G networks. The attractiveness of the technology is that it enables the possibility to achieve comparable data throughput to a similar MIMO system yet with as few as just one radio transceiver at each end. Otherwise, in conventional MIMO, several transceivers would be required ranging anything from 4 to 128 in next generation communication systems, which would be costly and energy inefficient. Therefore, SMTs are now reaching a matured level that they are integrated in this book to assist the research and development community in learning about the concepts. The book identifies and discusses in detail a number of emerging techniques for high data rate wireless communication systems. The book serves also as a motivating source for further research and development activities in SMT. The limitations of current approaches and challenges of emerging concepts are discussed. Furthermore, new directions of research and development are identified, hopefully providing fresh ideas and influential research topics to the interested readers.

SMTs provide unique method to convey information bits and require innovative thinking, which goes beyond existing theories. The book provides a comprehensive overview on the basic working principle of coherent and noncoherent SMTs. Practical system models with the minimum number of needed RF-chains at the transmitter are presented and discussed in terms of hardware cost, power efficiency, performance, and computational complexity. The advantages and disadvantages of each technique along with their detailed performance are discoursed. A general framework for analyzing the performance of these techniques is provided and used to provide detailed performance analysis over several generalized fading channels. In addition, capacity analysis of SMTs is provided and thoroughly discussed.

Amman, Jordan — *Raed Mesleh*
Benghazi, Libya, November 2017 — *Abdelhamid Alhassi*

1 Introduction

1.1 Wireless History

Wireless technology revolution started in 1896 when Guglielmo Marconi demonstrated a transmission of a signal through free space without placing a physical medium between the transmitter and the receiver [1, 2]. Based on the success of that experiment, several wireless applications were developed. Yet, it was widely believed that reliable communication over a noisy channel can be only achieved through either reducing data rate or increasing the transmitted signal power. In 1948, Claude Shannon characterizes the limits of reliable communication and showed that this belief is incorrect [3]. Alternatively, he demonstrated that through an intelligent coding of the information, communication at a strictly positive rate with small error probability can be achieved. There is, however, a maximal rate, called the channel capacity, for which this can be done. If communication is attempted beyond that rate, it is infeasible to drive the error probability to zero [4].

Since then, wireless technologies have experienced a preternatural growth. There are many systems in which wireless communication is applicable. Radio and television broadcasting along with satellite communication are perhaps some of the earliest successful common applications. However, the recent interest in wireless communication is perhaps inspired mostly by the establishment of the first-generation (1G) cellular phones in the early 1980s [5–7]. 1G wireless systems consider analog transmission and support voice services only. Second-generation (2G) cellular networks, introduced in the early 1990s, upgrade to digital technologies and cover services such as facsimile and low data rate (up to 9.6 kbps) in addition to voice [8, 9]. The enhanced versions of the second–generation (2G) systems, sometimes referred to as 2.5G systems, support more advanced services like medium-rate (up to 100 kbps) circuit- and packet-switched data [10–12]. Third-generation (3G) mobile systems were standardized around year 2000 to support high bit rate (144–384) kbps for fast-moving users and up to 2.048 Mbps for slow-moving users [13–15]. Following the third–generation (3G) concept, several enhanced technologies

Space Modulation Techniques, First Edition. Raed Mesleh and Abdelhamid Alhassi.

generally called 3.5G, such as high speed downlink packet access (HSPDA), which increases the downlink data rate up to 3.6 Mbps were proposed [16, 17]. Regardless of the huge developments in data rate from 1G to 3G and beyond systems, the demand for more data rate did not seem to layover at any point in near future. As such, much more enhanced techniques were developed leading to fourth-generation (4G) wireless standard. 4G systems promise data rates in the range of 1 Gbps and witnessed significant development and research interest since launched in 2013 [18]. However, a recent CISCO forecast [19] reported that global mobile data traffic grew 74% in 2015, where it reached 3.7 EB per month at the end of 2015, up from 2.1 EB per month at the end of 2014. As well, it is reported that mobile data traffic has grown 4000-fold over the past 10 years and almost 400-million-fold over the past 15 years. It is also anticipated in the same forecast that mobile data traffic will reach 30.6 EB by 2020, and the number of mobile-connected devices per capita will reach 1.5 [19]. With such huge demand for more data rates and better quality services, fifth-generation (5G) wireless standard is anticipated to be launched in 2020 and has been under intensive investigations in the past few years [20]. 5G standard is supposed to provide a downlink peak date rate of 20 Gbps and peak spectral efficiency of 30 b $(s/Hz)^{-1}$ [20]. Such huge data rate necessitates the need of new spectrum and more energy-efficient physical layer techniques [21].

1.2 MIMO Promise

Physical layer techniques such as millimeter-wave (mmWave) communications, cognitive and cooperative communications, visible light and free-space optical communications, and multiple-input multiple-output (MIMO) and massive MIMO techniques are under extensive investigations at the moment for possible deployments in 5G networks [21]. Among the set of existing technologies, MIMO systems promise a boost in the spectral efficiency by simultaneously transmitting data from multiple transmit antennas to the receiver [22–28].

In 1987, Jack Winters inspired by the work of Salz [23], investigated the fundamental limits on systems that exploit multipath propagation to allow multiple simultaneous transmission in the same bandwidth [29]. Later in 1991, Wittneben proposed the first bandwidth-efficient transmit diversity scheme in [30], where it was revealed that the diversity advantage of the proposed scheme is equal to the number of transmit antennas which is optimal [31]. Alamouti discovered a new and simple transmit diversity technique [24] that is generalized later by Tarokh et al. and given the name of space–time coding (STC) [32]. STC techniques achieve diversity gains by transmitting multiple, redundant copies of a data stream to the receiver in order to allow reliable decoding. Shortly after, Foschini introduced multilayered space–time architecture, called Bell Labs layered space time (BLAST), that uses spatial multiplexing to increase the data rate and not necessarily provides transmit diversity [27]. Capacity analysis of MIMO systems was reported by Telatar and shown that MIMO capacity increases linearly with the minimum number among the transmit and receiver antennas [25] as compared to a system with single transmit and receive antennas. However, spatial multiplexing (SMX) MIMO systems, as BLAST, suffer from several limitations that hinder their practical implementations. Simultaneous transmission of independent data from multiple transmit antennas creates high inter-channel interference (ICI) at the receiver input, which requires high computational complexity to be resolved. In addition, the presence of high ICI degrades the performance of SMX MIMO systems, significant performance degradations are reported for any channel imperfections [33, 34]. On the other hand, STC techniques alleviate SMX challenges at the cost of achievable data rate. In STCs, the maximum achievable spectral efficiency is one symbol per channel use and can be achieved only with two transmit antennas.

1.3 Introducing Space Modulation Techniques (SMTs)

Another group of MIMO techniques, called space modulation techniques (SMTs), consider an innovative approach to tackle previous challenges of MIMO systems. In SMTs, a new spatial constellation diagram is added and utilized to enhance the spectral efficiency while conserving energy resources and receiver computational complexity. The basic idea stems from [35] where a binary phase shift keying (BPSK) symbol is used to indicate an active antenna among the set of existing multiple antennas. The receiver estimates the transmitted BPSK symbol and the antenna that transmits this symbol. However, the first popular SMT was proposed by Mesleh et al. [36, 37] and called spatial modulation (SM), and all other SMTs are driven as spacial or generalized cases from SM. Opposite to traditional modulation schemes, SM conveys information by utilizing the multipath nature of the MIMO fading channel as an extra constellation diagram referred to as spatial constellation. The incoming data bits modulate the spatial constellation symbol, which represents the spatial position, or index, of one of the available transmit antennas that will be activated at this particular time to transmit a modulated carrier signal by a complex symbol drawn from an arbitrary constellation diagram. SM was the first scheme to define the concept of spatial constellation and proposes the use of modulating spatial symbols to convey information. It was shown that SM can achieve multiplexing gain while maintaining free ICI [37], reduced receiver computational complexity [38], enhances the bit error probability [39], and promises the use of single radio frequency (RF)-chain transmitter [40]. As such, the concept of SM attracted significant research interests, and different performance aspects were studied thoroughly in few years [41–88]. Hence, multiple variant schemes applying similar SM concept were proposed. In [89], space shift keying (SSK) system was proposed where only spatial symbols exist and no data symbol is transmitted. Generalized spatial modulation (GSM) where more than one transmit antenna is activated at each time instant to transmit identical data is proposed in [67]. Similarly, generalized space shift keying (GSSK) was proposed in [69]. In all these schemes, single-dimensional spatial constellation diagram was created and used to convey spatial bits. In [65, 70], an additional quadrature spatial constellation diagram is defined where the real part of the complex data symbol is transmitted from one spatial symbol and the imaginary part of the complex symbol is transmitted from another spatial symbol. As such, data rate enhancement of base two logarithm of the number of transmit antennas is achieved while maintaining all previous SM advantages. These schemes are called quadrature spatial modulation (QSM) and quadrature space shift keying (QSSK). In addition, their generalized parts can be defined as generalized quadrature spatial modulation (GQSM) and generalized quadrature space shift keying (GQSSK). These eight schemes are the basic SMTs and their working mechanism, performance and capacity analysis, limitations, and practical implementations will be the core of this book. Yet, there exist many other advanced techniques that were proposed utilizing the working mechanism of these techniques.

1.4 Advanced SMTs

1.4.1 Space–Time Shift Keying (STSK)

Space–time shift keying (STSK) is a generalization scheme that was developed based on the concept of SMTs [78–80, 90–93]. In space–time shift keying (STSK), and instead of activating a specific transmit antenna, dispersion matrices are designed to achieve certain performance metric, and incoming data bits activate one of the available dispersion matrices at each block time. It is shown that different MIMO configurations, including, STC, SMX, and SMTs, can be derived as special cases from STSK by properly designing the dispersion matrices. STSK and its system model will be discussed in Chapter 3.

1.4.2 Index Modulation (IM)

Index modulation (IM) is another interesting idea based on multicarrier communications, such as orthogonal frequency division multiplexing (OFDM), that has been proposed with inspiration from the concept of SMTs and can be applied to frequency domain without requiring multiple transmit antennas.

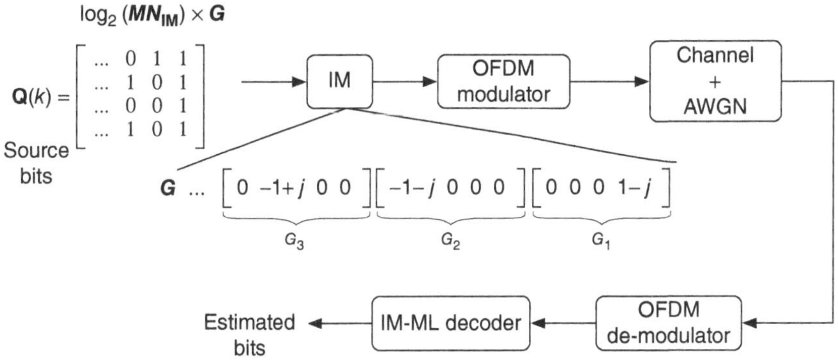

Figure 1.1 IM system model.

An illustration model for IM system is depicted in Figure 1.1. In IM, SMTs can be efficiently implemented for the OFDM subcarriers, where subcarriers are divided into groups, and certain subcarriers within each group are only activated. The index of such subcarriers is an extra information that can be utilized to convey additional data bits. In IM, the incoming bit stream is divided into two blocks, one of which modulates the index or indexes of active subcarriers and other block of bits modulates an ordinary constellation symbol to be transmitted on the activated subcarriers. In principle, this is very similar to the concept of SMTs but instead of having spatial symbols, index symbols are created now to modulate index bits [60, 94–96]. Very recently, a book was published entitled *Index Modulation for 5G Wireless Communications*, which covers the working principle and latest development of IM system. These techniques will not be covered in this book as they fall beyond the main scope of this book.

1.4.3 Differential SMTs

As will be discussed in this book for all SMTs, channel knowledge is mandatory at the receiver side to properly estimate the transmitted data. Such channel knowledge requires channel estimation algorithms through pilot symbols, which entails considerable overhead and not feasible in all applications. As such, significant interest in differential encoding and decoding of different SMTs led to the development of differential space modulation techniques (DSMTs) [63, 80, 97, 98]. In DSMTs, the receiver relies on the received signal at time t and the signal received at time $t-1$ to decode the message. Different differential schemes were proposed including differential spatial modulation (DSM) [63], differential space shift keying (DSSK), differential quadrature spatial modulation (DQSM) [97], differential quadrature space shift keying

(DQSSK), and differential space–time shift keying (DSTSK) [80]. These techniques will be discussed in detail in Chapter 3.

1.4.4 Optical Wireless SMTs

Another research area that benefited from SMTs is optical wireless communications (OWC) for both indoor and outdoor applications. Wireless transmission via optical carriers is motivated by the availability of a huge and unregulated spectrum at the optical frequencies. Traditional optical wireless communication was based on pulsed modulation since quadrature transmission of optical signals is not possible. However, OFDM was proposed for OWC by converting the time signal to real-unipolar signals through simple mathematical operations [99–105]. The use of OFDM for OWC facilitates the integration of SMTs, and optical spatial modulation (OSM) was proposed in [66]. Following similar concept of OSM, other SMTs can be considered as well for OWC.

SMTs found their way also in the application of outdoor optical wireless communication, generally called free-space optics (FSO). A direct application to FSO is foreseen through SSK scheme, where single transmitting laser among a set of available lasers is activated at each time instant to transmit unmodulated signal [106]. Other applications consider SM in conjunction with pulse amplitude modulation (PAM), pulse position modulation (PPM), or any other pulsed modulation scheme. Several recent books were published highlighting OWC techniques and include SMTs as promising techniques for OWC [107–110]. SMTs for OWC will not be addressed in this book.

1.5 Book Organization

The remaining of the book is organized as follows:

Chapter 2: MIMO System and Channel Models
MIMO system model is presented in this chapter along with different well-known channel models including Rayleigh, Nakagami-m, Rice, and generalized fading channel models such as $\eta-\mu$, $\kappa-\mu$, and $\alpha-\mu$. In addition, models for spatial correlation, mutual coupling, and channel estimation errors are discussed in this chapter.

Chapter 3: Space Modulation Transmission and Reception Techniques
This is the core chapter of the book presenting system models for the different SMTs, generalized space modulation techniques (GSMTs), and advanced SMTs. The working mechanism of each system, the maximum-likelihood (ML) receiver, the computational complexity, and a simplified low complexity sphere

decoder (SD) algorithm that are applicable for all SMTs, quadrature space modulation techniques (QSMTs), and GSMTs are presented as well. In addition, practical models for energy efficiency and power consumption and hardware implementation costs are presented and discussed.

Chapter 4: Average Bit Error Probability Analysis for SMTs
The derivation of the analytical error probability for the different SMTs, GSMTs, QSMTs, and DSMTs are presented in this chapter over different conventional and generalized fading channels. Results of comparative studies are presented and thoroughly discussed.

Chapter 5: Information Theoretic Treatments for SMTs
Capacity analysis and mutual information derivation for SMTs are presented in this chapter. Different examples and results are presented and discussed.

Chapter 6: Cooperative SMTs
SMTs for cooperative communication are discussed in this chapter. System models for different SMTs with different cooperative scenarios are presented and analyzed. Average bit error ratio (ABER) derivations for different system models and scenarios are presented and different numerical examples are illustrated.

Chapter 7: SMTs for Millimeter Wave Communications
mmWave communications is an emerging technology and one of the promising candidates for 5G wireless standard. Applying SMTs for mmWave systems is presented in this chapter along with detailed performance analysis, channel models, and numerical results.

Chapter 8: Summary and Future Directions
This chapter summarizes the entire book contents and provides directions for future research in the field of SMTs.

Appendix A: Matlab Codes
Matlab simulation codes for the different SMTs, GSMTs, and QSMTs are provided in the appendix. Also, Matlab codes for evaluating the derived analytical formulas are provided.

2

MIMO System and Channel Models

2.1 MIMO System Model

In wireless communications, a transmitter is communicating with a receiver through free-space medium. The transmitter is generally called the input as it transmits data into the communication link. The receiver is called the output as it receives transmitted data from the input. Depending on the number of antennas at both transmitter and receiver, several link configurations can be found. The simplest configuration is single-input single-output (SISO), where both transmitter and receiver each equipped with single antenna. If the transmitter has more than one antenna and communicates with single antenna receiver, a multiple-input single-output (MISO) configuration is conceived. If the receiver has more than one antenna and receives signal from a single antenna transmitter, a single-input multiple-output (SIMO) configuration is formed. Finally, if both transmitter and receiver have multiple antennas, a multiple-input multiple-output (MIMO) configuration is established.

In Figure 2.1, MIMO system model with N_t transmit antennas and N_r receive antennas is depicted.

This system can be represented by the following discrete time model:

$$\begin{bmatrix} y_1 \\ y_2 \\ \vdots \\ y_{N_r} \end{bmatrix} = \begin{bmatrix} h_{11} & h_{12} & \cdots & h_{1N_t} \\ h_{21} & h_{22} & \cdots & h_{2N_t} \\ \vdots & \vdots & \ddots & \vdots \\ h_{N_r1} & h_{N_r2} & \cdots & h_{N_rN_t} \end{bmatrix} \begin{bmatrix} x_t^1 \\ x_t^2 \\ \vdots \\ x_t^{N_t} \end{bmatrix} + \begin{bmatrix} n_1 \\ n_2 \\ \vdots \\ n_{N_r} \end{bmatrix}, \tag{2.1}$$

which can be simplified to

$$\mathbf{y} = \mathbf{H}\mathbf{x}_t + \mathbf{n}, \tag{2.2}$$

where $\mathbf{x}_t$ is the N_t-length transmitted vector, $\mathbf{n}$ is an N_r-length additive white Gaussian noise (AWGN) seen at the receiver input, $\mathbf{H}$ is an $N_r \times N_t$ MIMO

Space Modulation Techniques, First Edition. Raed Mesleh and Abdelhamid Alhassi.

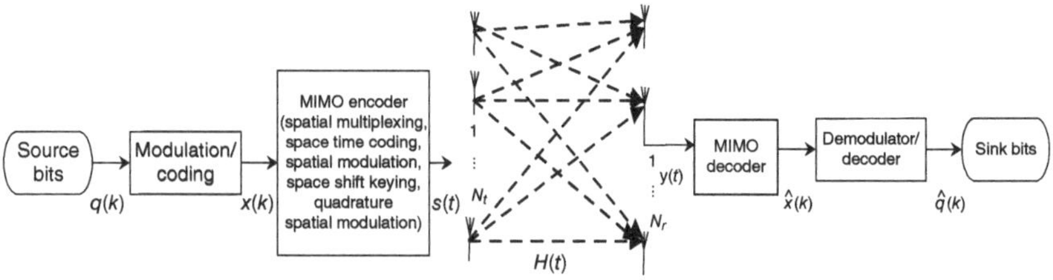

Figure 2.1 General MIMO system model with N_t transmit antennas and N_r receive antennas.

channel matrix representing the path gains $h_{n_r n_t}$ between transmit antenna n_t and receive antenna n_r, and $\mathbf{y}$ is the N_r-length vector received signal.

The transmitted vector $\mathbf{x}_t$ is created from the source data bit using a MIMO encoder, where arbitrary modulation techniques such as quadrature amplitude modulation (QAM), phase shift keying (PSK), or others are used by the MIMO encoder. The noise is generally modeled as a complex Gaussian noise that is temporally and spatially white with zero mean and a covariance matrix of $\sigma_n^2 \mathbf{I}_{N_r}$, where $\sigma_n^2 = N_0 B$, with N_0 denoting the noise power spectral density and B is the channel bandwidth. Also and when comparing systems with different configurations, the total transmit power from any number of transmit antennas is the same and for simplicity assumed to be 1. Therefore, the average signal-to-noise-ratio (SNR) at each receive antenna (n_r), under unity channel gain assumptions, is given by SNR $= 1/\sigma_n^2$. At the receiver, the optimum maximum-likelihood (ML) detector can be used to decode the transmitted messages as [111],

$$\hat{\mathbf{x}} = \arg \min_{\mathbf{x}_i \in \mathbf{X}} \left\| \mathbf{y} - \widetilde{\mathbf{H}} \mathbf{x}_i \right\|_F^2, \tag{2.3}$$

where $\hat{\mathbf{x}}$ denotes the estimated transmitted symbol, $\widetilde{\mathbf{H}}$ is the estimated channel matrix at the receiver, $\| \cdot \|_F$ is the frobenius norm, and $\mathbf{x}_i$ is a possible transmitted vector from $\mathbf{X}$, where $\mathcal{X}$ is a set containing all possible transmitted vectors combinations between transmit antennas and data symbols.[1]

The ML decoder in (2.3) searches the transmitted vectors space and selects the vector that is closest to the received signal vector $\mathbf{y}$ as the most probable transmitted vector. The closer the two vectors from the set $\mathcal{X}$ to each others, the higher the probability of error. Therefore, a better design is to place the vectors as far apart from each other as possible. This can be done also through proper design of the MIMO channel matrix $\mathbf{H}$. Also, the computational complexity of encoding and decoding should be practical, systems with higher complexity tends to perform better.

1 Several configurations for different space modulation techniques (SMTs) will be discussed in Chapter 3.

2.2 Spatial Multiplexing MIMO Systems

The first proposed spatial multiplexing (SMX) MIMO system was vertical Bell Labs layered space time (V-BLAST) system [112]. Later on, horizontal Bell Labs layered space time (H-BLAST) system was proposed [113]. In these systems, the input data stream is de-multiplexed into N_t parallel substreams. Each substream contains an independent data that will be transmitted from a single transmit antenna. In general wireless systems, channel coding and interleaving are generally applied. Based on the applied coding scheme, the different Bell Labs layered space time (BLAST) configurations are named. The H-BLAST took its name because the channel coding is applied horizontally on each substream. The earlier V-BLAST scheme called vertical since uncoded data symbol was viewed as one vector symbol. The transmitted streams from multiple transmit antennas are cochannel signals that share the same time and frequency slots. As such, the schemes mainly aim at decorrelating the received signals to retrieve the transmitted data. Each receive antenna observes a superposition of the transmitted signals, and the major task at the receiver is to resolve the inter-channel interference (ICI) between the transmitted symbols. The optimum solution is to use ML receiver as in (2.3). The ML compares the received signals with all possible transmitted signal vectors that are modified by the channel matrix and selects the optimal codeword. The problem of ML algorithm is the high complexity required to search over all possible combinations. Therefore, initial systems targeted other low complexity receiver such as sphere decoder (SD) algorithm [114] and the multiple variants of it proposed in [115]. The V-BLAST receiver was another low-complexity receiver that applies a successive interference cancellation technique with optimum ordering (OR-SIC). The optimal detection order is from the strongest symbol to the weakest one. The idea is to detect the strongest symbol first. Then, canceling the effect of this symbol from all received signals and detects the next strongest symbol and so on. The process is repeated until all symbols are detected. Details of this scheme can be found in [71, 112].

In this book, SMX MIMO systems will be used as a benchmark system for comparison purposes with SMTs. The ML optimum decoder from (2.3) will be considered.

2.3 MIMO Capacity

Before 1948, it was widely believed that the only way to reduce the probability of error of a wireless communication system was to reduce the transmission data rate for fixed power and bandwidth. In 1948, Shannon showed that this belief is incorrect, and lower probability of error can be achieved through intelligent coding of the information. However, there is a maximum limit of data rate, called the capacity of the channel, for which this can be done. If the transmission rate exceeds the channel capacity, it will be impossible to derive the probability of error to zero [3].

The channel capacity is, therefore, a measure of the maximum amount of information that can be transmitted over the channel and received with no errors at the receiver [25],

$$C = \max_{p_{\mathbf{X}}} I(\mathbf{X};\mathbf{Y}), \tag{2.4}$$

where $I(\mathbf{X};\mathbf{Y})$ is the mutual information between the transmitted vector space $\mathbf{X}$ and the received vector space $\mathbf{Y}$ and the maximization is carried over the choice of the probability distribution function (PDF) of $\mathbf{X}$.

In an AWGN channel and for SISO transmission of complex symbols, the channel capacity is given by [116],

$$C_{\text{AWGN}} = \log_2(1 + \text{SNR}). \tag{2.5}$$

The ergodic capacity of a SISO system over a slow fading random channel, assuming full channel state information (CSI) at the receiver side only, is given by [25, 116],

$$C = \text{E}_h\{\log_2(1 + \text{SNR} \times |h|^2)\}, \tag{2.6}$$

where $|h|^2$ is the squared magnitude of the channel coefficient, and $\text{E}\{\cdot\}$ is the expectation operator. As the number of receiver antenna increases, the statistics of capacity improves. The capacity, C in (2.6), is often referred to as the error-free spectral efficiency, or the data rate that can be sustained reliably over the link [4].

In a SIMO system with an N_r receiver antennas, there exist N_r various copies of the faded signal at the receiver. If these signals are, on average, the same amplitude, then they may be added coherently to produce an N_r^2 increase in signal power. Of course, there are N_r sets of noise that will add together as well. Fortunately, noise adds incoherently to create only an N_r-fold increase in the noise power. Thus, there is still a net overall increase in SNR by $N_r^2/(N_r N_0)$ compared to SISO systems. Following this, the ergodic channel capacity of this system is [117],

$$C = \text{E}_{\mathbf{h}}\{\log_2|\mathbf{I}_{N_r} + \text{SNR} \times \mathbf{h}\mathbf{h}^H|\}, \tag{2.7}$$

where in SIMO system, the channel matrix $\mathbf{H}$ can be reduced to an N_r-length channel vector $\mathbf{h}$, and $\{\cdot\}^H$ is the Hermitian operator.

In a MISO system, where the transmitter is equipped with multiple antennas, whereas the receiver has single antenna, a special design of the transmit signal needs to exist for any possible advantages. Without precoding of transmitted data, received data from the multiple antennas will interfere at the receiver input and the capacity will be zero. Special techniques such as space–time coding (STC), repetition coding, and others are used in such topologies. The aim is to create orthogonal transmitted data that can be decoded by the receiver under a total power constraint; i.e. the transmit power is divided among existing transmit antennas. With such precoding, orthogonal signals are transmitted and the channel capacity is [4],

$$C = \mathrm{E}_{\mathbf{h}} \left\{ \log_2 \left(1 + \mathrm{SNR} \times \frac{1}{N_t} \|\mathbf{h}\|_{\mathrm{F}}^2 \right) \right\}, \tag{2.8}$$

where in the case of MISO channels, $\mathbf{h}$ is $1 \times N_t$-length channel vector.

Having N_t antennas at the transmitter and N_r antennas at the receiver results in a MIMO configuration as discussed earlier. The ergodic capacity for a MIMO system over uncorrelated channel paths assuming equal total power transmission as in SISO systems is given by [25, 71, 116, 117],[2]

$$C = \mathrm{E}_{\mathbf{H}} \left\{ \log_2 \left| \mathbf{I}_{N_r} + \mathrm{SNR} \times \frac{1}{N_t} \mathbf{H}\, \mathbf{H}^H \right| \right\}. \tag{2.9}$$

In order to interpret (2.9), let $\mathbf{H} = \mathbf{U}_h \mathbf{D}_h \mathbf{V}_h^H$ be the singular value decomposition (SVD) of the channel matrix $\mathbf{H}$. $\mathbf{U}_h$ and $\mathbf{V}_h$ are unitary matrices. $\mathbf{D}_h$ is a diagonal matrix of (σ_i) of $\mathbf{Z} = \mathbf{H}^H\mathbf{H}$ with $\sigma_i \geq 0$ and $\sigma_i \geq \sigma_{i+1}$ being the positive eigenvalues of $\mathbf{Z}$. Rewriting (2.9) as

$$C = \mathrm{E}_{\mathbf{H}} \left\{ \log_2 \left| \mathbf{I}_{N_r} + \mathrm{SNR} \times \frac{1}{N_t} \mathbf{U}_h \mathbf{D}_h \mathbf{D}_h^H \mathbf{U}_h^H \right| \right\}. \tag{2.10}$$

The result of $\mathbf{D}_h\mathbf{D}_h^H$ is a diagonal matrix containing the positive eigenvalues of $\mathbf{H}\mathbf{H}^H$. The diagonal elements are given by $\lambda_i = \sigma_i^2, (i = 1 \cdots r)$, where $r = \min(N_t, N_r)$ is the rank of the channel matrix. Substituting this in (2.10) and using the identity $|\mathbf{I}_a + \mathbf{AB}| = |\mathbf{I}_b + \mathbf{BA}|$ for matrices $\mathbf{A}$ $(a \times b)$ and $\mathbf{B}$ $(b \times a)$ [111], and $\mathbf{U}_h\mathbf{U}_h^H = \mathbf{I}_{N_r}$ simplifies (2.10) to

$$C = \mathrm{E}_{\mathbf{H}} \left\{ \sum_{i=1}^{r} \log_2 \left(1 + \mathrm{SNR} \times \frac{1}{N_t} \lambda_i \right) \right\}. \tag{2.11}$$

It is shown in Figure 2.2 that using multiple antennas increases the ergodic capacity. The capacity increases with the increasing number of transmit antennas, receive antennas, or by increasing both of them at the same time.

2.4 MIMO Channel Models

Propagating signals from transmitter to receiver arrives from multipaths and suffer from multipath fading. The combined signals at the receiver are random in nature, and the received signal power changes over a period of time. The propagation channel consists of static or moving reflecting objects, and

2 The derivation of this equation can be found in Chapter 5.

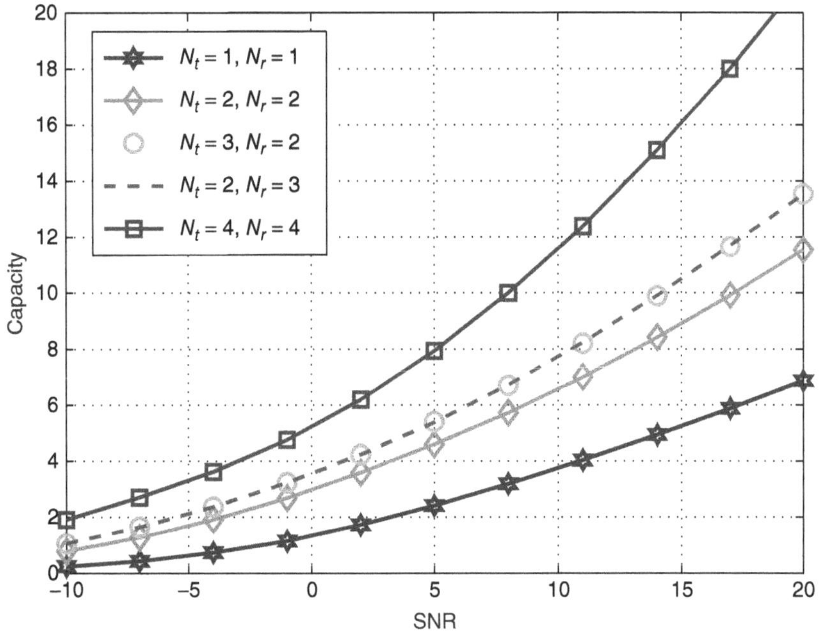

Figure 2.2 Ergodic MIMO capacity for different antenna configurations. Capacity improves with larger antenna configurations.

scatterers that create a randomly changing environment. If the channel has a constant gain and linear phase response over a bandwidth that is greater than the bandwidth of the transmitted signal, it is called flat fading or frequency non-selective fading channel [118]. This specific bandwidth is generally called the *Coherence bandwidth* and is a statistical measure of the range of frequencies over which the channel can be considered flat.

The movement of the transmitter, receiver, or the surrounding environment results in a random frequency modulation due to different Doppler shifts on each of the multipath components. Hence, a spectral broadening at the receiver side occurs and is measured by the Doppler spread, which is defined as the range of frequencies over which the received Doppler spectrum is not zero [119, 120].

Based on time and frequency statistics, fading channels can be classified into flat and frequency selective according to their time changes and slow and fast according to their frequency variations. These two phenomena are independent and result in the following fading types [121]:

- *Flat slow fading or frequency nonselective slow fading*: when the bandwidth of the signal is smaller than the coherence bandwidth of the channel and the signal duration is smaller than the coherence time of the channel. The coherence time is the duration of time in which the channel impulse response is effectively invariant.

- *Flat fast fading or frequency nonselective fast fading*: when the bandwidth of the signal is smaller than the coherence bandwidth of the channel and the signal duration is larger than the coherence time of the channel.
- *Frequency selective slow fading*: when the bandwidth of the signal is larger than the coherence bandwidth of the channel and the signal duration is smaller than the coherence time of the channel.
- *Frequency selective fast fading*: when the bandwidth of the signal is larger than the coherence bandwidth of the channel and the signal duration is larger than the coherence time of the channel.

The propagation environment plays a dominant role in determining the capacity of the MIMO channel. In what follows, several MIMO channel models are discussed.

2.4.1 Rayleigh Fading

The Rayleigh fading distribution is generally considered when the transmitter and receiver have no line-of-sight (LOS) [122, 123]. As such, the sum of all scattered and reflected components of the complex received signal is modeled as a zero mean complex Gaussian random process given by $h_{n_r n_t} \sim \mathcal{CN}(0, 1)$. Hence, the phase of the random process $h_{n_r n_t}$ takes an uniform distribution, and is given by

$$p_\Theta(\theta) = \frac{1}{2\pi} \coprod\nolimits_{[-\pi,\pi]}(\theta), \tag{2.12}$$

where $\coprod_B(b) = 1$ if $b \in B$ and zero otherwise. Furthermore, the amplitude takes a Rayleigh distribution given by

$$p_R(r) = \frac{r}{\sigma^2} \exp\left(-\frac{r^2}{2\sigma^2}\right) \coprod\nolimits_{\mathbb{R}^+}(r), \tag{2.13}$$

where $\mathbb{R}^+$ denotes the set of all positive real numbers.

2.4.2 Nakagami-*n* (Rician Fading)

If the transmitter and receiver can see each other through a LOS path, the channel amplitude gain is characterized by a Rician distribution, and the channel is said to exhibit Rician fading [120, 123, 124]. The Rician fading MIMO channel matrix can be modeled as the sum of a LOS matrix and a Rayleigh fading channel matrix as [123],

$$\mathbf{H}_{\text{Rician}} = \underbrace{\sqrt{\frac{K}{1+K}}\bar{\mathbf{H}}}_{\text{LOS component}} + \underbrace{\sqrt{\frac{1}{1+K}}\mathbf{H}}_{\text{Fading component}}, \tag{2.14}$$

where K is the Rician K-factor. The Rician K-factor is defined as the ratio of the LOS and the scatter power components. There are two contrasting prototypes of $\bar{\mathbf{H}}$ for a MIMO channel and unipolarized antennas. The first one is a matrix with all elements being one, which can be applied when the distance between the transmit antennas and the receive antennas is much larger than the spacing between the transmit antennas and the receive antennas. The second alternative is for the case when the distance between the transmit antennas and the receive antennas is comparable to the spacing between the transmit antennas and/or the receive antennas. The LOS component of the channel matrix, assuming $N_t = N_r = 2$ for instance, is then given by

$$\bar{\mathbf{H}}_2 = \begin{bmatrix} 1 & -1 \\ 1 & 1 \end{bmatrix}. \tag{2.15}$$

Perfect orthogonality of this channel matrix requires specific antenna locations and geometry. Therefore, $\bar{\mathbf{H}}_2$ is likely only in multibase operations when transmit (or receive) antennas are located at different base stations [111].

In Rician fading channel, the capacity of the MIMO system depends on the value of the Rician K-factor and on the channel geometry. When the value of K is low, the random matrix $\mathbf{H}$ has more influence than $\bar{\mathbf{H}}$, resulting in an expression of the MIMO capacity similar to (2.11). However, when the value of K is high, the LOS component of the channel matrix dominates and the capacity depends on the channel geometry of the LOS component. As discussed before, there exist two contrasting prototypes of $\bar{\mathbf{H}}$ for a MIMO channel. The second one, $\bar{\mathbf{H}}_2$, clearly outperforms the first channel with an increasing K-factor. This is because the second matrix is orthogonal while the first one is rank-deficient. Hence, the geometry of the LOS component of the channel matrix plays a critical rule in channel capacity at high Rician factor [125].

2.4.3 Nakagami-*m* Fading

Nakagami-m distribution is widely used to describe channels with severe to moderate fading [126–128]. The Nakagami-m channel is a generalized fading channel that includes the one-sided Gaussian ($m = 1/2$), the Rayleigh fading ($m = 1$), and if $m \to \infty$, the Nakagami-m fading channel converges to a non-fading AWGN channel. Furthermore, when $m < 1$, the Nakagami-m can closely approximate the Nakagami-q (Hoyt) distribution.

The entries of the Nakagami-m fading channels are modeled as [127]:

$$h_{n_r n_t} = \sqrt{\sum_{i=1}^{m} |X_i|^2} + j\sqrt{\sum_{i=1}^{m} |Y_i|^2}, \tag{2.16}$$

where X_i and Y_i are an identical and independently distributed (i.i.d.) Gaussian random variables with μ_X and μ_Y means and σ_X^2 and σ_Y^2 variances.

The joint envelope-phase distribution of the random variable $h_{n_r n_t}$ is given by [127],

$$p_{R,\Theta}(r,\theta) = \frac{m^m |\sin\theta\cos\theta|^{m-1} r^{2m-1} \exp\left(-\frac{mr^2}{\Omega}\right)}{\Omega^m \Gamma\left(\frac{1+p}{2}m\right)\Gamma\left(\frac{1-p}{2}m\right)|\tan\theta|^{pm}} \coprod_{\mathbb{R}^+}(r) \coprod_{[-\pi,\pi]}(\theta) \quad (2.17)$$

where $2\Omega = \sigma_X^2 + \sigma_Y^2$, $p = (\mu_X - \mu_Y)/(\mu_X + \mu_Y)$, and $\Gamma(\cdot)$ is the gamma function.

The envelope of the Nakagami-m channel is given by [127]

$$p_R(r) = \frac{2m^m r^{2m-1}}{\Omega^m \Gamma(m)} \exp\left(-\frac{mr^2}{\Omega}\right) \coprod_{\mathbb{R}^+}(r) \quad (2.18)$$

and the phase distribution is given by [127]

$$p_\Theta(\theta) = \frac{\Gamma(m)|\sin(2\theta)|^{m-1}}{2^m \Gamma\left(\frac{1+p}{2}m\right)\Gamma\left(\frac{1-p}{2}m\right)|\tan\theta|^{pm}} \coprod_{[-\pi,\pi]}(\theta). \quad (2.19)$$

Assuming $\mu_X = \mu_Y$ and $\sigma_X^2 = \sigma_Y^2 = \sigma_H^2$, the mean and the variance of the Nakagami-m channel are then given by

$$\mu_{\mathrm{H}} = \frac{\Gamma((m/2)+1/2)}{\Gamma(m/2)\sqrt{m/2}} \exp\left(j\frac{\pi}{4}\right), \quad (2.20)$$

$$\sigma_{\mathrm{H}}^2 = 1 - \frac{2}{m}\left(\frac{\Gamma((m/2)+1/2)}{\Gamma(m/2)}\right)^2. \quad (2.21)$$

The joint distribution $p_{R,\Theta}(r,\theta)$ for different values of m are depicted in Figure 2.3. As can be seen from the figure, when m increases, the Nakagami-m channel approaches Gaussian distribution, which increases the correlation between different channel paths from different transmit antennas. It can be also seen from (2.19) that the phase distribution of the Nakagami-m channel is uniform only if $m = 1$, which corresponds to Rayleigh distribution. The impact of varying the value of m on the performance of SMTs and other MIMO systems will be discussed in coming chapters.

2.4.4 The η–μ MIMO Channel

The η–μ channel is a generalized fading distribution that represents the small-scale variation of the signal in a nonline–of–sight (NLOS) environment [73, 129]. The previously discussed channels can be driven as special cases from the η–μ distribution. The Nakagami-m channel can be obtained by setting $\eta = 1$ and $\mu = m/2$. The Rayleigh fading channel is deduced when $\eta = 1$ and $\mu = 1/2$. The one-sided Gaussian distribution can be obtained by setting $\eta = 1$ and $\mu = 1/4$ and the Nakagami-q (Hoyt) distribution can be obtained when $\eta = q^2$ and $\mu = 1/2$.

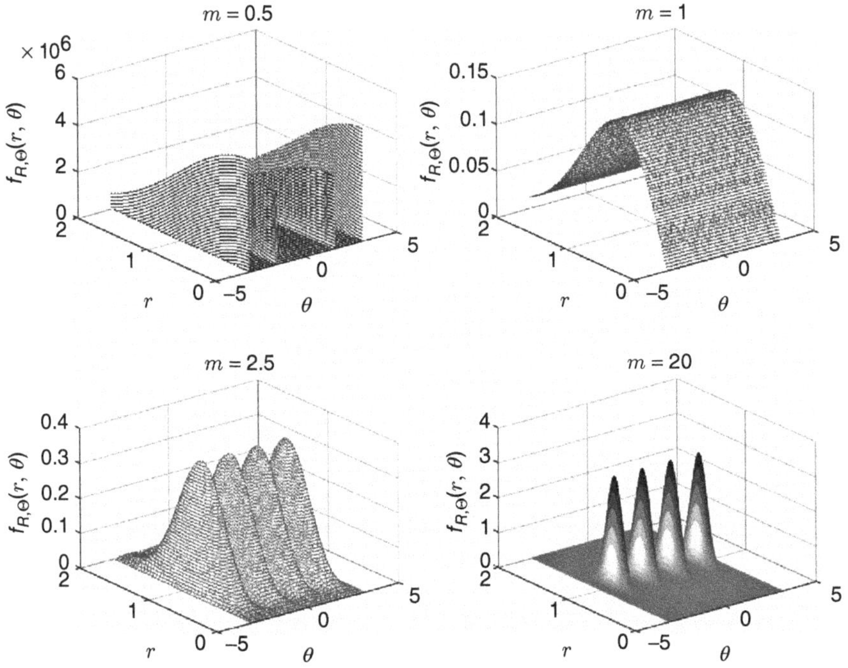

Figure 2.3 Nakagami-m joint envelope-phase pdf behavior for variable m values and $\sigma_X^2 = \sigma_Y^2 = 1/2m$ and $\mu_X = \mu_Y = 0$.

The complex $\eta-\mu$ fading channel coefficients can be numerically generated using the envelope and the phase distributions. The envelope R can be obtained through

$$R^2 = X^2 + Y^2, \tag{2.22}$$

where $X^2 = \sum_{i=1}^{2\mu} x_i^2$ and $Y^2 = \sum_{i=1}^{2\mu} y_i^2$ with x_i and y_i being mutually independent Gaussian processes with $\mathrm{E}\{x_i\} = \mathrm{E}\{y_i\} = 0$, $\mathrm{E}\{x_i^2\} = \sigma_X^2 = \eta\hat{r}^2/(2\mu(1+\eta))$, $\mathrm{E}\{y_i^2\} = \sigma_Y^2 = \hat{r}^2/(2\mu(1+\eta))$, and $\eta = \sigma_X^2/\sigma_Y^2$.

The phase Θ can be obtained via

$$\Theta = \arctan\left(\frac{Y}{X}\right). \tag{2.23}$$

The $\eta-\mu$ joint envelope-phase PDF, $p_{R,\Theta}(r,\theta)$ can be expressed as [129]

$$\begin{aligned} p_{R,\Theta}(r,\theta) =& \frac{2\mu^{2\mu}A^{2\mu}r^{4\mu-1}|\sin 2\theta|^{2\mu-1}}{(A^2-B^2)^{\mu}\hat{r}^{4\mu}\Gamma^2(\mu)} \\ &\times \exp\left(-\frac{2\mu A r^2}{\hat{r}^2(A^2-B^2)}(A+B\cos(2\theta))\right)\coprod_{\mathbb{R}^+}(r)\coprod_{[-\pi,\pi]}(\theta), \end{aligned} \tag{2.24}$$

where $\hat{r} = \sqrt{\mathrm{E}\{R^2\}}$ is the root mean square (*rms*) value of R, $B = \eta/(1-\eta^2)$ and $A = 1/(1-\eta^2)$. μ (where $\mu > 0$) represents the number of multipaths in each cluster and η (where $0 < \eta < \infty$) represents the scattered-wave power ratio between the in-phase and quadrature components of each cluster of multipath.

The PDF of the normalized envelope, after random variable transformation, is given as

$$p_R(r) = \frac{4\sqrt{\pi}\mu^{\mu+\frac{1}{2}}A^{\mu}}{\hat{r}\Gamma(\mu)B^{\mu-\frac{1}{2}}}\left(\frac{r}{\hat{r}}\right)^{2\mu}\exp\left(-2\mu A\left(\frac{r}{\hat{r}}\right)^{2}\right)I_{\mu-\frac{1}{2}}\left(2\mu B\left(\frac{r}{\hat{r}}\right)^{2}\right)\coprod_{\mathbb{R}^{+}}(r). \tag{2.25}$$

The phase distribution, $p_{\Theta}(\theta)$, is given as

$$p_{\Theta}(\theta) = \frac{(A^2-B^2)^{\mu}\Gamma(2\mu)|\sin(2\theta)|^{2\mu-1}}{2^{2\mu}\Gamma^2(\mu)(A+B\cos(2\theta))^{2\mu}}\coprod_{[-\pi,\pi]}(\theta). \tag{2.26}$$

The joint envelope-phase distributions of the $\eta-\mu$ channel for variable values of μ and η are shown in Figures 2.4 and 2.5, respectively. Increasing the value of μ has similar impact as increasing the value of m for the Nakagami-m channel.

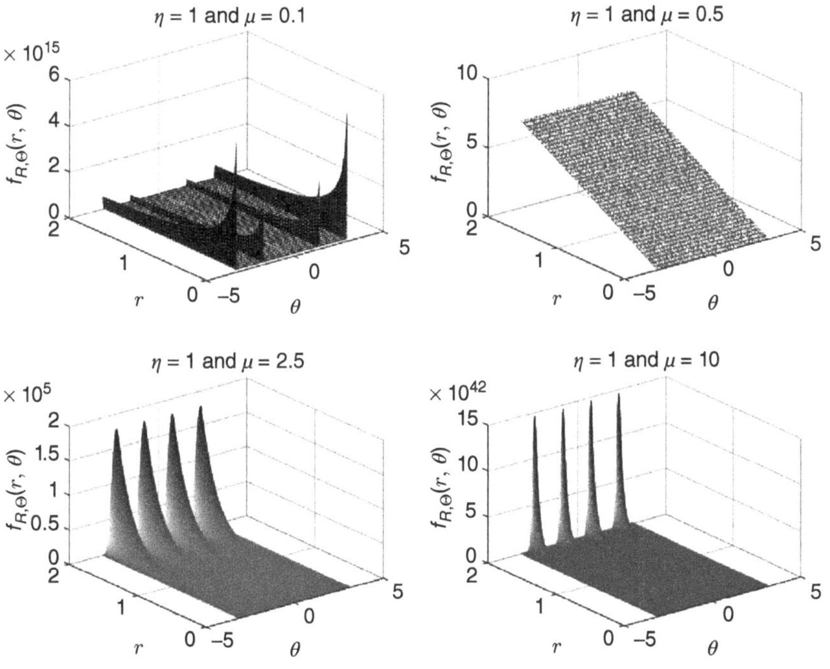

Figure 2.4 $\eta-\mu$ Joint PDF for fixed η and variable μ. η Value is fixed to 1, while μ takes the values of 0.1, 0.5, 2.5, and 10.

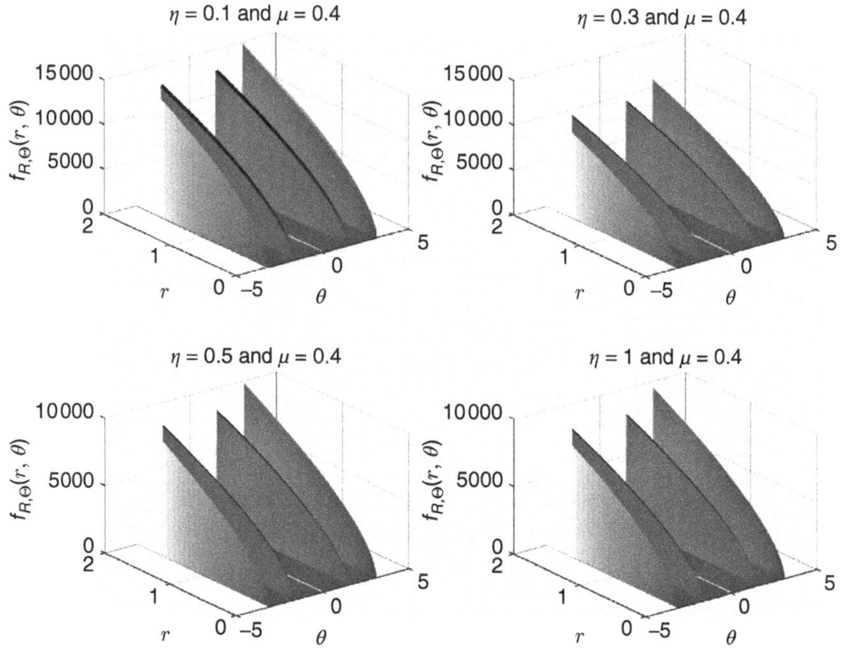

Figure 2.5 η–μ Joint PDF for fixed μ and variable η. η Value is varied as 0.1, 0.3, 0.5, and 1 while μ is fixed to 0.4.

However, increasing the value of η has almost no impact on the shape of the distribution but reduces the value of the envelope r.

2.4.5 The κ–μ Distribution

The κ–μ distribution is another general fading distribution that describes the small-scale variation of the fading signal in a LOS environment. The parameter $\kappa > 0$ represents the ratio between the total power of the dominant component and the total power of the scattered waves, and $\mu > 0$ is the number of the multipath clusters. As such, it includes other well-known fading distributions, such as

(1) The Nakagami-n distribution that is realized when $\kappa = n^2$ and $\mu = 1$.
(2) The Nakagami-m distribution that is obtained when $\kappa \to 0$ and $\mu = m$.

The complex κ–μ fading channel coefficients can be numerically generated using the envelope and the phase distributions. The envelope R can be obtained through

$$R^2 = X^2 + Y^2, \tag{2.27}$$

where $X^2 = \sum_{i=1}^{\mu} (x_i + p_i)^2$ and $Y^2 = \sum_{i=1}^{\mu} (y_i + q_i)^2$ with x_i and y_i being mutually independent Gaussian processes with $\mathrm{E}\{x_i\} = \mathrm{E}\{y_i\} = 0$, $\mathrm{E}\{x_i^2\} = \mathrm{E}\{y_i^2\} = \sigma^2$, and p_i and q_i, respectively, denote the mean values of the in-phase and the quadrature components of the multipath waves of cluster i. Let $p^2 = \sum_{i=1}^{\mu} p_i^2$ and $q^2 = \sum_{i=1}^{\mu} q_i^2$, then

$$\kappa = \frac{p^2 + q^2}{2\mu\sigma^2}. \tag{2.28}$$

Accordingly,

$$\sigma^2 = \frac{\Omega}{2\mu(1+\kappa)} \tag{2.29}$$

with $\Omega = \mathrm{E}\{R^2\}$. The phase Θ of the complex fading channel can be obtained via,

$$\Theta = \arctan\left(\frac{Y}{X}\right). \tag{2.30}$$

Define $\phi = \arg(p + jq)$ as a phase parameter, then for a fading signal with envelope R and $\hat{r} = \sqrt{\mathrm{E}\{R^2\}}$ being the *rms* value of R, the κ–μ joint phase-envelope distribution $p(r, \theta)$ is given by [130],

$$\begin{aligned} p_{R,\Theta}(r,\theta) &= \frac{r^{\mu+1}\mu^2(1+\kappa)^{\mu+1}|\sin(2\theta)|^{\mu}}{2\Omega^{\mu+1}\kappa^{\mu-1}|\sin(2\theta)|^{\mu-1}} \\ &\times \exp\left[-\mu\frac{(1+\kappa)r^2}{\Omega} - 2r\sqrt{\frac{\kappa(1+\kappa)}{\Omega}}\cos(\theta-\phi) + \kappa\right] \\ &\times \operatorname{sech}\left(2\mu r\sqrt{\frac{\kappa(1+\kappa)}{\Omega}}|\cos\theta\cos\phi|\right)\operatorname{sech}\left(2\mu r\sqrt{\frac{\kappa(1+\kappa)}{\Omega}}|\sin\theta\sin\phi|\right) \\ &\times I_{\mu-1}\left(2\mu r\sqrt{\frac{\kappa(1+\kappa)}{\Omega}}\cos\theta\cos\phi\right) I_{\mu-1}\left(2\mu r\sqrt{\frac{\kappa(1+\kappa)}{\Omega}}\cos\theta\cos\phi\right), \end{aligned} \tag{2.31}$$

where $0 \le r < \infty$, $-\pi \le \theta \le \pi$, and $-\pi \le \phi \le \pi$. The parameters p and q can be obtained as $p = \sqrt{\kappa/(1+\kappa)}\hat{r}\cos(\phi)$ and $q = \sqrt{\kappa/(1+\kappa)}\hat{r}\sin(\phi)$. The function $I_v(\cdot)$ denotes the modified Bessel function of the first kind and order v.

The κ–μ envelope PDF is then given by [131]

$$\begin{aligned} p_R(r) &= \frac{2\mu(1+\kappa)^{\frac{\mu+1}{2}}}{\hat{r}\kappa^{\frac{\mu+1}{2}}e^{\mu\kappa}}\left(\frac{r}{\hat{r}}\right)^{\mu}\exp\left(-\mu(1+\kappa)\left(\frac{r}{\hat{r}}\right)^2\right) \\ &\times I_{\mu-1}\left(2\mu\sqrt{\kappa(1+\kappa)}\frac{r}{\hat{r}}\right)\coprod_{\mathbb{R}^+}(r). \end{aligned} \tag{2.32}$$

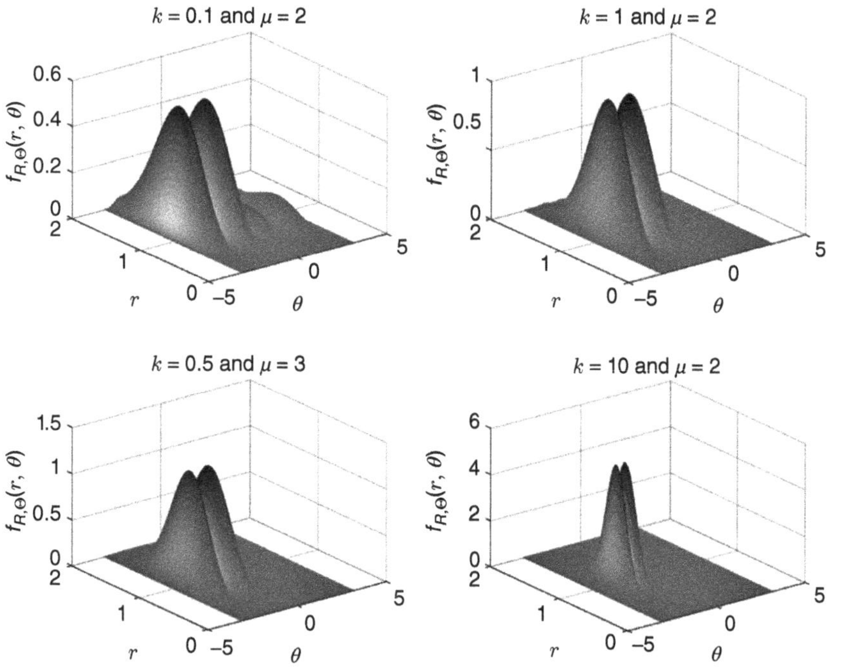

Figure 2.6 $\kappa{-}\mu$ Joint PDF for fixed μ and variable κ and for $\phi = -\pi/2$. μ-Value is set to 2 and κ varies from 0.1, 1, 1.5, and 10.

The PDF of the normalized envelope is

$$p_P(\rho) = \frac{2\mu(1+\kappa)^{\frac{\mu+1}{2}}}{\kappa^{\frac{\mu+1}{2}}\exp(\mu\kappa)}\rho^{\mu}\exp(-\mu(1+\kappa)\rho^2)$$

$$\times I_{\mu-1}\left(2\mu\sqrt{\kappa(1+\kappa)}\rho\right)\coprod_{\mathbb{R}^+}(p). \quad (2.33)$$

The ℓth moment, $\mathrm{E}\{P^{\ell}\}$, of P in (2.33) is given as

$$\mathrm{E}\{P^{\ell}\} = \frac{\Gamma(2\mu+\frac{\ell}{2})\exp(-\kappa\mu)}{[(1+\kappa)\mu]^{\frac{\ell}{2}}\Gamma(\mu)}1F_1\left(\mu+\frac{\ell}{2};\mu;\kappa\mu\right), \quad (2.34)$$

where ${}_1F_1(\cdot)$ is the confluent hypergeometric function [[132], Eq. (13.1.2)].

The $\kappa{-}\mu$ joint PDF for different values of κ, μ, and ϕ is numerically computed and depicted in Figures 2.6–2.8. Figure 2.6 demonstrates the impact of varying κ for fixed μ and $\phi = -\pi/2$. The impact of varying ϕ can be seen when comparing the results in Figure 2.6 with those in Figure 2.7. For the same values of κ and μ, a π change of ϕ leads to a PDF flip around the θ access. Large values

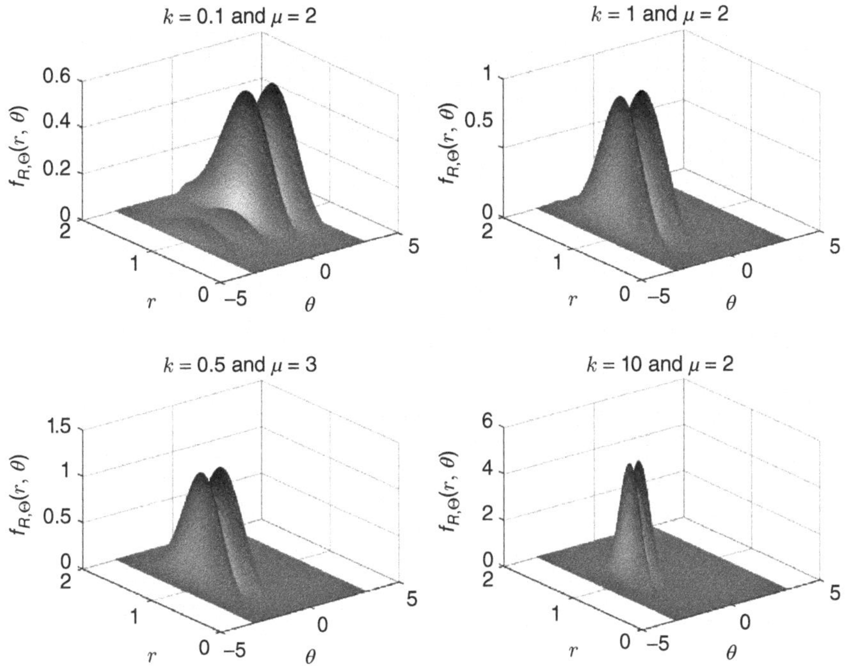

Figure 2.7 κ–μ Joint PDF for fixed μ and variable κ and for $\phi = \pi/2$. μ-Value is set to 2, and κ varies from 0.1, 1, 1.5, and 10.

of κ indicate stronger LOS path component. Varying μ has similar impact as discussed for η–μ channel as it has the same definition.

2.4.6 The α–μ Distribution

Another generalized fading distribution that describes the small-scale variation of the fading signal in a NLOS environment is called α–μ channel. The parameter $\alpha > 0$ denotes the nonlinearity of the propagation medium and $\mu \geq 1/2$ is the number of the multipath clusters. Hence, the α–μ distribution includes the Weibull and the Nakagami-m distributions as special cases. The Weibull distribution can be obtained when $\mu = 1$, whereas Nakagami-m is obtained when $\alpha = 2$ and $\mu = m$.

The envelope R and the phase Θ of the α–μ fading channel are given by

$$R = (X^2 + Y^2)^{\frac{1}{\alpha}}, \tag{2.35}$$

$$\Theta = \left(\frac{2}{\alpha}\right) \arctan\left(\frac{Y}{X}\right), \tag{2.36}$$

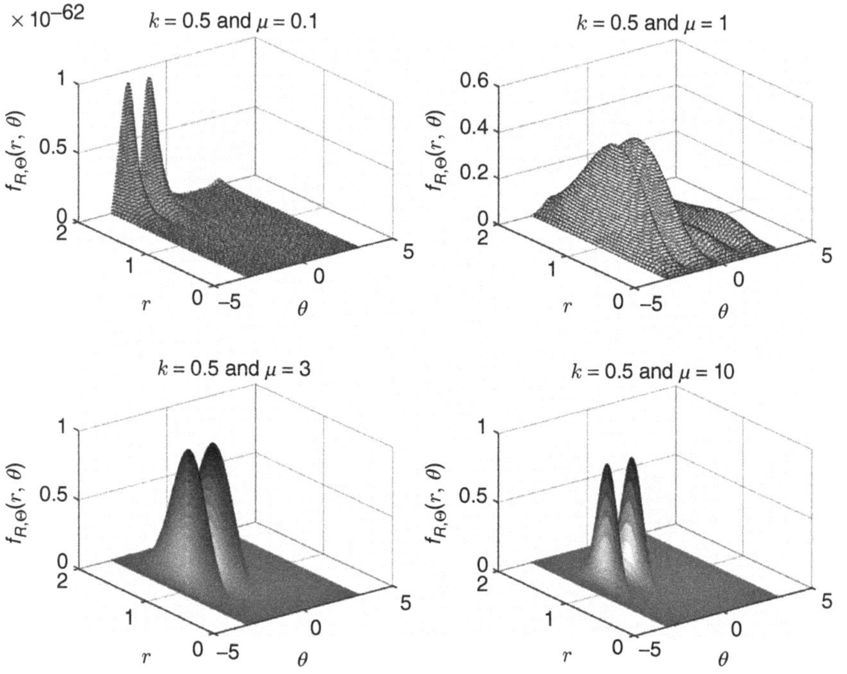

Figure 2.8 κ–μ Joint PDF for fixed κ and variable μ and for $\phi = -\pi/2$. μ-Value is varied from 0.1, 1, 3, and 10, and κ is fixed to 0.5.

where $X^2 = \sum_{i=1}^{\mu} x_i^2$ and $Y^2 = \sum_{i=1}^{\mu} y_i^2$ with x_i and y_i being mutually independent Gaussian processes with $\mathrm{E}\{x_i\} = \mathrm{E}\{y_i\} = 0$, and identical variances $\mathrm{E}\{x_i^2\} = \mathrm{E}\{Y_i^2\} = \sigma^2 = \hat{r}^\alpha/2\mu$.

For a fading signal with envelope r and $\hat{r} = \sqrt[\alpha]{\mathrm{E}\{R^\alpha\}}$ being the α-*rms* of R, the α–μ joint phase-envelope distribution $p(r,\theta)$ is given by [133]

$$p(r,\theta) = \frac{\alpha^2 \mu^\mu r^{\alpha\mu-1} \mid \sin(\alpha\theta)|^{\mu-1}}{2^{\mu+1}\Omega^\mu \Gamma^2(\frac{\mu}{2})} \exp\left(-\frac{\mu r^\alpha}{\Omega}\right) \coprod_{\mathbb{R}^+}(r) \coprod_{[-\pi,\pi]}(\theta), \tag{2.37}$$

where $0 \leq r < \infty$, $0 \leq \theta \leq 4\pi/\alpha$, and $\Omega = \hat{r}^\alpha$.

The α–μ PDF of envelope R, $p_R(r)$, is given by [134]

$$p_R(r) = \frac{\alpha\mu^\mu}{\hat{r}\Gamma(\mu)}\left(\frac{r}{\hat{r}}\right)^{\alpha\mu-1} \exp\left[-\mu\left(\frac{r}{\hat{r}}\right)^\alpha\right] \coprod_{\mathbb{R}^+}(r). \tag{2.38}$$

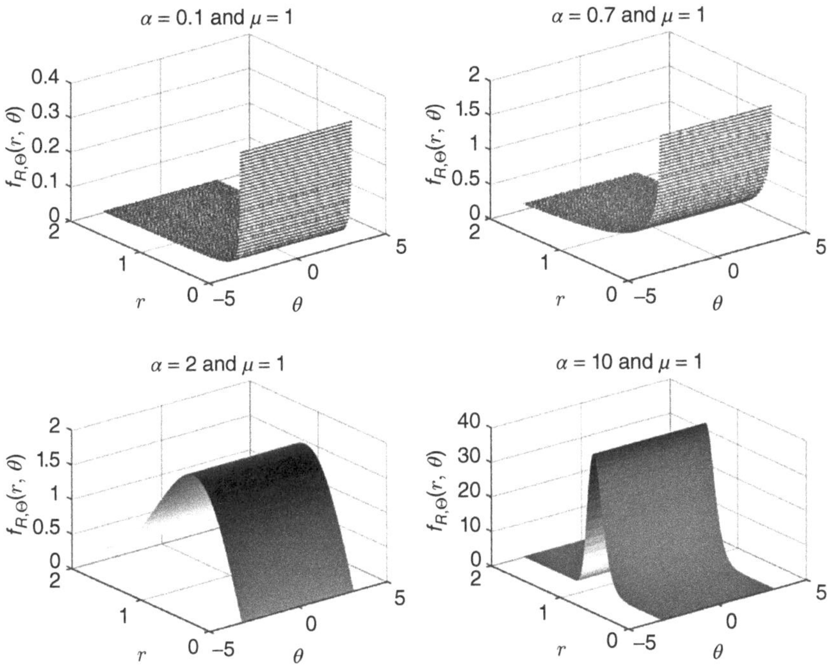

Figure 2.9 $\alpha-\mu$ Joint PDF for fixed μ and variable α. μ-Value is set to 1 and α varies from 0.1, 0.7, 2, and 10.

The PDF of the normalized envelope $\rho = r/\hat{r}$, $p_P(\rho)$, after random variable transformation is given as

$$p_P(\rho) = \frac{\alpha\mu^{\mu}}{\Gamma(\mu)}\rho^{\alpha\mu-1}\exp(-\mu\rho^{\alpha})\coprod\nolimits_{\mathbb{R}^+}(p). \tag{2.39}$$

The ℓth moment, $\mathrm{E}\{P^{\ell}\}$, of P in (2.39) is given as

$$\mathrm{E}\{P^{\ell}\} = \frac{\Gamma(\mu + \frac{\ell}{\alpha})}{\mu^{\frac{\ell}{\alpha}}\Gamma(\mu)}. \tag{2.40}$$

The PDF of the phase is given by

$$p_{\Theta}(\theta) = \frac{\alpha\Gamma(\mu)|\sin(\alpha\theta)|^{\mu-1}}{2^{\mu+1}\Gamma^2(\mu/2)}\coprod\nolimits_{[-\pi,\pi]}(\theta). \tag{2.41}$$

The joint PDF distribution for variable α and fixed μ is shown in Figure 2.9 and for fixed α and variable μ in Figure 2.10. Changing the value of α significantly changes the joint distribution PDF while changing μ has the same impact as discussed before for $\eta-\mu$ and $\kappa-\mu$ channels. The impact of varying these parameters on the performance of SMT and other MIMO systems will be discussed in detail in the coming chapters.

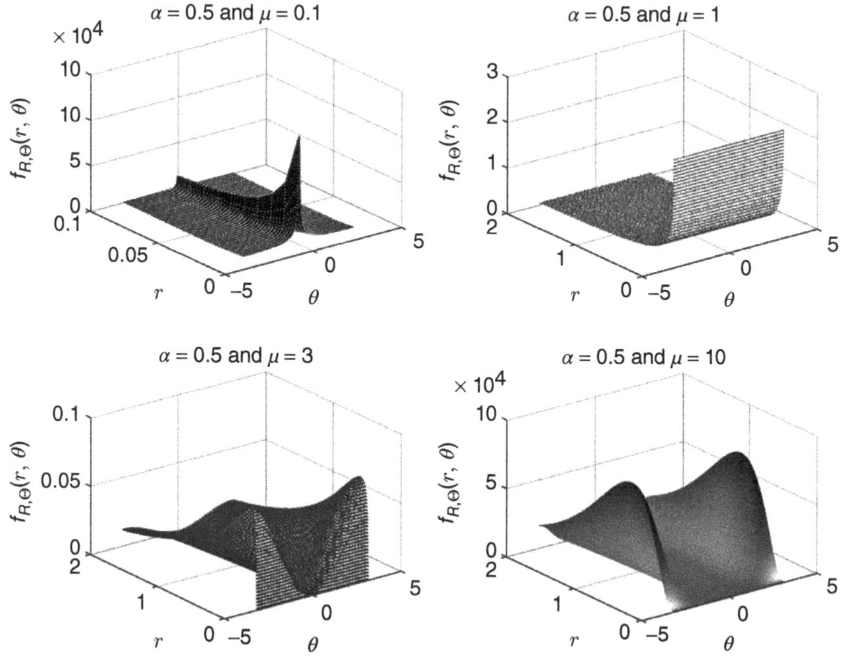

Figure 2.10 α–μ Joint PDF for fixed α and variable μ. μ-Value varies from 0.1, 1, 3, and 10 and α is fixed to 0.5.

2.5 Channel Imperfections

In this section, several channel impairments are considered and their impacts on the MIMO channel capacity are studied. In particular, spatial correlation (SC), multual coupling (MC), and imperfect channel estimation are studied.

2.5.1 Spatial Correlation

The channel correlation depends on both the environment and the spacing of the antenna elements. A terminal, surrounded by a large number of local scatterers, can achieve relatively low correlation values even if the antennas are only separated by half a wavelength [135, 136]. In outdoor base stations, the antennas are significantly higher than the scatterers, and sufficiently low correlation is likely to require more than 10 wavelengths between neighboring antenna elements. In indoor base stations, however, the required antenna separation is likely to be in between these two extremes [136].

The magnitude of correlation depends on the antenna spacing, angular values of the signals, power azimuth spectrum (PAS), and the radiation pattern [137]. Generally, it is fair to assume that correlations at the transmitter and the receiver arrays are independent of each other because the distance between the transmit and receive arrays is large compared to the antenna element spacing. All the elements in the transmit array illuminate the same scatterers in the environment. As a result, the signals at the receive array antennas will have the same PAS [137].

To incorporate the SC into the channel model, the correlation among channels at multiple elements needs to be calculated. The cross correlation $\varphi_{ij}^{\mathrm{Tx}}$ between the channel coefficients of the two antenna elements i and j at the transmitter array can be calculated as

$$\varphi_{ij}^{\mathrm{Tx}} = \left\langle |\mathbf{h}_i|^2, |\mathbf{h}_j|^2 \right\rangle, \tag{2.42}$$

where $\mathbf{h}_i$ is the channel vector between transmit antenna i and all receive antennas, and $\langle \cdot, \cdot \rangle$ is the inner product. In a similar way, the cross correlation $\varphi_{ij}^{\mathrm{Rx}}$ between the two antenna elements i and j at the receiver array can be computed. The transmit and receive correlation matrices ($\mathbf{R}_{\mathrm{Tx}}$ and $\mathbf{R}_{\mathrm{Rx}}$) contain information about how signals from each element at the transmitter and receiver are correlated with each other and they are given by

$$\mathbf{R}_{\mathrm{Tx}} = \begin{pmatrix} \varphi_{11}^{\mathrm{Tx}} & \varphi_{12}^{\mathrm{Tx}} & \cdots & \varphi_{1N_t}^{\mathrm{Tx}} \\ \varphi_{21}^{\mathrm{Tx}} & \varphi_{22}^{\mathrm{Tx}} & \cdots & \varphi_{2N_t}^{\mathrm{Tx}} \\ \vdots & \vdots & \ddots & \vdots \\ \varphi_{N_t1}^{\mathrm{Tx}} & \varphi_{N_t2}^{\mathrm{Tx}} & \cdots & \varphi_{N_tN_t}^{\mathrm{Tx}} \end{pmatrix}, \tag{2.43}$$

$$\mathbf{R}_{\mathrm{Rx}} = \begin{pmatrix} \varphi_{11}^{\mathrm{Rx}} & \varphi_{12}^{\mathrm{Rx}} & \cdots & \varphi_{1N_r}^{\mathrm{Rx}} \\ \varphi_{21}^{\mathrm{Rx}} & \varphi_{22}^{\mathrm{Rx}} & \cdots & \varphi_{2N_r}^{\mathrm{Rx}} \\ \vdots & \vdots & \ddots & \vdots \\ \varphi_{N_r1}^{\mathrm{Rx}} & \varphi_{N_r2}^{\mathrm{Rx}} & \cdots & \varphi_{N_rN_r}^{\mathrm{Rx}} \end{pmatrix}. \tag{2.44}$$

The correlated channel matrix is then obtained as

$$\mathbf{H}^{\mathrm{corr}} = \mathbf{R}_{\mathrm{Rx}}^{1/2} \mathbf{H} \mathbf{R}_{\mathrm{Tx}}^{1/2}. \tag{2.45}$$

The correlation matrices can be generated based on measurement data such as the spatial channel model (SCM) approach [138], or computed analytically based on the PAS distribution and array geometry [137]. The latter can be computed assuming a clustered channel model (as seen in Figure 2.11), in which groups of scatterers are modeled as clusters located around the transmit and receive antennas. The clustered channel model has been validated through measurements [139, 140] and adopted by various wireless system standard bodies such as the IEEE 802.11n technical group (TG) [141] and the 3GPP/3GPP2 technical specification group (TSG) [138].

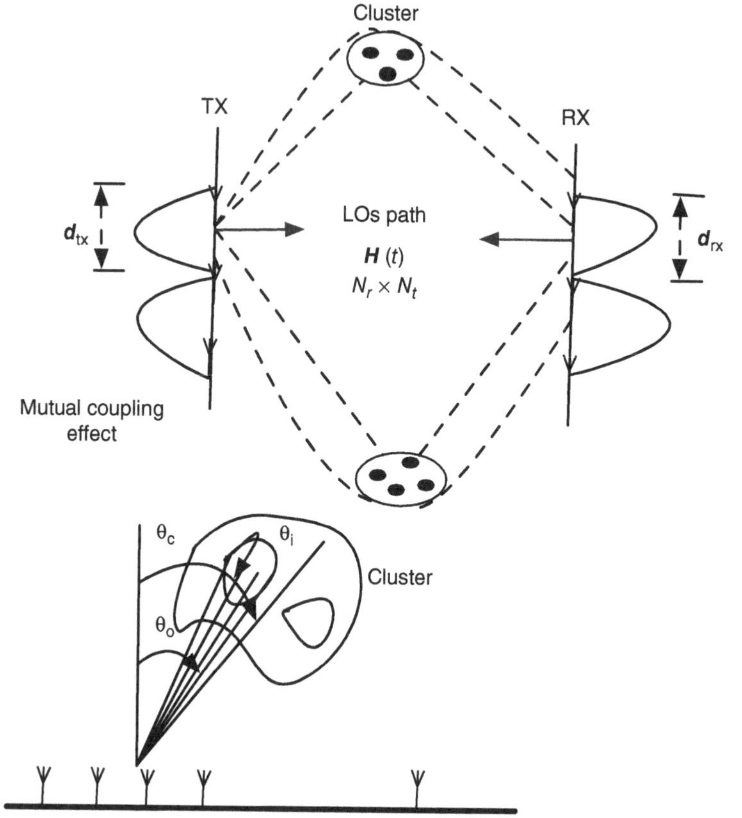

Figure 2.11 Geometry of cluster channel model – SC between transmit/receive signals. Angles θ_c, θ_0 and θ_i are the mean AOA of cluster, channel tap and the AOA offset of the channel tap.

In the clustered channel model, a group of scatterers are modeled as clusters located around the transmitter and the receiver antenna arrays. Each multipath resulting from the scattering is associated with a time delay and an angle of arrival (AOA). Multipaths are grouped to form clusters. In each cluster, the delay differences between the multipaths are not resolvable within the transmission signal bandwidth. The clustered model is characterized by multiple clusters with each cluster having a mean AOA of cluster and channel tap (θ_c and θ_0 in Figure 2.11). Multipaths within a cluster are generated with respect to a certain PDF that best fits the PAS of the channel. The standard deviation of each cluster PAS is a measure of the cluster angular spread (AS).

In the early 1970s, Lee modeled the PAS in outdoor scenarios as the nth power of a cosine function [142]. This model has been regarded as inconvenient [138], since it does not enable one to derive closed-form expressions. Therefore, two other distributions, a truncated Gaussian and a uniform one, have been introduced in [143, 144], respectively. Another model considers a truncated Laplacian distribution in [145] and is shown to best fits to the measurement results in urban and rural areas.

2.5.1.1 Simulating SC Matrix

The signal received at the n_rth antenna element, assuming noise-free transmission, is

$$y_{n_r}(t) = s(t)\frac{1}{\sqrt{N_{\mathrm{sp}}}}\sum_{i=1}^{N_{\mathrm{sp}}} h_i(t)e^{jDn_r \sin(\theta_0-\theta_i)}, \tag{2.46}$$

where $D = 2\pi d_a/\lambda$, N_{sp} is the number of subpaths per channel tap, $s(t)$ is the complex envelope of the transmitted signal, $h_i(t)$ is the complex fading channel, which can be Rayleigh for NLOS channel, Rician for LOS channels, or any other fading channel, d_a is the antenna element spacing denoted by d_{Tx} at the transmitter and d_{Rx} at the receiver, and finally from Figure 2.11, θ_i is the AoA offset compared to the mean AoA of the channel tab, θ_0.

Each channel tap is assumed to exhibit a truncated Laplacian PAS. Then, the random variable θ_i is distributed according to the Laplacian PDF with σ_{θ_i} standard deviation given by

$$P_{\theta_i}(\theta_i) = \frac{1}{\sqrt{2}\sigma_{\theta_i}}\exp\left(-\left|\frac{\sqrt{2}\theta_i}{\sigma_{\theta_i}}\right|\right)\coprod_{[-\pi,\pi]}(\theta). \tag{2.47}$$

The correlation between signals received at antenna n_r^1 and n_r^2 is calculated as [137]

$$\mathbf{R}_{n_r^1,\,n_r^2} = \mathrm{E}\{y_{n_r^1}y^*_{n_r^2}\}. \tag{2.48}$$

Substituting y_{n_r} in (2.46) in the previous equation results in

$$\mathbf{R}_{n_r^1,\,n_r^2} = \frac{1}{N_{\mathrm{sp}}}\mathrm{E}\left\{\left(s(t)\sum_{i=1}^{N_{\mathrm{sp}}} h_i(t)e^{jDn_r^1 \sin(\theta_0-\theta_i)}\right)\right.$$
$$\left.\left(s(t)\sum_{i=1}^{N_{\mathrm{sp}}} h_i(t)e^{jDn_r^2 \sin(\theta_0-\theta_i)}\right)^*\right\}. \tag{2.49}$$

Assume that the transmitted signal power is unity, i.e. $E_s = 1$ and the complex fading coefficients are i.i.d. over different rays and channel time delays,

$$\mathrm{E}\{h_i(t)h_k^*(t)\} = \delta_{ik}; \quad \delta_{ik} = \begin{cases} 1 & \text{if } i = k; \\ 0 & \text{otherwise.} \end{cases} \tag{2.50}$$

In addition, let θ_i be independent across different rays and the antenna gain pattern to be unity. Then, (2.49) can be reduced to the following

$$\mathbf{R}_{n_r^1,n_r^2 2} = \mathrm{E}_{\theta_i}\{e^{jD(n_r^1-n_r^2)\ \sin(\theta_0-\theta_i)}\}. \tag{2.51}$$

From calculus, $\sin(\theta_0 - \theta_i) = \sin(\theta_0)\cos(\theta_i) - \cos(\theta_0)\sin(\theta_i)$. Assume that the AOA offset compared to the mean AOA of the channel tab $\theta_i \approx 0$, then, $\sin(\theta_0 - \theta_i) = \sin(\theta_0) - \cos(\theta_0)\theta_i$.

Substituting this in (2.51), recalling that θ_i is a random variable (RV) with a PDF given in (2.47), and using the definition of first-order moment of a RV gives

$$\begin{aligned}\mathbf{R}_{n_r^1,n_r^2} &= \int_{-\pi}^{\pi} e^{jD(n_r^1-n_r^2)\ (\sin(\theta_0)-\cos(\theta_0)\theta_i)} P(\theta_i)\ d\theta_i \\ &= e^{jD(n_r^1-n_r^2)\ \sin(\theta_0)} \int_{-\pi}^{\pi} e^{-jD(n_r^1-n_r^2)\ \cos(\theta_0)\theta_i} P(\theta_i)\ d\theta_i.\end{aligned} \tag{2.52}$$

It was observed in [137, 141] that the Laplacian PAS distribution decays rapidly to zero within the range $(-\pi, \pi]$, also for high values of *rms* AS. Therefore, the integration of $P(\theta_i)$ truncated over $(-\pi, \pi]$ is equivalent to the one over infinite domain. Assuming $w = D(n_r^1 - n_r^2)\ \cos(\theta_0)$, (2.52) is written as follows:

$$\begin{aligned}\mathbf{R}_{n_r^1,n_r^2} &= e^{jD(n_r^1-n_r^2)\ \sin(\theta_0)} \int_{-\pi}^{\pi} P(\theta_i) e^{-jw\theta_i}\ d\theta_i \\ &= e^{jD(n_r^1-n_r^2)\ \sin(\theta_0)} \left[\sqrt{2\pi}\mathcal{F}_w\{P(\theta_i)\}\right].\end{aligned} \tag{2.53}$$

This is equivalent to computing the Fourier transform of the PAS distribution. The Fourier transform $\mathcal{F}_w$ of $e^{-a|t|}$ is given as [146]

$$\mathcal{F}_w(e^{-a|t|}) = \frac{1}{\sqrt{2\pi}} \int_{-\infty}^{\infty} e^{-a|t|} e^{-jwt} dt = \sqrt{\frac{2}{\pi}} \frac{1}{a^2 + w^2}. \tag{2.54}$$

From (2.47), $a = \sqrt{2}/\sigma_{\theta_i}$, then the $\mathcal{F}_w\{P(\theta_i)\}$ is given as

$$\mathcal{F}_w\{P(\theta_i)\} = \frac{1}{1 + \frac{\sigma_{\theta_i}^2}{2} w^2}, \tag{2.55}$$

replacing w from the above equation and substituting (2.55) in (2.52) results in

$$[\mathbf{R}(\theta_0, \sigma_{\theta_i})]_{n_r^1,n_r^2} \approx \overbrace{e^{jD(n_r^1-n_r^2)\sin(\theta_0)}}^{[\mathbf{a}(\theta_0)\cdot\mathbf{a}(\theta_0)^H]_{n_r^1,n_r^2}} \cdot \overbrace{\left(\frac{1}{1 + \frac{\sigma_{\theta_i}^2}{2} \cdot [D(n_r^1 - n_r^2)] \cos\theta_0}\right)}^{[\mathbf{B}(\theta_0,\sigma_{\theta_i})]_{n_r^1,n_r^2}}. \tag{2.56}$$

In general, the SC matrix for a receive array is given by [137]

$$\mathbf{R}_{\mathrm{Rx}}(\theta_0, \sigma_\theta) \approx [\mathbf{a}(\theta_0)\mathbf{a}^H(\theta_0)] \odot \mathbf{B}(\theta_0, \sigma_{\theta_i}), \tag{2.57}$$

where $\odot$ denotes the Schur–Hadamard (or elementwise) product and $\mathbf{B}(\theta_0, \sigma_{\theta_i})$ is the Fourier transform of the PAS PDF whose standard deviation is given by σ_{θ_i}. The column vector $\mathbf{a}(\theta_0) = [1\ e^{jD\sin\theta_0} \ldots e^{jD(N_r-1)\sin\theta_0}]^T$ is the array response vector for the signal. In a similar way, the transmits correlation matrix can be computed.

2.5.1.2 Effect of SC on MIMO Capacity

The correlated channel matrix is given in (2.45). Substituting the modified channel matrix in (2.9) results in

$$C = \mathrm{E}\left\{ \log_2 \left| \mathbf{I}_{N_r} + \mathrm{SNR} \times \frac{1}{N_t} \mathbf{R}_{\mathrm{Rx}}^{1/2} \mathbf{H} \mathbf{R}_{\mathrm{Tx}}^{1/2} \left(\mathbf{R}_{\mathrm{Rx}}^{1/2} \mathbf{H} \mathbf{R}_{\mathrm{Tx}}^{1/2} \right)^H \right| \right\}. \tag{2.58}$$

Without loss of generality, assume that $N_t = N_r = N$ and the receive and transmit correlation matrices are full rank. At high SNR, the capacity of the MIMO channel can be written as [111]

$$C \approx \mathrm{E}\left\{ \log_2 \left| \mathrm{SNR} \times \frac{1}{N} \mathbf{H}\mathbf{H}^H \right| + \log_2 |\mathbf{R}_{\mathrm{Rx}}| + \log_2 |\mathbf{R}_{\mathrm{Tx}}| \right\}. \tag{2.59}$$

From (2.59), it can be clearly seen that correlation at either transmitter or receiver has similar impact on the capacity of MIMO system. Let $e_i(\mathbf{R}_{\mathrm{Rx}})$; $(i = 1 : N)$ be the eigenvalues of the receiver correlation matrix, same can be done for transmitter correlation matrix, such that,

$$\sum_{i=1}^{N} e_i(\mathbf{R}_{\mathrm{Rx}}) = N. \tag{2.60}$$

It is shown in [111, 147] that $\prod_{i=1}^{N} e_i(\mathbf{R}_{\mathrm{Rx}}) \leq 1$ for any number of transmit and receive antennas. However, $|\mathbf{R}_{\mathrm{Rx}}| = \prod_{i=1}^{N} e_i(\mathbf{R}_{\mathrm{Rx}})$. This implies that $\log_2 |\mathbf{R}_{\mathrm{Rx}}| \leq 0$, and is zero only if all eigenvalues of $(\mathbf{R}_{\mathrm{Rx}})$ are equal, i.e. $e_i(\mathbf{R}_{\mathrm{Rx}}) = \mathbf{I}_N$, which is the case of no correlation. Therefore, SC will reduce the MIMO capacity at high SNR by $(\log_2 |\mathbf{R}_{\mathrm{Rx}}| + \log_2 |\mathbf{R}_{\mathrm{Tx}}|)$.

2.5.2 Mutual Coupling

A radio signal impinging upon an antenna element induces a current in that element which in turn radiates a field that generates a surface current on the surrounding antenna elements. This effect is known as *mutual coupling (MC)*. MC influences the radiation pattern and the antenna correlation. Parameters affecting MC are element separation, frequency, and array geometry [148]. It is shown in [149] that MC impacts the performance of adaptive arrays and for a relatively large number of antennas in a MIMO system, MC limits the effective degrees of freedom and reduces the ergodic capacity [150].

An N_t element antenna array can be regarded as a coupled N_t-port network with N_t terminals as seen in Figure 2.12. Let $\mathbf{s} = [s_1 \cdots s_{N_t}]^T$ and

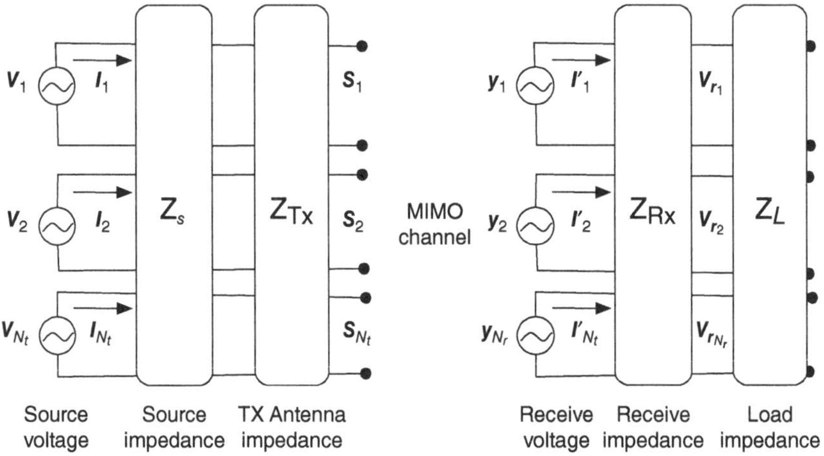

Figure 2.12 Mutual coupling in MIMO system – a network representation.

$\mathbf{V} = [V_1 \cdots V_{N_t}]$ be the vector of terminal voltages and source voltages at the transmit array, respectively. The two vectors are related as [150]

$$\mathbf{s} = (\mathbf{Z}_t/c_t)\mathbf{V}, \tag{2.61}$$

where $\mathbf{Z}_t = \mathbf{Z}_{\text{Tx}}(\mathbf{Z}_{\text{Tx}} + \mathbf{Z}_s)^{-1}$, $\mathbf{Z}_s$ is the source impedance diagonal matrix whose entries are equal to the conjugate of the diagonal entries of the transmitter mutual impedance matrix $\mathbf{Z}_{\text{Tx}}$, i.e. $(\mathbf{Z}_s)_{ii} = (\mathbf{Z}_{\text{Tx}})^*_{ii}$ and $c_t = (\mathbf{Z}_{\text{Tx}})_{11}/((\mathbf{Z}_{\text{Tx}})_{11} + (\mathbf{Z}_{\text{Tx}})^*_{11})$ is a normalization factor that guarantees $\mathbf{s} = \mathbf{V}$ for zero mutual coupling.

Similarly at the receiver, $\mathbf{y} = [y_1 \cdots y_{N_r}]^T$ are the open circuit induced voltages across the array and $\mathbf{v}_r = [v_{r1} \ldots v_{r_{N_r}}]^T$ are the voltages at the output of the array. They are related as

$$\mathbf{V}_r = (\mathbf{Z}_r/c_r)\mathbf{y}, \tag{2.62}$$

where $\mathbf{Z}_r = \mathbf{Z}_{\text{Rx}}(\mathbf{Z}_{\text{Rx}} + \mathbf{Z}_L)^{-1}$, $\mathbf{Z}_L$ is the load impedance diagonal matrix whose entries are equal to the conjugate of the diagonal entries of the receiver mutual impedance matrix $\mathbf{Z}_{\text{Rx}}$ to guarantee maximum power transfer, i.e. $(\mathbf{Z}_L)_{ii} = (\mathbf{Z}_{\text{Rx}})^*_{ii}$ and $c_r = (\mathbf{Z}_{\text{Rx}})_{11}/((\mathbf{Z}_{\text{Rx}})_{11} + (\mathbf{Z}_{\text{Rx}})^*_{11})$ is a normalization factor.

Let $I = [I_1 \cdots I_{N_t}]^T$ be the vector of terminal currents in the transmitter array. At the transmitting end, the circuit relations are

$$\mathbf{V} = (\mathbf{Z}_{\text{Tx}} + \mathbf{Z}_s)I, \tag{2.63}$$

$$\mathbf{s} = (\mathbf{Z}_{\text{Tx}}I). \tag{2.64}$$

Rearranging (2.63) and (2.64)

$$\mathbf{s} = \left(\mathbf{Z}_{\text{Tx}}(\mathbf{Z}_{\text{Tx}} + \mathbf{Z}_s)^{-1}\right)\mathbf{V} = (\mathbf{Z}_t/c_t)\mathbf{V}. \tag{2.65}$$

From (2.65), and knowing that the received signal without noise is $\mathbf{y} = \mathbf{Hs}$, the received signal $\mathbf{y}$ can be written as

$$\mathbf{y} = \mathbf{H}(\mathbf{Z}_t/c_t)\mathbf{V}. \tag{2.66}$$

Substituting (2.66) in (2.62),

$$\mathbf{V}_r = (\mathbf{Z}_r/c_r)\mathbf{H}(\mathbf{Z}_t/c_t)\mathbf{V}. \tag{2.67}$$

Using $\mathbf{C}_{\mathrm{Rx}} = \mathbf{Z}_r/c_r$ and $\mathbf{C}_{\mathrm{Tx}} = \mathbf{Z}_t/c_t$, the MIMO channel matrix in the presence of MC is modified as

$$\mathbf{H}^{\mathrm{coup}} = \mathbf{C}_{\mathrm{Rx}}\ \mathbf{H}\ \mathbf{C}_{\mathrm{Tx}}. \tag{2.68}$$

2.5.2.1 Effect of MC on MIMO Capacity

Replacing the modified MIMO channel matrix in the presence of MC in the general MIMO capacity equation in (2.9) gives

$$C = \mathrm{E}\left\{\log_2\left|\mathbf{I}_{N_r} + \mathrm{SNR}\times\frac{1}{N_t}\mathbf{C}_{\mathrm{Rx}}\mathbf{H}\mathbf{C}_{\mathrm{Tx}}\ (\mathbf{C}_{\mathrm{Rx}}\mathbf{H}\mathbf{C}_{\mathrm{Tx}})^H\right|\right\}. \tag{2.69}$$

At high SNR, the previous equation can be simplified to

$$C \approx \mathrm{E}\left\{\log_2\left|\mathrm{SNR}\times\frac{1}{N_t}\mathbf{C}_{\mathrm{Rx}}\mathbf{H}\mathbf{C}_{\mathrm{Tx}}\ (\mathbf{C}_{\mathrm{Rx}}\mathbf{H}\mathbf{C}_{\mathrm{Tx}})^H\right|\right\}. \tag{2.70}$$

It can be seen from (2.70) that MC affects both the channel correlation properties and the target average receive SNR. Assume that $N_t = N_r = N$ and the receive and transmit correlation matrices are full rank, then (2.70) can be written as

$$C = \mathrm{E}\left\{\log_2\left|\mathrm{SNR}\times\frac{1}{N_t}\mathbf{H}\mathbf{H}^H\right| + \log_2|\mathbf{C}_{\mathrm{Rx}}\mathbf{C}_{\mathrm{Rx}}^H| + \log_2|\mathbf{C}_{\mathrm{Tx}}\mathbf{C}_{\mathrm{Tx}}^H|\right\}, \tag{2.71}$$

which can be further simplified to

$$\begin{aligned} C = &\sum_{i=1}^{N}\log_2\left(\mathrm{SNR}\times\frac{1}{N_t}e_i(\mathbf{H}\mathbf{H}^H)\right) \\ &+\sum_{i=1}^{N}\log_2 e_i(\mathbf{C}_{\mathrm{Rx}}\mathbf{C}_{\mathrm{Rx}}^H) + \sum_{i=1}^{N}\log_2 e_i\left(\mathbf{C}_{\mathrm{Tx}}\mathbf{C}_{\mathrm{Tx}}^H\right). \end{aligned} \tag{2.72}$$

Comparing this result to (2.9), it can be noticed that the last two terms in (2.72) represent the effect of the two MC matrices. MC can enhance the capacity of the MIMO system if the following condition is satisfied,

$$\prod_{i=1}^{N} e_i(\mathbf{C}_{\mathrm{rx}}\mathbf{C}_{\mathrm{rx}}^H)e_i(\mathbf{C}_{\mathrm{tx}}\mathbf{C}_{\mathrm{tx}}^H) > 1. \tag{2.73}$$

In other words, the coupling effect will have a positive impact on the channel capacity if the product of the eigenvalues of the two ends MC correlation matrices is larger than one. This has been shown to be the case of closely spaced antennas [148, 150]. However, placing the antennas near to each others results in high SC, which degrades the performance and reduces the MIMO capacity. In the case that antennas were not very close to each other, the product of the eigenvalues of the two ends MC correlation matrices will be smaller than one and MC will then have negative impact on the channel capacity.

2.5.3 Channel Estimation Errors

The ML decoder for MIMO systems as given in (2.3) relies on the knowledge of the channel matrix $\mathbf{H}$ at the receiver. Practically, exact channel knowledge at the receiver is not possible due to the presence of AWGN. Therefore, channel estimation algorithm is generally used to obtain an estimate of the channel matrix $\tilde{\mathbf{H}}$ [120]. Assuming that $\tilde{\mathbf{H}}$ and $\mathbf{H}$ are jointly ergodic and stationary processes and assuming that the estimation channel and the estimation error are orthogonal yields

$$\mathbf{H} = \widetilde{\mathbf{H}} + \mathbf{e}, \tag{2.74}$$

where $\mathbf{e}$ denotes the channel estimation errors (CSEs) with complex Gaussian entries $\mathcal{CN}(0, \sigma_{\mathbf{e}}^2)$. Note that $\sigma_{\mathbf{e}}^2$ captures the quality of the channel estimation and can be chosen depending on the channel dynamics and estimation methods. In practical MIMO systems, interpolation techniques are generally considered for channel estimation methods. In such methods, the channel is estimated at a specific time or frequency and suitable interpolation methods are used to determine the channel at other points based on channel statistics [151].

2.5.3.1 Impact of Channel Estimation Error on the MIMO Capacity

The impact of channel estimation error on the capacity of MIMO systems is discussed in what follows. The MIMO capacity in the presence of channel estimation error can be derived by substituting $\mathbf{H} = \widetilde{\mathbf{H}} + \mathbf{e}$ and maximizing the mutual information given $\widetilde{\mathbf{H}}$, which yields with the lower bound [152]

$$C > \mathrm{E}\left\{ \log_2 \left| \mathbf{I}_{N_r} + \mathrm{SNR} \times \frac{1}{N_t} \frac{1}{1 + \mathrm{SNR} \times \sigma_{\mathbf{e}}^2} \mathbf{H}\, \mathbf{H}^H \right| \right\}. \tag{2.75}$$

Comparing (2.75) and (2.9) clearly highlights the negative impact of channel estimation error on the MIMO capacity. The SNR decays by a factor of $(1 + \mathrm{SNR} \times \sigma_{\mathbf{e}}^2)$. For small values of $\sigma_{\mathbf{e}}^2$ the impact of CSE is negligible. But for large values, the channel estimation error could deteriorate the achievable capacity significantly.[3]

3 More elaboration on this will be discussed in Chapter 5.

3

Space Modulation Transmission and Reception Techniques

In this chapter, the different space modulation techniques (SMTs) available in literature will be discussed. These include space shift keying (SSK) [89], generalized space shift keying (GSSK) [69], spatial modulation (SM) [37], generalized spatial modulation (GSM) [67], quadrature space shift keying (QSSK) [65], quadrature spatial modulation (QSM) [65], generalized quadrature space shift keying (GQSSK), generalized quadrature spatial modulation (GQSM) and the advanced SMTs including differential space shift keying (DSSK), differential spatial modulation (DSM) [63], differential quadrature spatial modulation (DQSM) [153], space–time shift keying (STSK) [78], and trellis coded spatial modulation (TCSM) [154] systems.

SMTs are unique multiple-input multiple-output (MIMO) transmission schemes that utilize the differences among different channel paths to convey additional information bits. In such systems, one or more of the available transmit antennas at the transmitter are activated at one particular time instant and all other antennas are turned off. The channel path from each transmit antenna to all receive antennas denotes a spatial constellation symbol denoted by $(\mathcal{H}_{\ell_t} \in \mathcal{H})$ or ℓ_t in this book, where $\mathcal{H}$ denotes the spatial constellation diagram generated for the $N_r \times N_t$ channel matrix $\mathbf{H}$, and N_t and N_r are the number of transmit and receive antennas, respectively. At each time instance, the active transmit antennas transmit a modulated or unmodulated radio frequency (RF) signal. A common advantage for such systems is the ability to design MIMO transmitters with single RF-chain, which promises implementation cost reduction, low computational complexity, and high-energy efficiency. This is unlike conventional MIMO systems, such as spatial multiplexing (SMX), where each transmit antenna is driven by one RF-chain, and independent data streams are transmitted from the available antennas. In such MIMO systems, the Euclidean difference among different channel vectors is utilized to transmit cochannel signals to increase the data rate. However, all antennas must be active, and the receiver needs to resolve the cochannel interference (CCI) to correctly estimate the transmitted data.

Space Modulation Techniques, First Edition. Raed Mesleh and Abdelhamid Alhassi.

In SMTs, the transmitted data are either modulated by a complex signal symbol, $\mathcal{S}_{\iota_t}$, drawn from an arbitrary constellation diagram, $\mathcal{S}_{\iota_t} \in \mathcal{S}$, such as quadrature amplitude modulation (QAM), or phase shift keying (PSK), or unmodulated RF signals. Hence, the received signal at the input of the receive antennas is given by

$$\mathbf{y} = \mathbf{H}\mathbf{x}_t + \mathbf{n} = \mathcal{H}_{\ell_t}\mathcal{S}_{\iota_t} + \mathbf{n}, \tag{3.1}$$

where $\mathbf{x}_t$ is the N_t-length transmitted vector, $\mathbf{n}$ is an N_r-dimensional additive white Gaussian noise (AWGN) with zero mean and covariance matrix of $\sigma_n^2\mathbf{I}_{N_r}$. As such, the signal-to-noise ratio (SNR) at the receiver input, assuming normalized channel $E_s = \mathrm{E}\{\|\mathbf{H}\mathbf{x}_t\|_\mathrm{F}^2\} = N_r$, is given by $\mathrm{SNR} = E_s/N_0 = 1/\sigma_n^2$.

The received signal is then processed by a maximum-likelihood (ML) decoder to jointly estimate the spatial symbol, $\hat{\mathcal{H}}_\ell$, and the signal symbol, $\hat{\mathcal{S}}_\iota$, as

$$[\hat{\mathcal{H}}_\ell, \hat{\mathcal{S}}_\iota] = \underset{\substack{\mathcal{H}_\ell \in \mathcal{H} \\ \mathcal{S}_\iota \in \mathcal{S}}}{\arg\min} \|\mathbf{y} - \mathcal{H}_\ell \mathcal{S}_\iota\|_\mathrm{F}^2, \tag{3.2}$$

where $\|\cdot\|_\mathrm{F}$ is the Frobenius norm. The estimated spatial and constellation symbols are then used to retrieve transmitted data bits by inversing the mapping procedure considered at the transmitter.

3.1 Space Shift Keying (SSK)

SSK is the simplest form of the family of SMTs [89] even though it was proposed after SM [37]. In the SSK system, the data are transmitted through spatial symbols only, and the transmitted signal is unmodulated RF signal considered to indicate the spatial index of the active transmit antenna. At each time instant, $\eta_t = \log_2(N_t)$ bits modulate a transmit antenna with an index, $\ell \in 1, 2, \ldots, N_t$, among the set of existing N_t transmit antennas. Only that particular antenna is active and transmits a fixed unmodulated RF signal. In Figure 3.1, the cosine part of the RF carrier, $x_t(t) = \cos(2\pi f_c t)$, is considered and any other fixed signal can be utilized. Therefore, SSK scheme requires no RF-chain at the transmitter, and the transmitter can be entirely designed through RF switches [40]. Since no information is modulated on the carrier signal, it can be generated once and stored for further use in all other transmissions. An RF digital to analog converter (DAC) with an internal memory [155] can be utilized to store the RF signal and continuously transmits it at each symbol time. However, the RF DAC board generally produces low output power, and power amplifier (PA) will be needed to boost the signal output power before transmitted by the antennas.

As such, SSK transmitter is an RF switch with single input and N_t RF outputs. The incoming $\eta = \log_2(N_t)$ bits control the RF switch and determine the active port at each particular time instant. An illustration of the spatial symbols

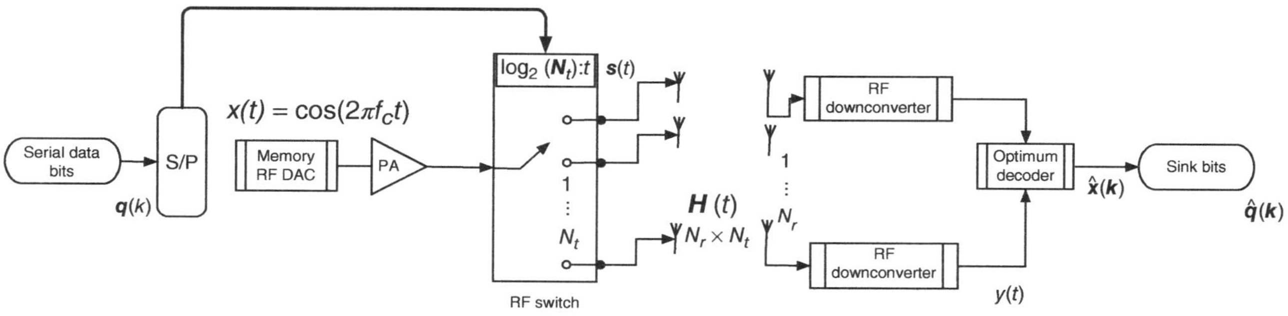

Figure 3.1 SSK system model with single RF-chain and with N_t transmit and N_r receive antennas.

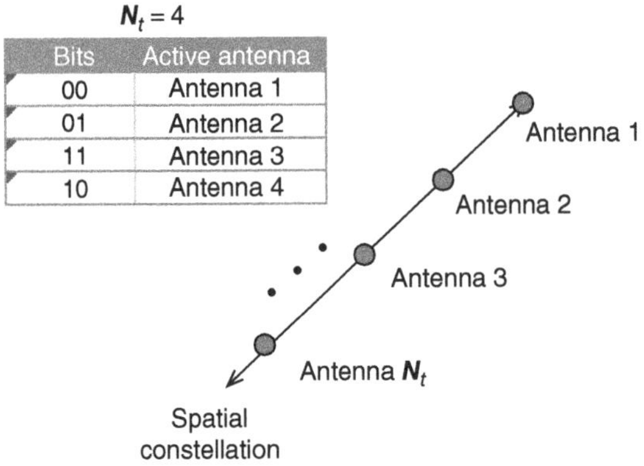

Figure 3.2 An example of SSK constellation diagram with the mapping table for $N_t = 4$.

and the mapping table for $N_t = 4$ is shown in Figure 3.2. Such RF switches are generally very cheap and cost roughly around 1–3 US$. The RF switching time including the rise and fall times of a pulse, T_{sw}, plays a major role in determining the maximum data rate of the SSK scheme. The maximum data rate that can be transmitted in SSK scheme is given by

$$R_b = \eta / T_{\text{sw}} \quad \text{bit (s Hz)}^{-1}. \tag{3.3}$$

Hence, a slow switching time degrades the spectral efficiency of SSK scheme, whereas a fast switching time achieves increased data rates. A study in [156] investigated the impact of RF switches on the achievable data rate of SM system. Different RF switches are available commercially with various switching times ranging from about 20 ns to a few microseconds. It should also be noted that there exists several RF switches that can support a different number of transmit antennas; $N_t = 2$, $N_t = 4$, $N_t = 8$, and $N_t = 16$, and can be obtained easily with very low cost [157, 158]. However, the switching time depends on the transistor technology and number of output ports and generally increases with increasing the number of output ports for the same technology. In some cases, like the 16 output switch, the decoder bit information has to be fed through a serial communication protocol such as serial peripheral interface (SPI). Therefore, the time overhead introduced by SPI has to be added to the RF switching time [40].

To demonstrate the working mechanism of SSK-MIMO system, an example is discussed in what follows. Assume that the incoming data bits $\mathbf{q} = [1 \;\; 0]$ bits are to be transmitted at one time instant from $N_t = 4$ transmit antennas. Considering the mapping table in Figure 3.2, the incoming bits, $\mathbf{q}$, activate the fourth transmit antenna, $\ell_t = 4$, and transmit the carrier signal through the RF switch. Therefore, the transmitted RF signal vector is given by

$$\mathbf{x}_t^{\text{RF}}(t) = \begin{bmatrix} 0 & 0 & 0 & \cos(w_c t) \end{bmatrix}^T. \tag{3.4}$$

Hence,

1. there is no constellation symbol, $\mathcal{S} = \{1\}$;
2. all incoming data bits are modulated in the spatial domain, where the spatial constellation diagram is $\mathcal{H} = \{\mathbf{h}_1, \mathbf{h}_2, \ldots, \mathbf{h}_{N_t}\}$, and $\mathbf{h}_\ell$ is the ℓth N_r-length column of $\mathbf{H}$.

3.2 Generalized Space Shift Keying (GSSK)

GSSK [69] generalizes the SSK scheme by activating more than one transmit antenna at the same time. The activated antennas transmit the same data symbol and the transmitted energy is divided among them. Hence, the spatial and signal constellations diagrams for GSSK are as follows:

1. The spatial diagram is

$$\mathcal{H} = \frac{1}{n_u}\left\{\sum_{i\in\ell_1}\mathbf{h}_i, \sum_{i\in\ell_2}\mathbf{h}_i, \ldots, \sum_{i\in\ell_{2^\eta}}\mathbf{h}_i\right\}, \tag{3.5}$$

   where ℓ_l is the lth combination of active antennas in the space $\eth$.
2. The cardinality of the signal constellation diagram is one, $\mathcal{S} = \{1\}$. That is similar to SSK scheme, where data are conveyed solely through spatial symbols.

One of the major advantages of such generalization is that it allows for an arbitrary number of transmit antennas. It is important to note that SSK scheme can work only for N_t being a power of two integers. In GSSK, however, any number of transmit antennas can be considered.

A system model for GSSK scheme with $N_t = 6$ antennas and arbitrary N_r receive antennas is depicted in Figure 3.3. The system model is similar to SSK with the only difference in the RF switch part. In GSSK scheme considering the mapping table shown in Figure 3.3, two RF switches are needed. The first switch with two outputs selects an antenna based on the most significant bit, b_3. The other switch with four outputs selects an antenna based on the other bits, b_2 and b_1. As such, two transmit antennas are activated at one time instant in the considered example. In general, n_u antennas among the available N_t antennas can be activated, and the system model can be designed based on the mapping table. In principle, an RF switch with 2^η outputs can be considered or multiple RF switches can be used to support the selections from the mapping table.

The number of data bits that can be transmitted at any particular time instant for GSSK is given by $\eta = \lfloor \log_2 \binom{N_t}{n_u} \rfloor$ bits. Please note that for $\eta = 3$ bits, as the example considered in Figure 3.3, $N_t = 5$ and $n_u = 2$ can support such spectral efficiency as illustrated in the mapping table shown in Figure 3.4. However, such

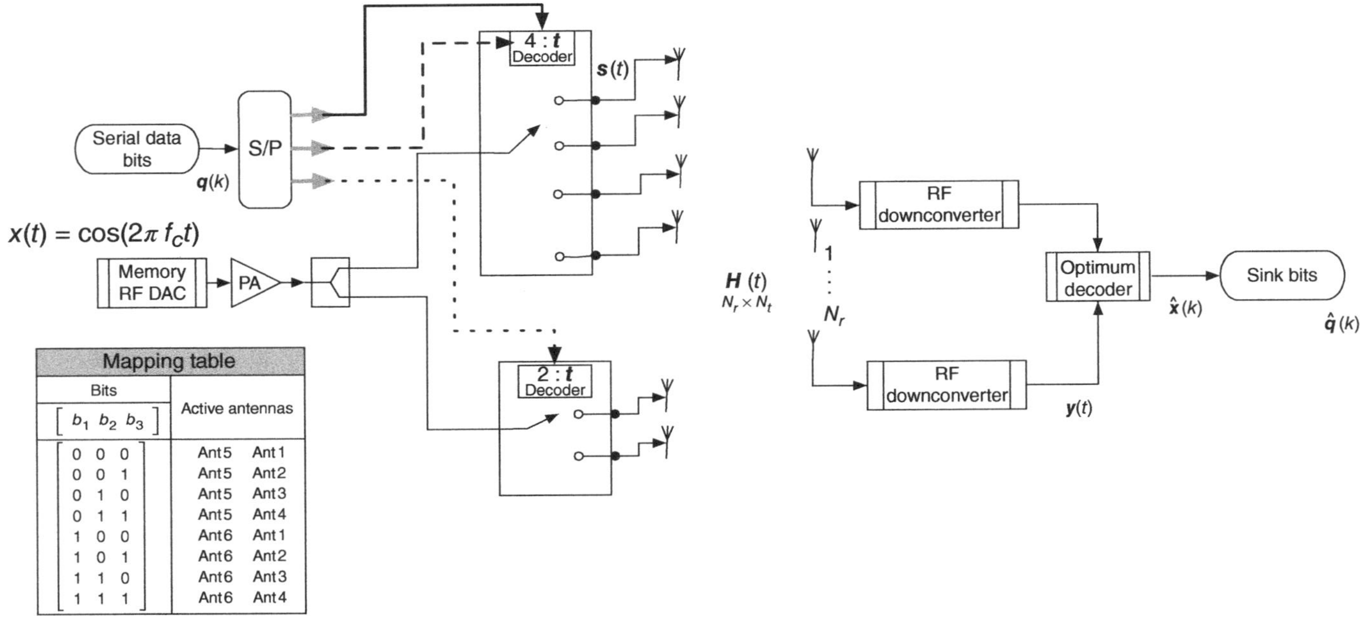

Bits $[b_1\ b_2\ b_3]$	Active antennas
0 0 0	Ant5 Ant1
0 0 1	Ant5 Ant2
0 1 0	Ant5 Ant3
0 1 1	Ant5 Ant4
1 0 0	Ant6 Ant1
1 0 1	Ant6 Ant2
1 1 0	Ant6 Ant3
1 1 1	Ant6 Ant4

Figure 3.3 GSSK system model with single RF-chain and with $N_t = 6$ transmit and N_r receive antennas.

Figure 3.4 An example of GSSK constellation diagram with the mapping table for $N_t = 5$ and $n_u = 2$.

$N_t = 5$ and $n_u = 2$

Bits	Active antenna
000	Antenna 1 & Antenna 2
001	Antenna 1 & Antenna 3
010	Antenna 1 & Antenna 4
011	Antenna 1 & Antenna 5
100	Antenna 2 & Antenna 3
101	Antenna 2 & Antenna 4
110	Antenna 2 & Antenna 5
111	Antenna 3 & Antenna 4
Not used	Antenna 3 & Antenna 5
Not used	Antenna 4 & Antenna 5

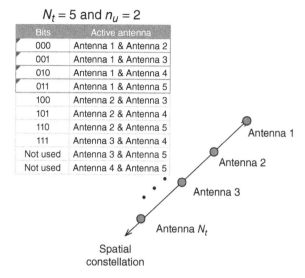

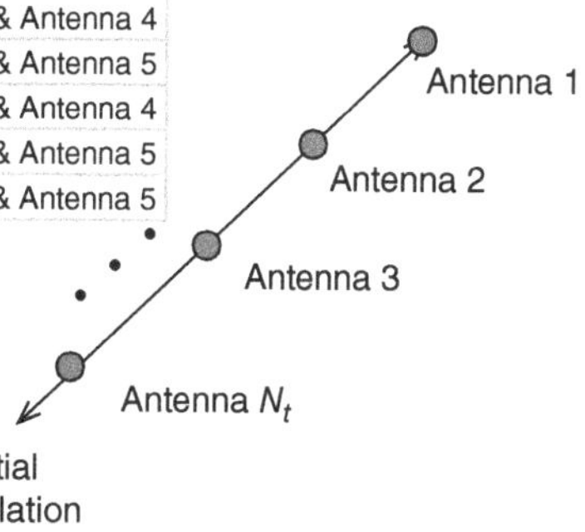

mapping table requires sophisticated RF switching circuits that can be simplified by considering $N_t = 6$ antennas instead of $N_t = 5$, as shown in Figure 3.3.

As multiple antennas transmit at the same time instant, transmit antenna synchronization is required. This is a drawback for GSSK where it increases the hardware complexity of the transmitter. Such synchronization is not required for SSK scheme since only one antenna is active at a time. Again, the unmodulated RF signal is generated once and stored in an RF DAC memory to be used regularly. The output from the RF memory is splitted through an RF splitter with n_u outputs. The splitter can be thought of as a power divider of the input signal by the number of output ports of the RF splitter. The splitter outputs are then transmitted from n_u antennas determined by the RF switches.

To demonstrate the working mechanism of GSSK system, let $\mathbf{q} = [1\ 0\ 1]$ be the data bits to be transmitted at one time instant from $N_t = 6$ transmit antennas, while activating $n_u = 2$ antennas at a time. Considering the mapping table shown in Figure 3.3, antennas $\ell_t^1 = 2$ and $\ell_t^2 = 6$ will be activated and the transmitted RF signal vector is given by $\mathbf{x}_t^{\mathrm{RF}}(t) = [0\ \frac{\cos(w_c t)}{\sqrt{2}}\ 0\ 0\ 0\ \frac{\cos(w_c t)}{\sqrt{2}}]^T$, where $S_{l_t} = 1$ and $\mathcal{H}_{\ell_t=\{1,6\}} = \frac{1}{\sqrt{2}}\sum_{\ell_t^i \in \{1,6\}} \mathbf{h}_{\ell_t^i} = \frac{1}{\sqrt{2}}(\mathbf{h}_1 + \mathbf{h}_6)$.

3.3 Spatial Modulation (SM)

SM is the first proposed technique among the set of SMTs and most existing methods are derived as special or generalized cases from it [37]. However, prior work in [159] caught the name of SSK, but it works totally different than the discussed SSK scheme above. In [159], two antennas exist at the transmitter, where one antenna is active for bit “0” and both antennas are active for bit “1.”

The idea is extended such that quadrature phase shift keying (QPSK) and binary phase shift keying (BPSK) signals can be transmitted to increase the data rate. This is a typical MIMO system that aims at enhancing the diversity by applying a repetition coding among antennas. The idea to modulate data bits in the spatial index of transmit antennas where suggested for the first time when proposing SM [37, 71].

An SM system model with single RF chain and RF switch is shown in Figure 3.5. Let $\mathbf{q}$ denote the data bits to be transmitted at one particular time instant. In SM, $\eta = \log_2(N_t) + \log_2(M)$ data bits can be transmitted at any particular time instant. The incoming serial data bits are converted to parallel data bits through serial/parallel shift register and grouped into two groups. The first group contains $\log_2(N_t)$ bits and activates one antenna among the set of N_t antennas using the RF switch. The second group with $\log_2(M)$ bits modulates a signal constellation symbol from arbitrary M–QAM/PSK or any other constellation diagram. Hence, as illustrated in Figure 3.6, the spatial and signal constellation diagrams are

(1) The spatial diagram is $\mathcal{H} = \{\mathbf{h}_1, \mathbf{h}_2, \dots, \mathbf{h}_{N_t}\}$.
(2) The signal constellation space is $\mathcal{S} = \{s_1, s_2, \dots, s_M\}$, where s_i is the ith symbol drawn from the considered M-QAM/PSK diagram.

The modulated complex symbol is processed by an IQ modulator to generate the RF carrier signal as

$$x_{\text{RF}} = \text{Re}\{\mathcal{S}_\iota\} \cos(w_c t) + \text{Im}\{\mathcal{S}_\iota) \sin(w_c t\}, \quad (3.6)$$

which is then transmitted from the active antenna ℓ_t.

To better explain this, an example is given in what follows. Assume that $\mathbf{q} = [0\ 1\ 1\ 0]^T$ input data bits are to be transmitted at a particular time instant using SM. The first group of data bits [0 1] determines the active antenna index $\ell = 2$, i.e. $\mathcal{H}_2 = \mathbf{h}_2$. The second group of data bits [1 0] selects the symbol $\mathcal{S}_3 = -1 - j \in \mathcal{S}$. A mapping table for SM with $M = 4$ and $N_t = 4$ is given in Table 3.1. The resultant symbol vector after the RF switch can be written as

$$\mathbf{x}_t^{\text{RF}}(t) = \begin{bmatrix} 0 & -\cos(w_c t) - \sin(w_c t) & 0 & 0 \end{bmatrix}^T. \quad (3.7)$$

Compared to other MIMO techniques, SM is shown to have many advantages that are

(1) interchannel interference (ICI) is totally avoided by SM, since only one antenna is active at a time;
(2) transmit antenna synchronization is not required;
(3) single RF-chain can be used similar to other SMTs. Therefore, transmitter complexity and cost are reduced significantly;
(4) the ML receiver complexity is much less than other MIMO techniques such as SMX as will be discussed later;

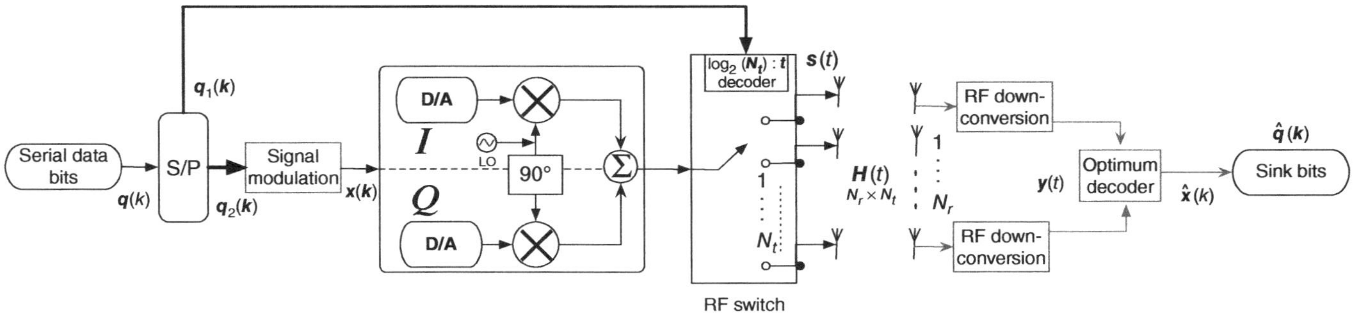

Figure 3.5 SM system model with single RF-chain and with N_t transmit and N_r receive antennas.

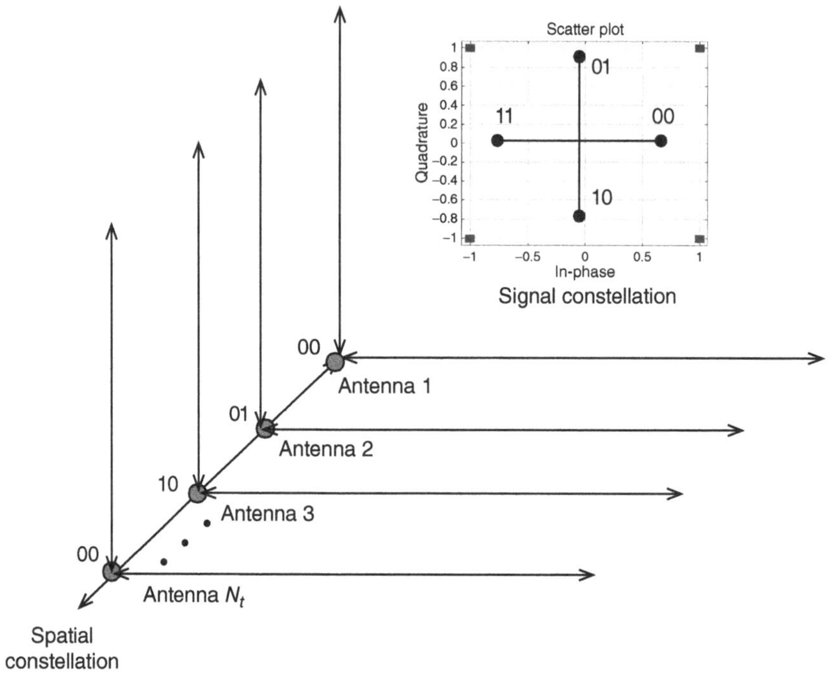

Figure 3.6 An example of SM constellation diagram with $N_t = 4$ and 4-QAM modulation.

(5) bit error performance, which will be discussed in Chapter 4, demonstrates that SM can achieve better error performance as compared to SMX MIMO system.

3.4 Generalized Spatial Modulation (GSM)

GSM is an expansion to SM similar to GSSK scheme [67]. In GSM, as shown in Figure 3.7, a group of transmit antennas, two or more, are activated at any particular time instant and transmit the same signal. Hence, the spatial constellation diagram is the same as in GSSK, and the signal constellation diagrams are the same as in SM. Thus, an overall spectral efficiency equal to $\eta = \lfloor \log_2 \binom{N_t}{n_u} \rfloor + \log_2 M$ is achieved by GSM scheme.

A mapping table for $N_t = 5$, $n_u = 2$, and BPSK modulation achieving $\eta = 4$ bits is shown in Table 3.2. Activating more than one antenna at a time reduces the required number of transmit antennas for a specific spectral efficiency and allows the use of N_t being a number that is not necessarily a power of 2. However, transmitted energy should be divided among all active

Table 3.1 SM mapping table for $N_t = 4$ and 4-QAM modulation.

Bits	Symbol bits	Symbols	Spatial bits	Antenna index
0000	00	+1+j	00	1
0001	01	−1+j	00	1
0010	10	−1−j	00	1
0011	11	+1−j	00	1
0100	00	+1+j	01	2
0101	01	−1+j	01	2
0110	10	−1−j	01	2
0111	11	+1−j	01	2
1000	00	+1+j	10	3
1001	01	−1+j	10	3
1010	10	−1−j	10	3
1011	11	+1−j	10	3
1100	00	+1+j	11	4
1101	01	−1+j	11	4
1110	10	−1−j	11	4
1111	11	+1−j	11	4

antennas, and transmit antennas need to be synchronized. Similar to GSSK, the number of transmit antennas might slightly increase to simplify the RF switching circuits as discussed before.

Another scheme called variable generalized spatial modulation (VGSM) proposed in [72] where the number of activated antennas is not fixed, i.e. depending on the bits to modulate in the spatial domain the number of activated antennas can vary from only one active to all antennas are active and transmitting the same signal symbol. As such, VGSM can further reduce the number of required transmit antennas for a specific spectral efficiency, where the number of bits that can be modulated in the spatial domain is $\eta = N_t - 1$. Hence, VGSM can achieve similar data rate of $\eta = 4$, as discussed for GSM system, with only $N_t = 4$ antennas as illustrated in Table 3.3.

3.5 Quadrature Space Shift Keying (QSSK)

QSSK was proposed to enhance the spectral efficiency of SSK scheme [65, 70]. As discussed at the beginning of this chapter, in SSK system, either the cosine part or the sine part of the carrier signal is transmitted. However, QSSK idea is

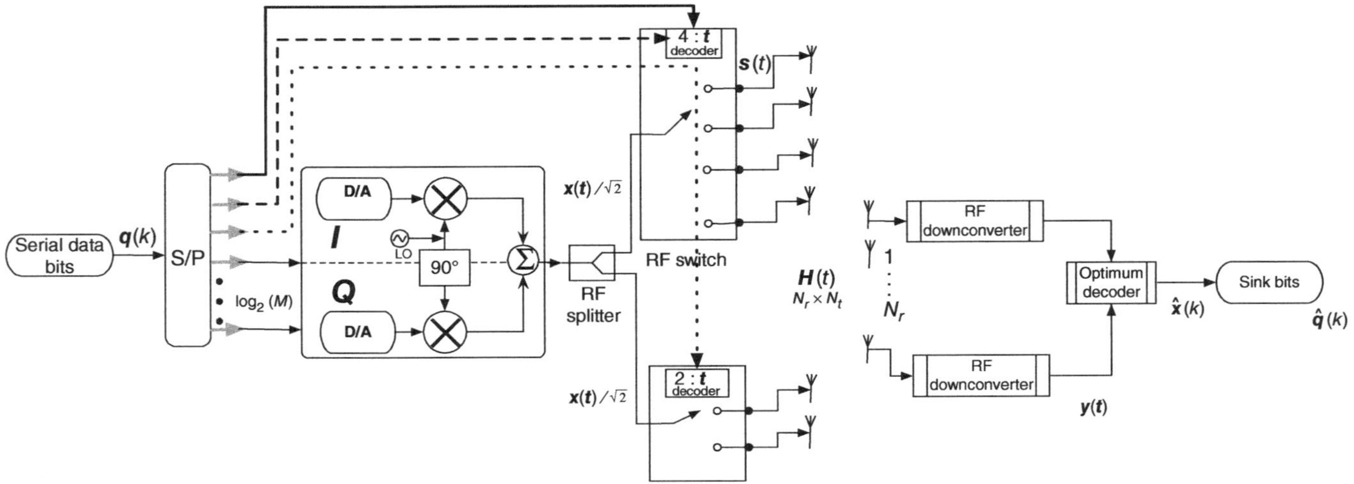

Figure 3.7 GSM system model with single RF-chain, multiple RF switches and with $N_t = 6$ transmit antennas, $n_u = 2$ active antennas at a time and N_r receive antennas.

Table 3.2 GSM mapping table with $N_t = 5$, $n_u = 2$ and BPSK modulation.

Bits	Symbol bits	Symbols	Spatial bits	Active antenna index
0000	0	+1	000	1,2
0001	1	−1	000	1,2
0010	0	+1	001	1,3
0011	1	−1	001	1,3
0100	0	+1	010	1,4
0101	1	−1	010	1,4
0110	0	+1	011	1,5
0111	1	−1	011	1,5
1000	0	+1	100	2,3
1001	1	−1	100	2,3
1010	0	+1	101	2,4
1011	1	−1	101	2,4
1100	0	+1	110	2,5
1101	1	−1	110	2,5
1110	0	+1	111	3,4
1111	1	−1	111	3,4

Table 3.3 VGSM with $N_t = 4$ and BPSK modulation.

Bits	Symbol bits	Symbols	Spatial bits	Antenna index
0000	0	+1	000	1
0001	1	−1	000	1
0010	0	+1	001	2
0011	1	−1	001	2
0100	0	+1	010	3
0101	1	−1	010	3
0110	0	+1	011	4
0111	1	−1	011	4
1000	0	+1	100	1,2
1001	1	−1	100	1,2
1010	0	+1	101	1,3
1011	1	−1	101	1,3
1100	0	+1	110	1,4
1101	1	−1	110	1,4
1110	0	+1	111	2,3
1111	1	−1	111	2,3

to utilize both parts to increase the data rate and enhance the performance of SSK scheme. This is done by transmitting the cosine part of the carrier from one antenna $\ell_t^{\Re}$ and the sine part from another or the same antenna $\ell_t^{\Im}$. Incoming data bits determine the active antennas. Hence, the spectral efficiency of QSSK is given by $\eta = 2\log_2(N_t)$.

A system model for QSSK technique is shown in Figure 3.8, and a mapping table with $N_t = 4$ is given in Table 3.4. Let $\mathbf{q}(k) = [1\ \ 0\ \ 1\ \ 1]^T$ denote the data bits to be transmitted at a particular time instant using QSSK scheme with $N_t = 4$. The incoming bits sequence is divided into two groups each with $\log_2(N_t) = 2$ bits. The first group $[1\ \ 0]^T$ will activate the antenna index $\ell_t^{\Re} = 3$ to transmit the cosine part of the carrier. The second group $[1\ \ 1]^T$ activates the antenna index, $\ell_t^{\Im} = 4$, which transmits the sine part of the carrier. Hence, the spatial and signal constellation diagrams for QSSK system are defined as

(1) The spatial diagram is $\mathcal{H} = \{[\mathbf{h}_1, \mathbf{h}_1], [\mathbf{h}_1, \mathbf{h}_2], \dots, [\mathbf{h}_2, \mathbf{h}_1], \dots, [\mathbf{h}_{N_t}, \mathbf{h}_{N_t}]\}$.
(2) The signal diagram is $\mathcal{S} = \{[1, j]^T\}$.

It is important to note that the cardinality of the signal diagram set is one and no data is transmitted in the signal domain. Similar to SSK and GSSK, data are transmitted exclusively in the spatial domain. As such, the transmitted vector for QSSK system in the previous example is given by,

$$\mathbf{x}_t^{\mathrm{RF}}(t) = \begin{bmatrix}0 & 0 & \cos(w_c t) & \sin(w_c t)\end{bmatrix}^T. \tag{3.8}$$

Please note that the cosine and the sine parts of the carrier signal are orthogonal and transmitting them simultaneously causes no ICI similar to SSK and SM algorithms. Also and even though two transmit antennas might be active at a time, no RF chain is needed as in SSK scheme. Hence, all inherent advantages of SSK scheme are retained but with an additional $\log_2(N_t)$ bits that can be transmitted. However, the transmit antennas must be synchronized to start the transmission simultaneously. Again, RF signals are stored in an RF memory and repeatedly used for transmission. Two RF DAC memories are needed for QSSK scheme. One memory storing the in-phase component of the carrier signal while the other one stores the quadrature component of the carrier signal. In addition, in QSSK scheme, there is a possibility that in-phase and quadrature bits will modulate the same transmit antennas as shown in Table 3.4. Hence, an RF combiner is needed before each transmit antenna connecting identical outputs from the RF switches as illustrated in Figure 3.8.

3.6 Quadrature Spatial Modulation (QSM)

QSM can be thought of as an amendment to SM system by utilizing the quadrature spatial dimension similar to QSSK [48, 65, 70, 73]. However, in QSM, the

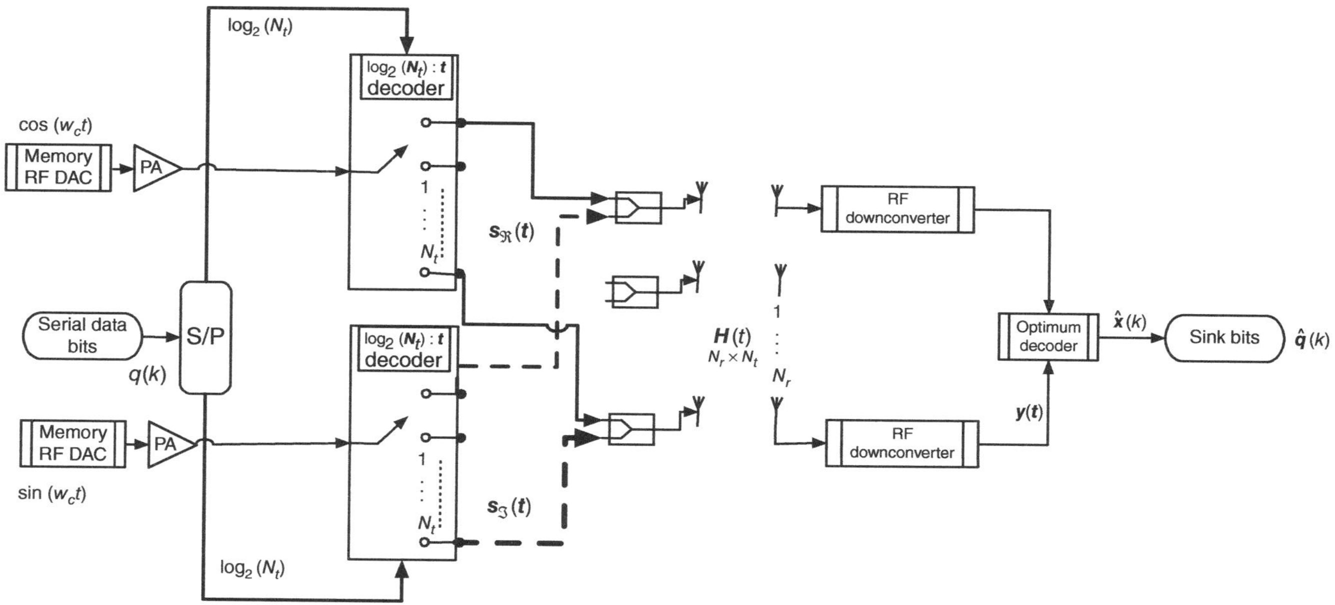

Figure 3.8 QSSK system model.

Table 3.4 QSSK mapping table for $N_t = 4$.

Bits	In-phase bits	In-phase antenna	Quadrature bits	Quadrature antenna
0000	00	1	00	1
0001	00	1	01	2
0010	00	1	10	3
0011	00	1	11	4
0100	01	2	00	1
0101	01	2	01	2
0110	01	2	10	3
0111	01	2	11	4
1000	10	3	00	1
1001	10	3	01	2
1010	10	3	10	3
1011	10	3	11	4
1100	11	4	00	1
1101	11	4	01	2
1110	11	4	10	3
1111	11	4	11	4

transmitted symbol is utilized to convey information bits and can be obtained from an arbitrary complex signal constellation diagram. Thus, the spatial constellation diagram is the same as for QSSK, while the signal constellation diagram is $\mathcal{S} = \{[\text{Re}\{s_1\}, \text{Im}\{s_1\}]^T, [\text{Re}\{s_2\}, \text{Im}\{s_2\}]^T, \dots, [\text{Re}\{s_M\}, \text{Im}\{s_M\}]^T\}$.

A system model for QSM is depicted in Figure 3.9. Similar to SM, QSM can be designed with a single RF-chain even though two antennas might be active at one time instant. The incoming data bits, $\mathbf{q}(k)$ with $\eta = 2\log_2 N_t + \log_2 M$ bits, are to be transmitted in one time slot using QSM system. The incoming bits are grouped into three groups. The first one contains $\log_2(M)$ bits, which is used to choose the signal symbol $\mathcal{S}_\iota \in \mathcal{S}$. The other two $\log_2(N_t)$ bits determine the indexes of the two antennas to activate, $\ell_t^{\Re}$ and $\ell_t^{\Im}$, resulting in spatial symbol $\mathcal{H}_{\ell_t} \in \mathcal{H}$. A constellation illustration for QSM system is shown in Figure 3.10. The first antenna index, $\ell_t^{\Re}$, will transmit the modulated in-phase part of the RF carrier by the real part of complex symbol $\mathcal{S}_\iota$. Whereas the second antenna will be transmitting the quadrature part of the carrier signal modulated by imaginary part of the complex symbol $\mathcal{S}_\iota$. The output from the RF-chain is given by

$$x_{\text{RF}} = \text{Re}\{\mathcal{S}_\iota\}\cos(w_c t) + \text{Im}\{\mathcal{S}_\iota\}\sin(w_c t). \tag{3.9}$$

The modulated cosine part of the carrier by $\text{Re}\{\mathcal{S}_\iota\}$ will be transmitted from antenna $\ell_t^{\Re}$ through the first RF switch, and the sine part of the carrier

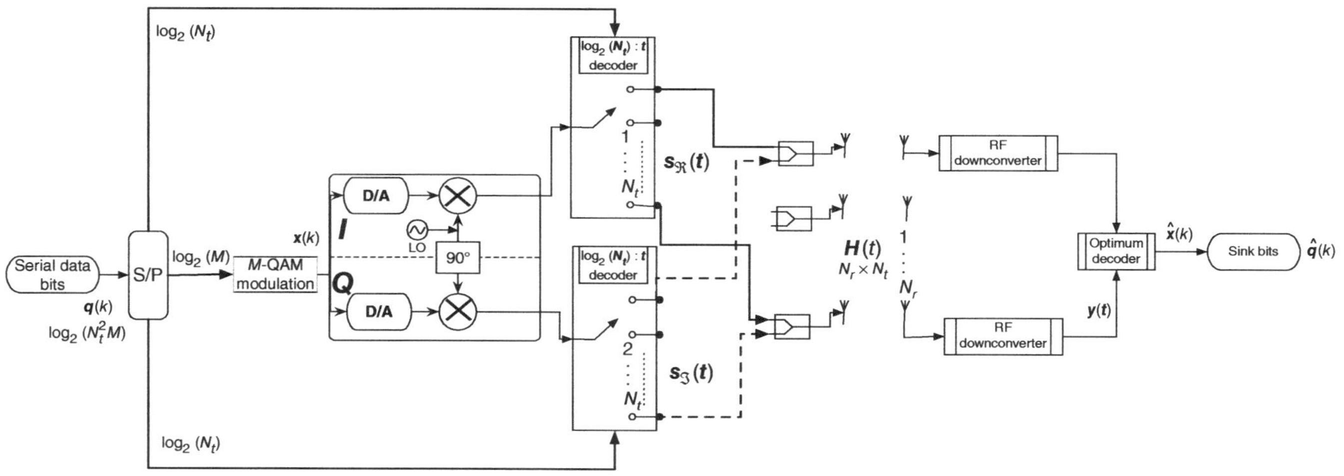

Figure 3.9 QSM system model with single RF-chain, two RF switches, N_t transmit antennas and N_r receive antennas.

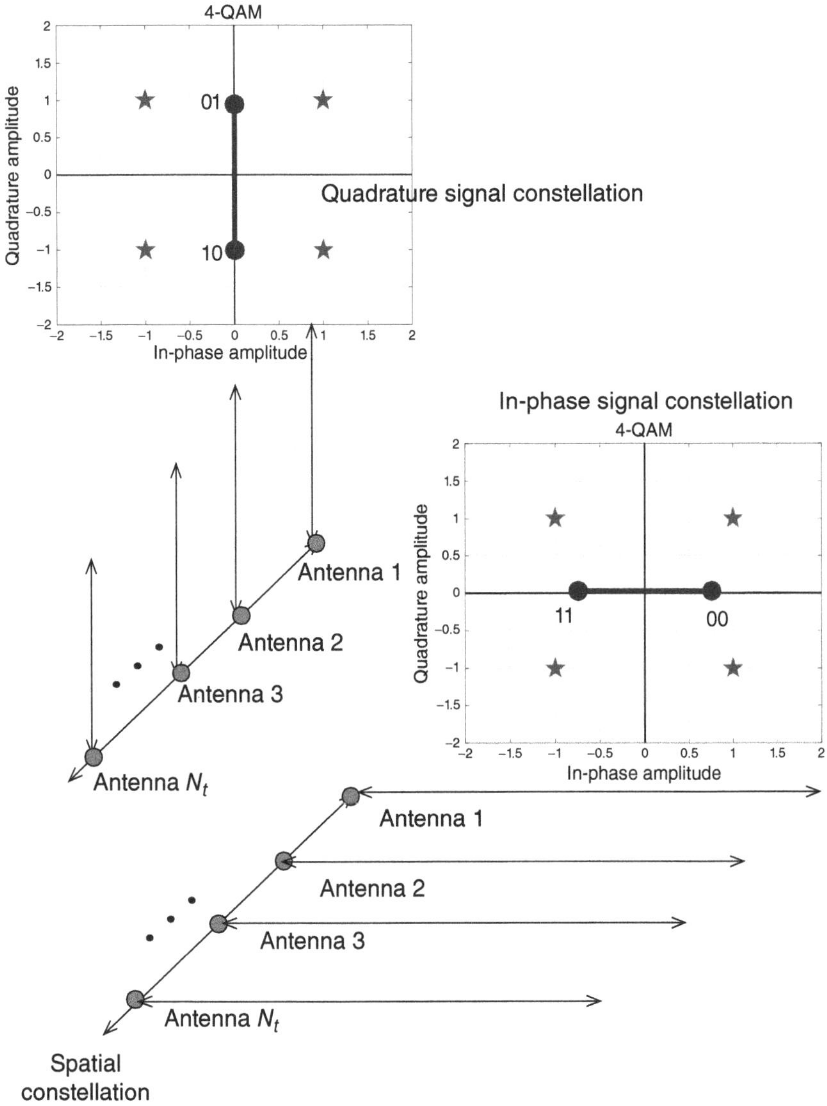

Figure 3.10 An illustration for QSM signal and spatial constellation diagrams.

modulated by $\mathrm{Im}\{S_t\}$ is transmitted from antenna $\ell_t^{\Im}$ through the second RF switch.

Consider an incoming sequence of bits given by $\mathbf{q}(k) = [1\ 0\ 0\ 1\ 1\ 1]$ is to be transmitted from $N_t = 4$ antennas and 4-QAM modulation. The first two data bits [1 0] modulate a 4-QAM symbol, $s_3 = 1 - j$. The second group,

[0 1] modulates the antenna index $\ell_t^{\Re} = 2$ used to transmit $\mathrm{Re}\{s_3\} = 1$, resulting in $\mathbf{x}_t^{\Re}(t) = [0\ \cos(w_c t)\ 0\ 0]$. The last group $[1\ 1]^T$ indicates that the transmit antenna $\ell_t^{\Im} = 4$ will be used to transmit $\mathrm{Im}\{s_3\} = -1$ and resulting in $\mathbf{x}_t^{\Im}(t) = [0\ 0\ 0\ -\sin(w_c t)]^T$. Hence, the spatial and signal symbols are $\mathcal{H}_{10} = [\mathbf{h}_2, \mathbf{h}_4]$ and $\mathcal{S}_3 = [1, -j]^T$, respectively, and the resultant RF vector at the transmit antennas is given by

$$\mathbf{x}_t^{\mathrm{RF}}(t) = \begin{bmatrix} 0 & \cos(w_c t) & 0 & -\sin(w_c t) \end{bmatrix}^T. \tag{3.10}$$

It is important to note that it is possible to have $\ell_t^{\Re} = \ell_t^{\Im}$ if identical spatial bits are to be transmitted as discussed above for QSSK system. Hence, one transmit antenna might be active at one time instant. To facilitate this, RF combiners are needed to connect the identical outputs from the RF switches to the corresponding antenna as shown in Figure 3.9.

3.7 Generalized QSSK (GQSSK)

A system model for GQSSK is depicted in Figure 3.11 with the mapping table for $N_t = 6$ and $n_u = 2$ and the way they can be connected to the RF switches. Following similar concept as in GSSK, a subset of transmit antennas can be activated at a time to transmit the in-phase part of the carrier and another subset to transmit the quadrature part of the carrier in GQSSK. Therefore, the number of data bits that can be transmitted in GQSSK scheme is $\eta = 2\lfloor \log_2 \binom{N_t}{n_u} \rfloor$, and the spatial and signal constellation diagrams are

(1) The spatial diagram is

$$\mathcal{H} = \frac{1}{n_u} \left\{ \left[\sum_{i \in \ell_1^{\Re}} \mathbf{h}_i, \sum_{i \in \ell_1^{\Im}} \mathbf{h}_i \right], \ldots, \left[\sum_{i \in \ell_{\wp}^{\Re}} \mathbf{h}_i, \sum_{i \in \ell_{\wp}^{\Im}} \mathbf{h}_i \right] \right\}, \tag{3.11}$$

where ℓ_l is the lth active antennas combination in the space $\eth$ containing all used antenna combinations.

(2) The signal diagram for GQSSK contains only one symbol $\mathcal{S} = \{[1, j]^T\}$.

To illustrate the working principle of this system, an example is given in what follows. Consider the mapping table in Figure 3.11 and assume $N_t = 6$ and $n_u = 2$. The number of data bits that can be transmitted using GQSSK at one time instant is $\eta = 6$ bits, which is double the number of bits in GSSK system. Assume that $\mathbf{q}(k) = [1\ 0\ 0\ 1\ 1\ 1]$ bits are to be transmitted at one time instant. The sequence of bits is divided into half. The first half [1 0 0] indicates that the second antenna combination, $\ell_t^{\Re_1} = 1$ and $\ell_t^{\Re_2} = 6$, will be transmitting the real part of the carrier. Hence, the real transmitting

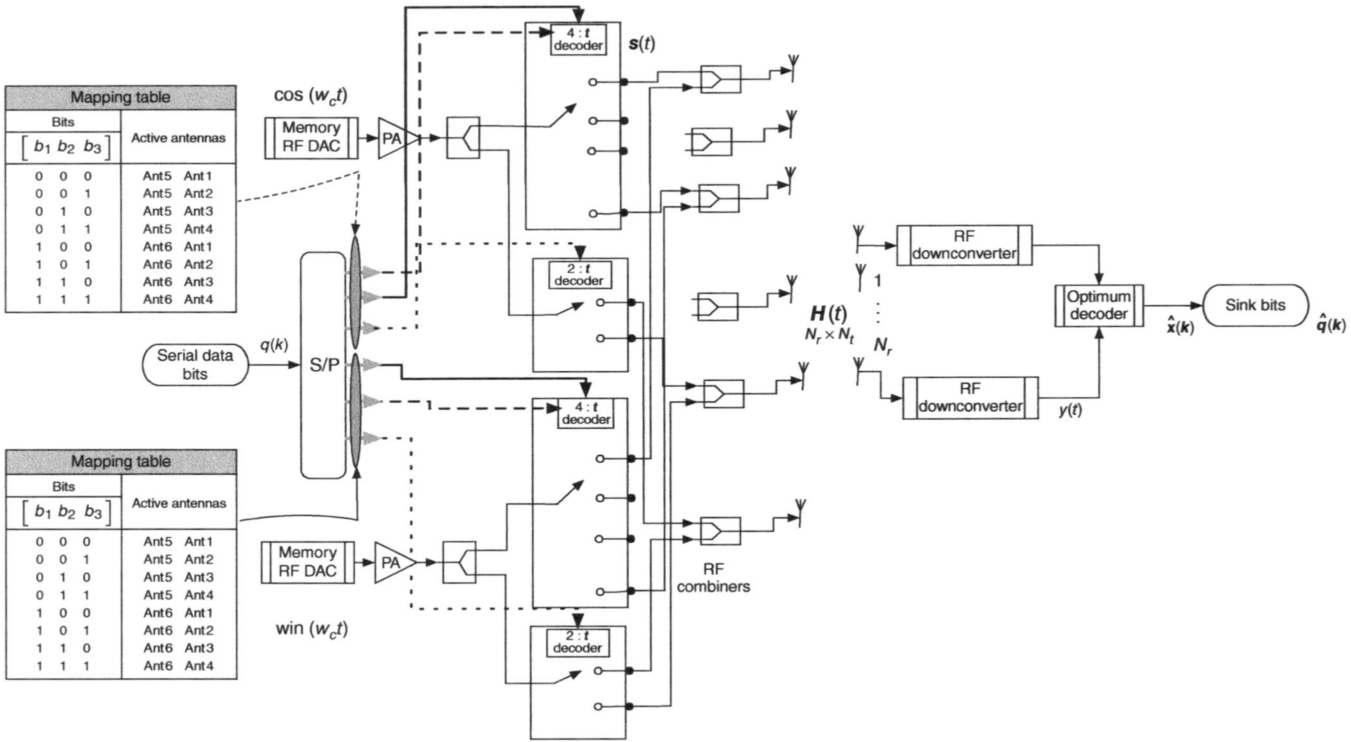

Figure 3.11 GQSSK system model with illustration for $N_t = 6$ an $n_u = 2$ achieving $\eta = 6$ bits.

vector is $\mathbf{x}_t^{\Re}(t) = [\frac{\cos(w_c t)}{\sqrt{2}}\ 0\ 0\ 0\ 0\ \frac{\cos(w_c t)}{\sqrt{2}}]^T$. Similarly, the second half [1 1 1] indicates that the fourth antenna combination, $\ell_t^{\Im_1} = 4$ and $\ell_t^{\Im_2} = 6$, will be transmitting the quadrature part of the carrier resulting in the transmitted vector $\mathbf{x}_t^{\Im}(t) = [0\ 0\ 0\ \frac{\sin(w_c t)}{\sqrt{2}}\ 0\ \frac{\sin(w_c t)}{\sqrt{2}}]^T$. The real and imaginary vectors are added coherently and the resultant RF vector at the input of transmit antennas is

$$\mathbf{x}_t^{\text{RF}}(t) = \left[\frac{\cos(w_c t)}{\sqrt{2}}\quad 0\quad 0\quad \frac{\sin(w_c t)}{\sqrt{2}}\quad 0\quad \frac{\cos(w_c t) + \sin(w_c t)}{\sqrt{2}}\right]^T \tag{3.12}$$

which is transmitted over the MIMO channel. Note, the spatial symbol for this example is $\mathcal{H}_{40} = \frac{1}{\sqrt{2}}[\sum_{i\in\{1,6\}}\mathbf{h}_i, \sum_{i\in\{4,6\}}\mathbf{h}_i]$.

3.8 Generalized QSM (GQSM)

Modulating the RF carrier in GQSSK system by an arbitrary complex symbol drawn from a signal constellation diagram will lead to GQSM as shown in Figure 3.12. In GQSM, a subset of transmit antennas is considered at each time instant to separately transmit the real and the imaginary parts of complex symbol, $\text{Re}\{s_t\}$ and $\text{Im}\{s_t\}$. As in QSM, the real part modulates the in-phase part of the carrier, whereas the imaginary part modulates the quadrature component of the carrier signal. The spectral efficiency of GQSM is then given by $\eta = 2\lfloor\log_2\binom{N_t}{n_u}\rfloor + \log_2(M)$. The discussion of GQSM system is similar to GQSSK except that the signal constellation diagram is the same as in QSM where it conveys $\log_2 M$ bits. Mapping tables for $N_t = 6$, $n_u = 2$, and $M = 8$-QAM is given in Figure 3.12.

3.9 Advanced SMTs

3.9.1 Differential Space Shift Keying (DSSK)

In all previous discussions for optimum receiver of the different presented SMTs so far, the MIMO channel matrix should be perfectly known at the receiver. The perfect knowledge of the channel matrix is idealistic, as discussed in the previous chapter, and channel estimation techniques should be used to obtain an estimate for $\mathbf{H}$. However, a scheme called DSSK is proposed in [63] aimed at alleviating this condition, where the requirement for channel knowledge at the receiver in SSK is totally avoided in DSSK. The idea is that the receiver will rely on the received signal block at time t, $\mathbf{Y}_t$, and the signal block received at time $t-1$, $\mathbf{Y}_{t-1}$, to decode the message.

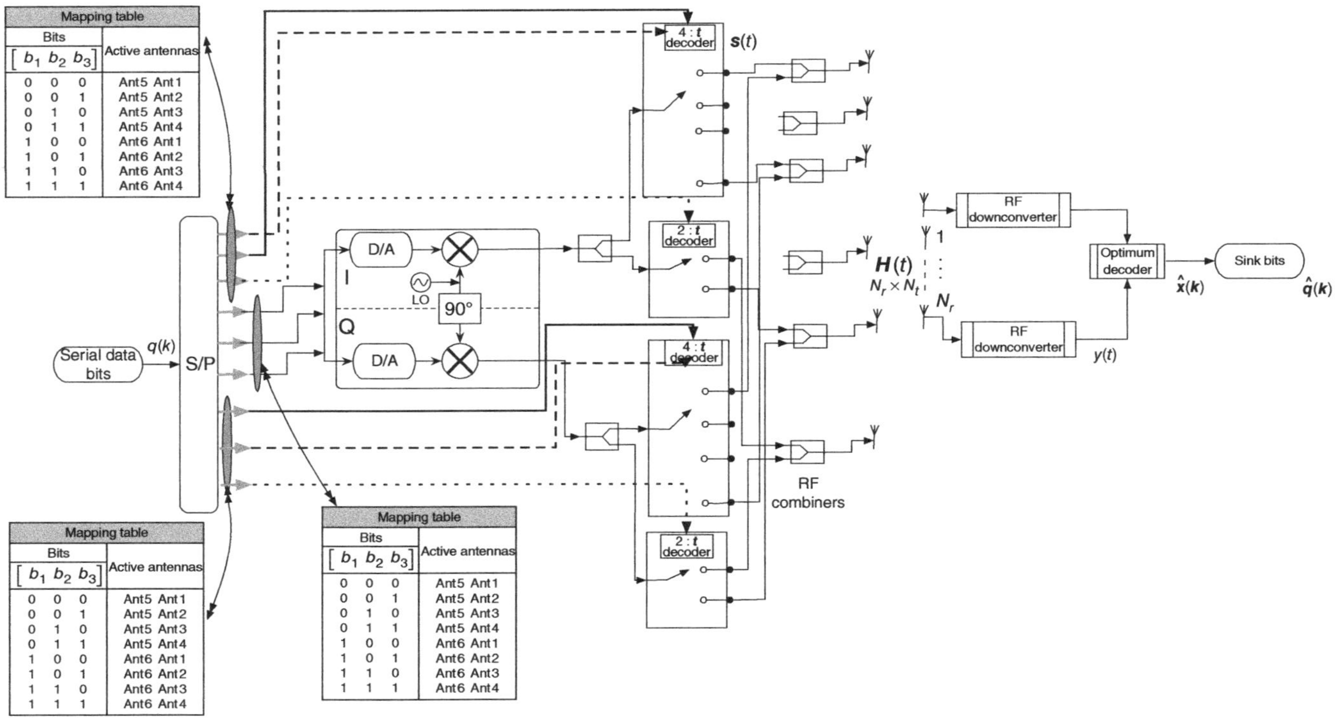

Figure 3.12 GQSM system model with illustration for $N_t = 6$, $n_u = 2$ and $M = 8$-QAM achieving 9 bits $(\mathrm{s\,Hz})^{-1}$.

A system model for DSSK with the a mapping table for $N_t = 2$ is shown in Figure 3.13.[1] Another mapping table for $N_t = 3$ is shown in Table 3.5. The mapping table is designed with the following conditions:

1. All columns should have only one nonzero element. This means that only one transmit antenna will be active at one particular time instant similar to SSK scheme. As such, ICI is totally avoided, and single RF-chain can be used with proper RF switching design.
2. All rows should not have the same symbol more than once. This also indicates that during the transmission time block, each symbol is transmitted only once.

It is shown in [63] that the spectral efficiency of DSSK is smaller than that of SSK scheme for the same N_t value and equal only for the case of $N_t = 2$, where DSSK transmits $\lfloor \log_2(N_t!) \rfloor$ bits each N_t time slots. As such, the spectral efficiency of DSSK is $\eta = \frac{1}{N_t} \lfloor \log_2(N_t!) \rfloor$ bits. However, in DSSK, the number of transmit antennas that can be used for communication is flexible and the power of two requirement as in SSK scheme is alleviated. For instance, with $N_t = 3$, 2 bits can be transmitted on 3 time slots using DSSK, whereas 3 bits can be transmitted on the three time slots when using SSK scheme with only two transmit antennas. The spectral efficiency of DSSK decays further for larger number of transmit antennas.

As in SSK, DSSK also does not have signal constellation symbols. However, its spatial constellation diagram contains a $2^{\lfloor \log_2(N_t!) \rfloor}$ spatial symbols, where each symbol is an $N_t \times N_t$ square matrix containing a different permutation of the channel matrix. For example, for $N_t = 3$, the spatial constellation diagram is

$$\mathcal{H} = \left\{ \overbrace{\begin{matrix} [\mathbf{h}_1 & \mathbf{h}_2 & \mathbf{h}_3] \\ t=0 & t=1 & t=2 \end{matrix}}^{\mathcal{H}_1}, \overbrace{[\mathbf{h}_2, \mathbf{h}_1, \mathbf{h}_3]}^{\mathcal{H}_2}, \overbrace{[\mathbf{h}_2, \mathbf{h}_3, \mathbf{h}_1]}^{\mathcal{H}_3}, \overbrace{[\mathbf{h}_3, \mathbf{h}_2, \mathbf{h}_1]}^{\mathcal{H}_4} \right\}. \tag{3.13}$$

In DSSK, the transmission begins with known symbol (bits), which maps to specific transmission matrix $\mathcal{H}_{\ell_t}(t)$. The next transmitted symbol is generated by multiplying the chosen spatial symbols $\mathcal{H}_{\ell_t}(t)$ with a delayed version of the transmitted signal $\mathbf{S}_t$, such that the next transmitted signal is

$$\mathbf{S}_t = \mathbf{S}_{t-1}\mathcal{H}_{\ell_t}(t). \tag{3.14}$$

Note, for simplicity, $\mathbf{S}_0 = \mathbf{I}_{N_t}$ is assumed.

1 The RF-chain at the transmitter and the RF down conversion at the receiver are not shown in Figure 3.13 for the sake of simplifying the drawing and focusing on the working mechanism of DSSK.

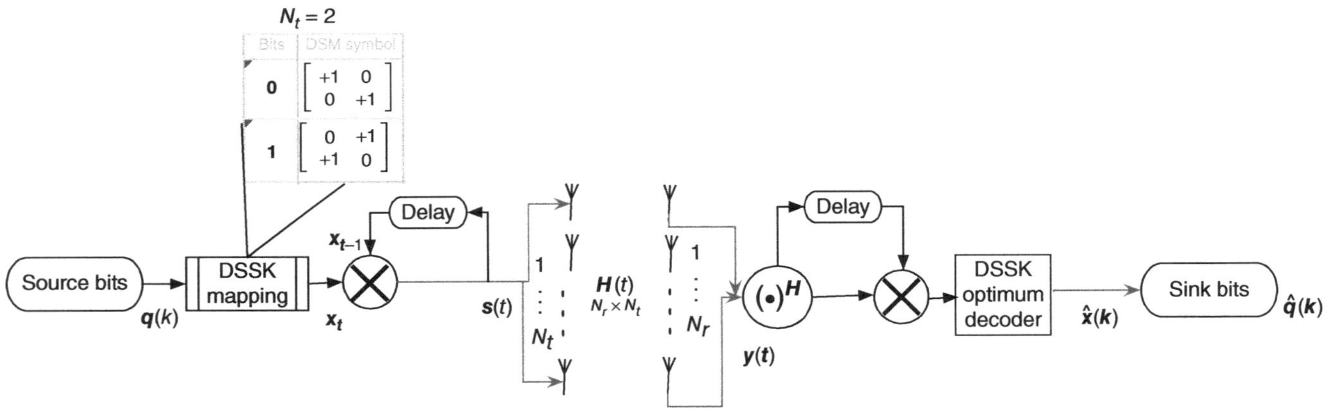

Figure 3.13 Differential space shift keying system model with the mapping table for $N_t = 2$.

Table 3.5 DSSK mapping table for $N_t = 3$ achieving a spectral efficiency of $\eta = 2/3$ bits Hz^{-1}.

Bits	DSSK symbol
00	$\begin{bmatrix} 1 & 0 & 0 \\ 0 & 1 & 0 \\ 0 & 0 & 1 \end{bmatrix}$
01	$\begin{bmatrix} 0 & 1 & 0 \\ 1 & 0 & 0 \\ 0 & 0 & 1 \end{bmatrix}$
11	$\begin{bmatrix} 0 & 0 & 1 \\ 1 & 0 & 0 \\ 0 & 1 & 0 \end{bmatrix}$
10	$\begin{bmatrix} 0 & 0 & 1 \\ 0 & 1 & 0 \\ 1 & 0 & 0 \end{bmatrix}$
Not used	$\begin{bmatrix} 1 & 0 & 0 \\ 0 & 0 & 1 \\ 0 & 1 & 0 \end{bmatrix}$
Not used	$\begin{bmatrix} 0 & 1 & 0 \\ 0 & 0 & 1 \\ 1 & 0 & 0 \end{bmatrix}$

The main idea behind DSSK is that the channel state information (CSI) at the receiver is not needed. The symbol $\mathbf{X}_{\ell_t}$ is an $N_t \times N_t$ square matrix indicating which antenna is active at time instance (t) as shown in Table 3.5, where $\mathcal{H}_{\ell_t} = \mathbf{H}\mathbf{X}_{\ell_t}$. As in (3.14), the transmitted symbol is

$$\dot{\mathbf{X}}_t = \dot{\mathbf{X}}_{t-1} \times \mathbf{X}_{\ell_t}. \tag{3.15}$$

Assuming that the channel is quasi-static such that $\mathbf{H}(t) = \mathbf{H}(t-1)$, which is generally assumed for space–time systems, the received signals at t and $t-1$ time slots are

$$\mathbf{Y}_{t-1} = \mathbf{H}\dot{\mathbf{X}}_{t-1} + \mathbf{N}_{t-1} \tag{3.16}$$

and

$$\mathbf{Y}_t = \mathbf{H}\dot{\mathbf{X}}_t + \mathbf{N}_t$$

$$= \mathbf{H} \times \dot{\mathbf{X}}_{t-1} \times \mathbf{X}_t + \mathbf{N}_t$$

$$= \mathbf{H} \times \overbrace{(\mathbf{H}^{-1}(\mathbf{Y}_{t-1} - \mathbf{N}_{t-1}))}^{\dot{\mathbf{X}}_{t-1}} \times \mathbf{X}_t + \mathbf{N}_t$$

$$= \mathbf{Y}_{t-1}\mathbf{X}_t - \mathbf{N}_{t-1}\mathbf{X}_t + \mathbf{N}_t. \tag{3.17}$$

Hence, the ML decoder for DSSK scheme is given by

$$\hat{\mathbf{X}}_t = \arg \min_{\mathbf{X}_i \in \mathbf{X}^{\mathrm{D}}} \|\mathbf{Y}_t - \mathbf{Y}_{t-1}\mathbf{X}_i\|_{\mathrm{F}}^2, \tag{3.18}$$

where $\mathbf{X}^{\mathrm{D}}$ is a space containing all possible symbols of $\mathbf{X}_i$. Note that the ML for DSSK in (3.18) does not require any knowledge of the MIMO channel.

3.9.2 Differential Spatial Modulation (DSM)

DSM is very similar to DSSK with the difference that transmitted symbols are now modulated [63]. Hence, the achievable spectral efficiency for DSM is given by $\eta = \frac{1}{N_t}\lfloor \log_2(N_t!) \rfloor + \log_2(M)$. DSM has the same spatial constellation diagram as DSSK and has a signal constellation diagram containing all possible signal symbols permutations. For instance, assuming BPSK constellation diagram and $N_t = 2$, the signal constellation diagram is

$$\mathcal{S} = \left\{ \begin{bmatrix} +1 \\ +1 \end{bmatrix}, \begin{bmatrix} +1 \\ -1 \end{bmatrix}, \begin{bmatrix} -1 \\ +1 \end{bmatrix}, \begin{bmatrix} -1 \\ -1 \end{bmatrix} \right\}. \tag{3.19}$$

Note, in general, it is required that the signal constellation must have equal unit energy, such as M-PSK constellations. This is required to maintain the closure property where the multiplication of any two transmitted vectors results in another vector from the existing set. Similar receiver as discussed for DSSK can be considered here as well, where $\mathbf{X}_i$ is for example as showing in the mapping table for $N_t = 2$ and $M = 2$-PSK is given in Table 3.6.

3.9.3 Differential Quadrature Spatial Modulation (DQSM)

The extension of DSM to the QSM technique is proposed recently in [97]. A system model for DQSM is depicted in Figure 3.14. The incoming data bits are partitioned into three groups. The first group containing $N_t \log_2 M$ bits modulates an N_t M-QAM symbols, $\mathbf{s}_t = [s_t^1 \ s_t^2 \ \cdots \ s_t^{N_t}]^T$, to be transmitted in N_t time instants. The other two groups each with $\lfloor \log_2(N_t!) \rfloor$ bits modulate two sets of active antennas, which will transmit the real and imaginary parts of the signal symbol, respectively.

The first spatial symbol, $\ell_t^{\Re}$, represents the permutation matrix, $\mathbf{P}^{\ell_t^{\Re}}$, used to generate the transmitter block that transmits the real part of the constellation symbol, $\mathbf{s}_t^{\Re} = \mathrm{Re}\{\mathbf{s}_t\}$. The other spatial symbol, $\ell_t^{\Im}$, denotes

Table 3.6 DSM mapping table for $N-t=2$ and $M=2$-PSK modulation achieving a spectral efficiency of $m=1.5$ bits Hz^{-1}.

Bits	DSM symbol
000	$\begin{bmatrix} +1 & 0 \\ 0 & +1 \end{bmatrix}$
001	$\begin{bmatrix} -1 & 0 \\ 0 & +1 \end{bmatrix}$
010	$\begin{bmatrix} +1 & 0 \\ 0 & -1 \end{bmatrix}$
011	$\begin{bmatrix} -1 & 0 \\ 0 & -1 \end{bmatrix}$
100	$\begin{bmatrix} 0 & +1 \\ +1 & 0 \end{bmatrix}$
101	$\begin{bmatrix} 0 & -1 \\ +1 & 0 \end{bmatrix}$
110	$\begin{bmatrix} 0 & +1 \\ -1 & 0 \end{bmatrix}$
111	$\begin{bmatrix} 0 & -1 \\ -1 & 0 \end{bmatrix}$

the permutation matrix, $\mathbf{P}^{\ell_t^{\Im}}$, that will generate the transmitter block which transmits the imaginary part of the constellation symbol, $\mathbf{s}_t^{\Im} = \mathrm{Im}\{\mathbf{s}_t\}$. The real symbol vector, $\mathbf{s}_t^{\Re}$, and the imaginary symbol vector, $\mathbf{s}_t^{\Im}$, are, respectively, multiplied with $\mathbf{P}^{\ell_t^{\Re}}$ and $\mathbf{P}^{\ell_t^{\Im}}$, where each element of the symbol vectors is multiplied by the corresponding column vector of the permutation matrix to generate $\mathbf{X}_t^{\Re}$ and $\mathbf{X}_t^{\Im}$. It should be mentioned, though, that there are $2^{N_t!}$ possible permutation matrices and only $2^{\lfloor \log_2(N_t!) \rfloor}$ are considered. It is also assumed that $\mathrm{E}\{|s_i^{\Re}|^2\} = \mathrm{E}\{|s_i^{\Im}|^2\} = 1, \ \forall i \in \{1 : N_t\}$. Therefore, $\forall\ \mathbf{X}_t^{\Re}, \mathbf{X}_{t-1}^{\Re} \in \mathcal{P}$, $\mathbf{X}_t^{\Re}\mathbf{X}_{t-1}^{\Re} \in \mathcal{P}$, where $\mathcal{P}$ denoting a set of all possible permutation matrices. Similarly, $\forall\ \mathbf{X}_t^{\Im}, \mathbf{X}_{t-1}^{\Im} \in \mathcal{P}, \mathbf{X}_t^{\Im}\mathbf{X}_{t-1}^{\Im} \in \mathcal{P}$.

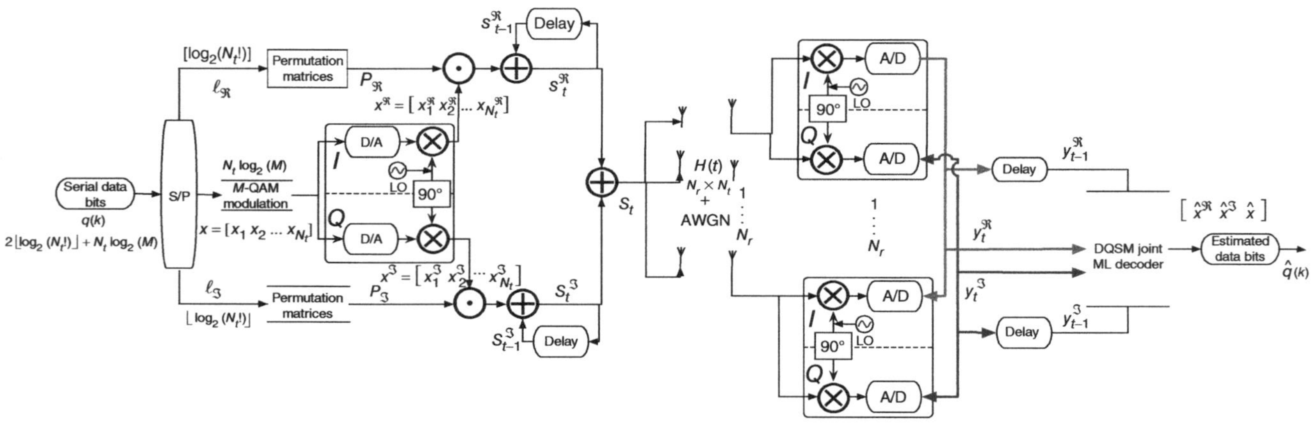

Figure 3.14 DQSM system model with arbitrary number of transmit, N_t, and receive, N_r, antennas and specific modulation order, M, utilizing single RF-chain transmitter.

To maintain the inherent advantages of QSM and similar to DSSK and DSM systems, transmitter blocks with $N_t \times N_t$ dimension are designed in such away that

(1) At each time instant and to maintain single RF-chain transmitter, a single transmit antenna is active to transmit the real part of the complex symbol and another or the same antenna to transmit the imaginary part of the data symbol. As such, each column of the transmitter block contains only one nonzero real value and one nonzero imaginary value.
(2) A transmit antenna is activated only once during the duration of one transmitter block.
(3) The closure property of the transmitted blocks is maintained, as discussed above, to facilitate differential modulation and demodulation.

To better explain the transmitter procedure of DQSM, an example is given in what follows considering $N_t = 3$ and $M = 4$-QAM modulation. The possible permutation matrices with $N_t = 3$ antennas along with the mapping bits are illustrated in Table 3.7. Let the incoming data bits to be transmitted at N_t time slots using DQSM be $\mathbf{q}(k) = [0\ 0\ 1\ 1\ 1\ 0\ 1\ 0\ 0\ 0]^T$. The first group with $N_t \log_2(M)$ data bits, $\mathbf{q}_1(k) = [0\ 0\ 1\ 1\ 1\ 0]^T$, modulates N_t 4-QAM symbols as,

$$\mathbf{x}_t = \begin{bmatrix} \underbrace{1+j}_{00} & \underbrace{-1-j}_{11} & \underbrace{1-j}_{10} \end{bmatrix}^T. \tag{3.20}$$

Table 3.7 DQSM bits mapping and permutation matrices for $N_t = 3$ transmit antennas.

Bits	Permutations	$\mathbf{S}^{\Re}$	$\mathbf{S}^{\Im}$
00	$\begin{bmatrix}1 & 2 & 3\end{bmatrix}$	$\begin{bmatrix} x_1^{\Re} & 0 & 0 \\ 0 & x_2^{\Re} & 0 \\ 0 & 0 & x_3^{\Re} \end{bmatrix}$	$\begin{bmatrix} x_1^{\Im} & 0 & 0 \\ 0 & x_2^{\Im} & 0 \\ 0 & 0 & x_3^{\Im} \end{bmatrix}$
01	$\begin{bmatrix}1 & 3 & 2\end{bmatrix}$	$\begin{bmatrix} x_1^{\Re} & 0 & 0 \\ 0 & 0 & x_3^{\Re} \\ 0 & x_2^{\Re} & 0 \end{bmatrix}$	$\begin{bmatrix} x_1^{\Im} & 0 & 0 \\ 0 & 0 & x_3^{\Im} \\ 0 & x_2^{\Im} & 0 \end{bmatrix}$
10	$\begin{bmatrix}2 & 1 & 3\end{bmatrix}$	$\begin{bmatrix} 0 & x_2^{\Re} & 0 \\ x_1^{\Re} & 0 & 0 \\ 0 & 0 & x_3^{\Re} \end{bmatrix}$	$\begin{bmatrix} 0 & x_2^{\Im} & 0 \\ x_1^{\Im} & 0 & 0 \\ 0 & 0 & x_3^{\Im} \end{bmatrix}$
11	$\begin{bmatrix}2 & 3 & 1\end{bmatrix}$	$\begin{bmatrix} 0 & 0 & x_3^{\Re} \\ x_1^{\Re} & 0 & 0 \\ 0 & x_2^{\Re} & 0 \end{bmatrix}$	$\begin{bmatrix} 0 & 0 & x_3^{\Im} \\ x_1^{\Im} & 0 & 0 \\ 0 & x_2^{\Im} & 0 \end{bmatrix}$

The other two groups each with $\lfloor \log_2(3!) \rfloor = 2$ bits, [10] and [00], modulate two spatial indexes, $\ell_t^{\Re} = 3$ and $\ell_t^{\Im} = 1$, respectively. The first index, $\ell_t^{\Re}$, denotes the permutation matrix that will be used to transmit the real parts of the transmitted symbols,

$$\mathbf{P}^{\ell_t^{\Re}} = \begin{bmatrix} 0 & 1 & 0 \\ 1 & 0 & 0 \\ 0 & 0 & 1 \end{bmatrix}. \tag{3.21}$$

The second index, $\ell_t^{\Im}$, denotes the permutation matrix that will be used to transmit the imaginary parts of the transmitted symbols as

$$\mathbf{P}^{\ell_t^{\Im}} = \begin{bmatrix} 1 & 0 & 0 \\ 0 & 1 & 0 \\ 0 & 0 & 1 \end{bmatrix}. \tag{3.22}$$

The real symbols vector, $\mathbf{s}_t^{\Re}$, and the imaginary symbols ,vector $\mathbf{s}_t^{\Im}$, are, respectively, multiplied with $\mathbf{P}^{\ell_t^{\Re}}$ and $\mathbf{P}^{\ell_t^{\Im}}$, where each element in the symbol vectors is multiplied by the corresponding column vector of the permutation matrix to generate

$$\mathbf{X}_t^{\Re} = \begin{bmatrix} 0 & -1 & 0 \\ +1 & 0 & 0 \\ 0 & 0 & +1 \end{bmatrix} \tag{3.23}$$

and

$$\mathbf{X}_t^{\Im} = \begin{bmatrix} +j & 0 & 0 \\ 0 & -j & 0 \\ 0 & 0 & -j \end{bmatrix}. \tag{3.24}$$

To facilitate differential demodulation, each transmitted block, $\mathbf{X}_t^{\Re}$, is multiplied by the previously transmitted block, $\mathbf{X}_{t-1}^{\Re}$, and the generated real and imaginary blocks are coherently added and transmitted over the MIMO channel matrix and suffer from an AWGN at the receiver inputs as shown in Figure 3.14.

The received signal for the tth block is then given by

$$\mathbf{Y}_t = \mathbf{H}_t\mathbf{X}_t + \mathbf{N}_t. \tag{3.25}$$

At the receiver, the received signals at each receive antenna are first demodulated through an IQ demodulator. Then, the obtained real, $\mathbf{Y}_t^{\Re} = \text{Re}\{\mathbf{Y}_t\}$, and imaginary, $\mathbf{Y}_t^{\Im} = \text{Im}\{\mathbf{Y}_t\}$, signals are differentially demodulated to retrieve the transmitted bits as will be discussed in what follows.

The received real signal at time $(t-1)$ is

$$\mathbf{Y}_{t-1}^{\Re} = \mathbf{H}_{t-1}\mathbf{X}_{t-1}^{\Re} + \mathbf{N}_{t-1}. \tag{3.26}$$

Assuming quasi-static channel where $\mathbf{H}_{t-1} = \mathbf{H}_t$, the received real part of the signal can be written as

$$\mathbf{Y}_t^{\Re} = \mathbf{Y}_{t-1}^{\Re}\mathbf{X}_t^{\Re} - \mathbf{Z}_{t-1}\mathbf{S}_t^{\Re} + \mathbf{N}_t. \tag{3.27}$$

Similarly, the received imaginary part can be obtained.

The optimum joint ML differential detector is given by

$$\left[\hat{\mathbf{X}}_t^{\Re}, \hat{\mathbf{X}}_t^{\Im}\right] = \underset{\substack{\mathbf{X}_t^{\Re} \in \mathbf{X}^{\Re} \\ \mathbf{X}_t^{\Im} \in \mathbf{X}^{\Im}}}{\arg\min} \left\| \mathbf{Y}_t^{\Re} - \mathbf{Y}_{t-1}^{\Re}\mathbf{X}_t^{\Re} + \mathbf{Y}_t^{\Im} - \mathbf{Y}_{t-1}^{\Im}\mathbf{X}_t^{\Im} \right\|_F^2, \tag{3.28}$$

where $\hat{\mathbf{X}}_t^{\Re}$ and $\hat{\mathbf{X}}_t^{\Im}$, respectively, denote the detected real and imaginary matrices, and $\mathbf{X}^{\Re}$ and $\mathbf{X}^{\Im}$ denote a set with $2^{\lfloor \log_2(N_t!) \rfloor + N_t \log_2(M)}$ dimension containing, respectively, all possible real and imaginary transmission matrices. The estimated matrices are used to retrieve the original information bits through an inverse mapping procedure considering the same mapping rules applied at the transmitter.

3.9.4 Space–Time Shift Keying (STSK)

STSK [79, 90, 92, 160] is another generalization MIMO transmission scheme that is based on the concept of SMTs. In STSK, incoming data bits activate a dispersion matrix to be transmitted from multiple transmit antennas. Different designs for the dispersion matrices are reported in literature. It is shown in [79] that different MIMO schemes can be obtained as special cases from STSK with proper design of the dispersion matrices. The spectral efficiency of STSK systems is

$$\eta = \frac{\log_2 \left\lfloor \binom{D}{\alpha} \right\rfloor + \alpha \log_2(M)}{T}, \tag{3.29}$$

where D is the number of total dispersion matrices, α is the number of used dispersion matrices for each transmitted block, and T is the time slots needed to transmit one dispersion matrix. The dispersion matrices D_i, $i \in \{1 : D\}$ can be designed to achieve any of the previously discussed modulation schemes. For instance and assuming $D = 4$, $N_t = 4$, $T = \alpha = 1$, the following dispersion matrices can be designed: $D_1 = [1\ 0\ 0\ 0]^T, D_2 = [0\ 1\ 0\ 0]^T, D_3 = [0\ 0\ 1\ 0]^T$, and $D_4 = [0\ 0\ 0\ 1]^T$. Now if SSK is targeted, $M = 1$, and a fixed transmitted symbol, $s = 1$, is transmitted at each time instant. The spectral efficiency is then $\eta = 2$ bits. Assume that the incoming data bits at one time instant are $\left[1\ 0\right]$, which modulate D_3 to be transmitted at this particular time. If SM is to be configured, M-QAM /PSK symbols are then modulated by another $\log_2(M)$ bits, and the modulated complex symbol is multiplied by the corresponding modulated dispersion matrix. Similarly, QSM and QSSK can be configured where the real part of the complex symbol is multiplied by a dispersion matrix and the imaginary part is multiplied by another dispersion matrix, i.e. $D = D_i + jD_u$, $\forall\{i, u\} \in 1 : \mathcal{D}$.

Also, other MIMO schemes can be designed. Consider, for instance, the following dispersion matrices:

$$D_1 = \frac{1}{\sqrt{2}} \begin{bmatrix} 1 & 0 \\ 0 & 1 \end{bmatrix}, \tag{3.30}$$

$$D_2 = \frac{1}{\sqrt{2}} \begin{bmatrix} 1 & 0 \\ 0 & -1 \end{bmatrix}, \tag{3.31}$$

$$D_3 = \frac{1}{\sqrt{2}} \begin{bmatrix} 0 & 1 \\ -1 & 0 \end{bmatrix}, \tag{3.32}$$

$$D_4 = \frac{1}{\sqrt{2}} \begin{bmatrix} 0 & 1 \\ 1 & 0 \end{bmatrix}, \tag{3.33}$$

where $N_t = 2$, $\alpha = 2$, and $T = 2$. Using these dispersion matrices, orthogonal space–time coding techniques, such as Alamouti code, combined with SMTs can be configured [121, 161–168].

A mapping table for $\eta = 4$ bits spectral efficiency, $N_t = 4$, $\mathcal{D} = 4$, where $\alpha = 2$ dispersion matrices are selected at each time $T = 2$ and $M = 2$ (BPSK) modulation is shown in Table 3.8. Please note that the number of possible combination in this configuration is six and only four combinations are considered as in generalized space modulation techniques (GSMTs). Two incoming data bits determine the active combination of transmitted matrices and two other bits determine the BPSK symbols to be transmitted over the two time slots. Each symbol is multiplied by the corresponding dispersion matrix and the resultant matrices are added coherently and then transmitted. The receiver task is to determine the set of active matrices and an estimate of the possible transmitted symbols.

3.9.5 Trellis Coded-Spatial Modulation (TCSM)

The last scheme that will be discussed in this chapter is different than all previous schemes since it includes channel coding techniques [169]. TCSM attracted significant interest in literature and many variant schemes have been developed [154, 162, 170–172]. The idea of TCSM is to apply trellis coded modulation (TCM) to the spatial domain [173, 174]. TCM is an efficient modulation technique that conserves bandwidth through convolutional coding by doubling the number of constellation points of a signal. In TCM, the η incoming bits are mapped to $\eta + 1$ bits using a convolutional encoder as illustrated in Figure 3.15. The basic idea is to use set partitioning to allow certain transitions among consecutive bits. An example of set partitioning for 8-PSK constellation diagram

Table 3.8 STSK mapping table for $N_t = 4$, $|\mathcal{D}| = 4$ and BPSK modulation.

Bits	Dispersion matrices	BPSK symbols	STSK codeword
0000	D_1,D_2	$+1, +1$	D_1+D_2
0001	D_1,D_2	$+1, -1$	D_1-D_2
0010	D_1,D_2	$-1, +1$	$-D_1+D_2$
0011	D_1,D_2	$-1, -1$	$-D_1-D_2$
0100	D_1,D_3	$+1, +1$	D_1+D_3
0101	D_1,D_3	$+1, -1$	D_1-D_3
0110	D_1,D_3	$-1, +1$	$-D_1+D_3$
0111	D_1,D_3	$-1, -1$	$-D_1-D_3$
1000	D_1,D_4	$+1, +1$	D_1+D_4
1001	D_1,D_4	$+1, -1$	D_1-D_4
1010	D_1,D_4	$-1, +1$	$-D_1+D_4$
1011	D_1,D_4	$-1, -1$	$-D_1-D_4$
1100	D_2,D_3	$+1, +1$	D_2+D_3
1101	D_2,D_3	$+1, -1$	D_2-D_3
1110	D_2,D_3	$-1, +1$	$-D_2+D_3$
1111	D_2,D_3	$-1, -1$	$-D_2-D_3$

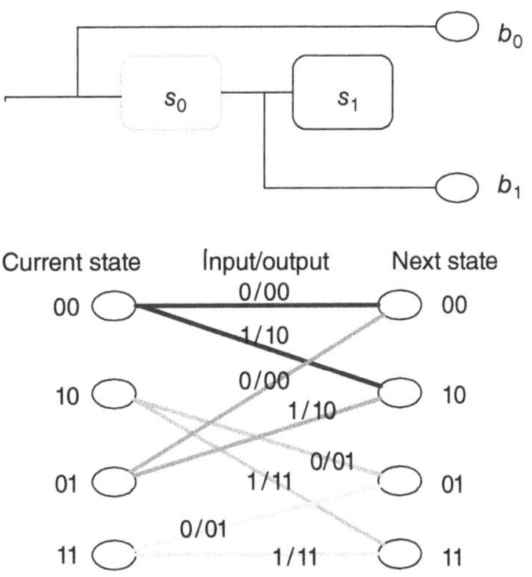

Figure 3.15 An example of rate 1/2 TCM encoder with the state diagram and convolutional encoder.

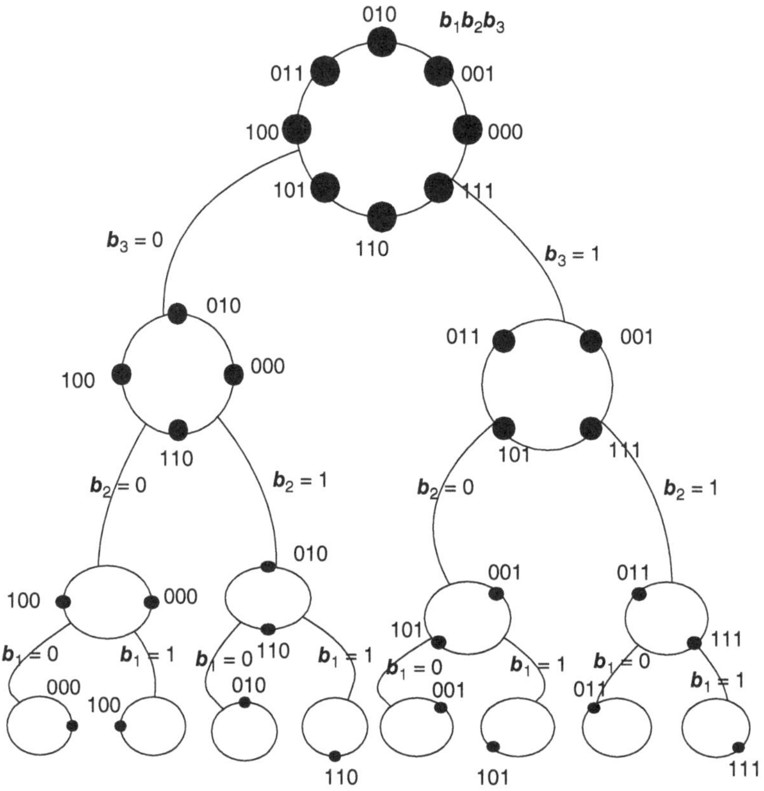

Figure 3.16 An example of TCM set partitioning for 8-PSK constellation diagram.

is depicted in Figure 3.16. Designing the sets such that they have maximum possible Euclidean distances among all possible symbol transitions in the set is shown to significantly enhance the performance [173].

In TCSM, the similar concept is applied to the spatial constellation symbols. The possible transition states for $N_t = 4$ antennas constellation along with the considered convolutional encoder are shown in Figure 3.17. As shown in figure, the antennas are partitioned in two sets, where Ant 1 and Ant 3 form a set and Ant 2 and Ant 4 form the other set. There is no possible transition between Ant 1 and Ant 2. Similarly, there is no transition between Ant 3 and Ant 4. Assuming that the antennas are horizontally aligned, the spacing between Ant 1 and Ant 3 is much larger than the spacing between Ant 1 and Ant 2. Therefore, the probability of correlation among each set elements is lower, which enhances the performance.

The idea of applying TCM in the spatial domain can be applied to all previously discussed SMTs. Spatial constellation symbols can be grouped in

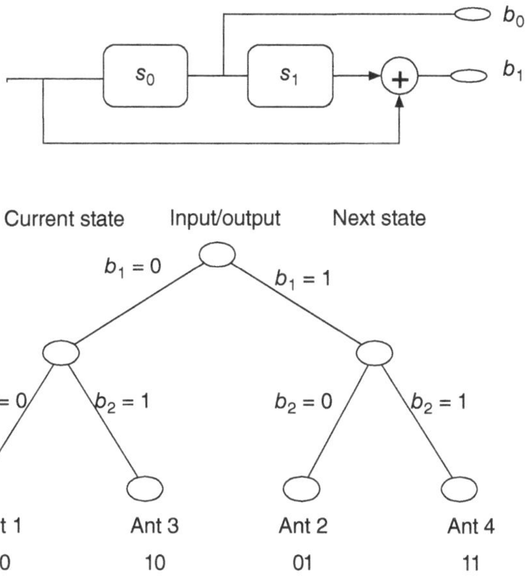

Figure 3.17 TCSM possible transition states for $N_t = 4$ antennas along with the considered convolutional encoder.

sets and special conditions among set elements can be guaranteed through the convolutional encoder.

3.10 Complexity Analysis of SMTs

3.10.1 Computational Complexity of the ML Decoder

One of the main advantages of SMTs is that they allow for simple receiver architecture with reduced complexity as compared to SMX and other MIMO systems. The receiver complexity is computed as the number of real multiplication and division operations needed by each algorithm [175]. Considering the SMT-ML receiver in (3.2), the computational complexity is calculated as

$$C_{\text{SMT}} = \begin{cases} 4N_r \times 2^\eta & \text{SMTs with no signal symbols,} \\ 8N_r \times 2^\eta & \text{the rest of SMTs.} \end{cases} \tag{3.34}$$

Note that each complex multiplication is a four real multiplications $(a + jb) \times (c + jd) = (a \times c - b \times d) + j(b \times c + a \times d)$. For SM, QSM, and other similar systems, (3.2) can be written as $\sum_{j=1}^{N_r} |y(j) - H(j, \ell)x_\ell|^2$, where the multiplication, $H(j, \ell)x_\ell$, requires four real multiplications and evaluating the square needs another four operations. These operations are done N_r times and over the cardinality of the set $\mathbf{X}$, which is 2^η. Therefore, QSM and SM requires $8N_r2^\eta$ operations. For SSK and similar systems, the first multiplication does not exist and only the square operation need to be evaluated. As such, the number of needed operations is $4N_r2^\eta$. Similarly, for SMX system, the multiplication $\mathbf{H}(j,:)\mathbf{x}_i$ requires $4N_tN_r$ operations and the square operations needs $4N_r$ operations done over 2^η possible symbols. Thereby, the number of required operations is $4N_r(N_t+1)2^\eta$.

3.10.2 Low-Complexity Sphere Decoder Receiver for SMTs

In this section, two sphere decoders (SDs) tailored for SMTs are considered. The first scheme called SMT-Rx and the second called SMT-Tx.

First, for ease of derivation, the real-valued equivalent of the complex-valued model in (3.2) is described as [176],

$$[\hat{\mathcal{H}}, \hat{\mathcal{S}}] = \underset{\substack{\mathcal{H}_\ell \in \mathcal{H} \\ \mathcal{S}_\iota \in \mathcal{S}}}{\arg\min} \left\{ \|\bar{\mathbf{y}} - \bar{\mathcal{H}}_\ell \bar{\mathcal{S}}_\iota\|_{\mathrm{F}}^2 \right\}, \tag{3.35}$$

where

$$\bar{\mathbf{y}} = \left[\mathrm{Re}\{\mathbf{y}^T\}, \mathrm{Im}\{\mathbf{y}^T\}\right]^T, \tag{3.36}$$

$$\bar{\mathcal{H}}_\ell = \begin{bmatrix} \mathrm{Re}\{\mathcal{H}_\ell\} & -\mathrm{Im}\{\mathcal{H}_\ell\} \\ \mathrm{Im}\{\mathcal{H}_\ell\} & \mathrm{Re}\{\mathcal{H}_\ell\} \end{bmatrix}, \tag{3.37}$$

$$\bar{\mathcal{S}}_\iota = \begin{bmatrix} \mathrm{Re}\{\mathcal{S}_\iota\} \\ \mathrm{Im}\{\mathcal{S}_\iota\} \end{bmatrix}. \tag{3.38}$$

3.10.2.1 SMT-Rx Detector

The SMT-Rx is a reduced-complexity and close-to-optimum average bit error ratio (ABER)-achieving decoder, which aims at reducing the receive search space. The detector can formally be written as [86],

$$\left[\hat{\mathcal{H}}_\ell, \hat{\mathcal{S}}_\iota\right] = \underset{\substack{\mathcal{H}_\ell \in \mathcal{H} \\ \mathcal{S}_\iota \in \mathcal{S}}}{\arg\min} \left\{ \sum_{r=1}^{\tilde{N}_r(\ell,\iota)} |\bar{y}_r - \bar{\mathcal{H}}_\ell^r \bar{\mathcal{S}}_\iota|^2 \right\}, \tag{3.39}$$

where $\bar{\mathcal{H}}_\ell^r$ is the rth row of $\bar{\mathcal{H}}_\ell$, $\bar{y}_r$ is the rth element of $\bar{\mathbf{y}}$, and

$$\tilde{N}_r(\ell,\iota) = \max_{n \in \{1,2,\ldots,2N_r\}} \left\{ n \,\middle|\, \sum_{r=1}^{n} |\bar{y}_r - \bar{\mathcal{H}}_\ell^r \bar{\mathcal{S}}_\iota|^2 \le R^2 \right\}, \tag{3.40}$$

where $\tilde{N}_r(\cdot,\cdot) = 2N_r$.

The idea behind SMT-Rx is that it keeps combining the received signals as long as the Euclidean distance in (3.39) is less or equal to the radius R. Whenever a point is found to be inside the sphere, the radius, R, is updated with the Euclidean distance of that point. The point with the minimum Euclidean distance is considered to be the solution.

3.10.2.2 SMT-Tx Detector

Conventional SD is designed for SMX, where all antennas are active at each time instance and transmitting different symbols [114, 177–179]. But, in SMT, there is none or only one constellation symbol transmitted from the active transmit antenna(s) depending on the used SMT. In [38, 87], a modified SD algorithm designed for SM only was presented. In this section, a generalized SD named SMT-Tx tailored to any SMTs system is described.

Similar to conventional SDs, the SMT-Tx scheme searches for points that lie inside a sphere with a radius R centered at the received point. Every time a point is found inside the sphere, the radius is decreased until only one point is left inside the sphere.

Th SD in (3.35) can be thought of as an inequality described by

$$\begin{aligned} R^2 &\geq \left\{\|\bar{\mathbf{y}} - \bar{\mathcal{H}}_\ell \bar{\mathcal{S}}_\iota\|_F^2\right\} \\ &\geq \left\{\bar{\mathbf{y}}^H\bar{\mathbf{y}} - \bar{\mathbf{y}}^H\bar{\mathcal{H}}_\ell\bar{\mathcal{S}}_\iota - \bar{\mathcal{S}}_\iota^H\bar{\mathcal{H}}_\ell^H\bar{\mathbf{y}} + \bar{\mathcal{S}}_\iota^H\bar{\mathcal{H}}_\ell^H\bar{\mathcal{H}}_\ell\bar{\mathcal{S}}_\iota\right\}. \end{aligned} \tag{3.41}$$

Let $\bar{\mathbf{G}}_\ell = \bar{\mathcal{H}}_\ell^H\bar{\mathcal{H}}_\ell$ be Cholesky factorized as $\bar{\mathbf{G}}_\ell = \bar{\mathbf{D}}_\ell^H\bar{\mathbf{D}}_\ell$, where $\bar{\mathbf{D}}_\ell$ is an upper triangular matrix. Define $\bar{\rho}_\ell = \bar{\mathbf{G}}_\ell^{-1}\bar{\mathcal{H}}_\ell^H\bar{\mathbf{y}}$ and $\bar{\mathbf{z}}_\ell = \bar{\mathbf{D}}_\ell\bar{\rho}_\ell$, then add $\bar{\mathbf{z}}_\ell^H\bar{\mathbf{z}}_\ell$ to the both sides of (3.41), which yields,

$$\left\{R^2 + \bar{\mathbf{z}}_\ell^H\bar{\mathbf{z}}_\ell - \bar{\mathbf{y}}^H\bar{\mathbf{y}}\right\} \geq \left\{\bar{\mathbf{z}}_\ell^H\bar{\mathbf{z}}_\ell - \bar{\mathbf{y}}^H\bar{\mathcal{H}}_\ell\bar{\mathcal{S}}_\iota - \bar{\mathcal{S}}_\iota^H\bar{\mathcal{H}}_\ell^H\bar{\mathbf{y}} + \bar{\mathcal{S}}_\iota^H\bar{\mathbf{G}}_\ell\bar{\mathcal{S}}_\iota\right\}. \tag{3.42}$$

Let $\bar{R}_\ell^2 = R^2 + \bar{\rho}^H\bar{\mathbf{D}}_\ell^H\bar{\mathbf{D}}_\ell\bar{\rho}_\ell - \bar{\mathbf{y}}^H\bar{\mathbf{y}}$,

$$\begin{aligned} \bar{R}_\ell^2 &\geq \left\{\bar{\mathbf{z}}_\ell^H\bar{\mathbf{z}}_\ell - \bar{\mathbf{y}}^H\bar{\mathcal{H}}_\ell\bar{\mathcal{S}}_\iota - \bar{\mathcal{S}}_\iota^H\bar{\mathcal{H}}_\ell^H\bar{\mathbf{y}} + \bar{\mathcal{S}}_\iota^H\bar{\mathbf{G}}_\ell\bar{\mathcal{S}}_\iota\right\} \\ &\geq \left\{\bar{\mathbf{z}}_\ell^H\bar{\mathbf{z}}_\ell - \bar{\mathbf{y}}^H\bar{\mathcal{H}}_\ell\bar{\mathbf{G}}_\ell^{-1}\bar{\mathbf{G}}_\ell\bar{\mathcal{S}}_\iota - \bar{\mathcal{S}}_\iota^H\bar{\mathbf{G}}_\ell\bar{\mathbf{G}}_\ell^{-1}\bar{\mathcal{H}}_\ell^H\bar{\mathbf{y}} + \bar{\mathcal{S}}_\iota^H\bar{\mathbf{G}}_\ell\bar{\mathcal{S}}_\iota\right\} \\ &\geq \left\{\bar{\mathbf{z}}_\ell^H\bar{\mathbf{z}}_\ell + \bar{\mathbf{z}}_\ell^H\bar{\mathbf{D}}_\ell\bar{\mathcal{S}}_\iota - \bar{\mathcal{S}}_\iota^H\bar{\mathbf{D}}_\ell^H\bar{\mathbf{z}}_\ell + \bar{\mathcal{S}}_\iota^H\bar{\mathbf{D}}_\ell^H\bar{\mathbf{D}}_\ell\bar{\mathcal{S}}_\iota\right\} \\ &\geq \left\{\|\bar{\mathbf{z}}_\ell - \bar{\mathbf{D}}_\ell\bar{\mathcal{S}}_\iota\|_F^2\right\}. \end{aligned} \tag{3.43}$$

3.10.2.3 Single Spatial Symbol SMTs (SS-SMTs)

For nonquadrature SMTs with only one spatial symbol, such as SM and GSM, $\bar{\mathbf{z}}_\ell$ is a two elements length vector, $\bar{\mathbf{D}}_\ell = \sqrt{\|\mathcal{H}_\ell\|_F^2} \times \mathbf{I}_2$ and $\bar{z}_\ell = \bar{\mathcal{H}}_\ell^H\bar{\mathbf{y}}$, where $\mathbf{I}_N$ is an $N \times N$ square identity matrix.

Thus, (3.43) can be written as

$$\bar{R}_\ell^2 \geq \left\{\left(\bar{z}_\ell^1 - \sqrt{\|\mathcal{H}_\ell\|_F^2}\mathrm{Re}\{\bar{\mathcal{S}}_\iota\}\right)^2 + \left(\bar{z}_\ell^2 - \sqrt{\|\mathcal{H}_\ell\|_F^2}\mathrm{Im}\{\bar{\mathcal{S}}_\iota\}\right)^2\right\}, \tag{3.44}$$

where $\bar{z}_\ell^i$ is the ith element of $\bar{\mathbf{z}}_\ell$.

The necessary conditions for the point $(\ell, \imath)$ to lie inside the sphere are

$$\bar{R}_\ell^2 \geq \left\{ \left(\bar{z}_\ell^2 - \sqrt{\|\mathcal{H}_\ell\|_F^2} \mathrm{Im}\{\bar{\mathcal{S}}_\imath\} \right)^2 \right\}, \tag{3.45}$$

$$\bar{R}_\ell^2 \geq \left\{ \left(\bar{z}_\ell^1 - \sqrt{\|\mathcal{H}_\ell\|_F^2} \mathrm{Re}\{\bar{\mathcal{S}}_\imath\} \right)^2 + \left(\bar{z}_\ell^2 - \sqrt{\|\mathcal{H}_\ell\|_F^2} \mathrm{Im}\{\bar{\mathcal{S}}_\imath\} \right)^2 \right\}. \tag{3.46}$$

Solving (3.45) and (3.46) gives the bounds

$$\frac{-\bar{R}_\ell + \bar{z}_\ell^2}{\sqrt{\|\mathcal{H}_\ell\|_F^2}} \leq \mathrm{Im}\{\bar{\mathcal{S}}_\imath\} \leq \frac{\bar{R}_\ell + \bar{z}_\ell^2}{\sqrt{\|\mathcal{H}_\ell\|_F^2}}, \tag{3.47}$$

$$\frac{-\sqrt{\bar{R}_\ell^2 - \Sigma_2^2} + \bar{z}_\ell^1}{\sqrt{\|\mathcal{H}_\ell\|_F^2}} \leq \mathrm{Re}\{\bar{\mathcal{S}}_\imath\} \leq \frac{\sqrt{\bar{R}_\ell^2 - \Sigma_2^2} + \bar{z}_\ell^1}{\sqrt{\|\mathcal{H}_\ell\|_F^2}}, \tag{3.48}$$

where $\Sigma_l^i = \left(\bar{z}_\ell^i - \bar{D}_\ell^{(i,l)} \mathrm{Im}\{\bar{\mathcal{S}}_\imath\} \right)^2$.

Every time a point is found inside the sphere, the radius R is updated with

$$R^2 = \left\{ \left(\bar{z}_\ell^1 - \sqrt{\|\mathcal{H}_\ell\|_F^2} \mathrm{Re}\{\bar{\mathcal{S}}_\imath\} \right)^2 + \left(\bar{z}_\ell^2 - \sqrt{\|\mathcal{H}_\ell\|_F^2} \mathrm{Im}\{\bar{\mathcal{S}}_\imath\} \right)^2 \right\} - \|\bar{\mathbf{z}}_\ell\|_F^2. \tag{3.49}$$

The point with the smallest radius is the solution; hence, the last point is found inside the sphere.

3.10.2.4 Double Spatial Symbols SMTs (DS-SMTs)

For quadrature SMTs with two spatial symbols, such as QSM, $\bar{\mathbf{D}}$ is an upper-triangular matrix. Hence, (3.43) can be rewritten as

$$\bar{R}_\ell^2 \geq \left\{ \sum_{i=2}^{4} \Sigma_4^i + \left(\bar{z}_\ell^1 - \left(\bar{D}_\ell^{(1,1)} \mathrm{Re}\{\bar{\mathcal{S}}_\imath\} + \bar{D}_\ell^{(1,4)} \mathrm{Im}\{\bar{\mathcal{S}}_\imath\} \right) \right)^2 \right\}. \tag{3.50}$$

From (3.50), the necessary conditions for the point $(\ell, \imath)$ to lie inside the sphere is

$$\bar{R}_\ell^2 \geq \left(\bar{z}_\ell^4 - \bar{D}_\ell^{(4,4)} \mathrm{Im}\{\bar{\mathcal{S}}_\imath\} \right)^2, \tag{3.51}$$

$$\bar{R}_\ell^2 \geq \left\{ \sum_{i=2}^{4} \Sigma_4^i + \left(\bar{z}_\ell^1 - \left(\bar{D}_\ell^{(1,1)} \mathrm{Re}\{\bar{\mathcal{S}}_\imath\} + \bar{D}_\ell^{(1,4)} \mathrm{Im}\{\bar{\mathcal{S}}_\imath\} \right) \right)^2 \right\}. \tag{3.52}$$

Solving (3.51) and (3.52) results in the following the bounds:

$$\frac{-\bar{R}_\ell + \bar{z}_\ell^4}{\bar{D}_\ell^{(4,4)}} \leq \quad \text{Im}\{\bar{S}_\iota\} \leq \frac{\bar{R}_\ell + \bar{z}_\ell^4}{\bar{D}_\ell^{(4,4)}}, \tag{3.53}$$

$$\frac{-\sqrt{\bar{R}_\ell^2 - \sum_{i=2}^{4} \Sigma_4^i + \sqrt{\Sigma_4^1}}}{\bar{D}_\ell^{(1,1)}} \leq \quad \text{Re}\{\bar{S}_\iota\} \frac{\sqrt{\bar{R}_\ell^2 - \sum_{i=2}^{4} \Sigma_4^i + \sqrt{\Sigma_4^1}}}{\bar{D}_\ell^{(1,1)}}. \tag{3.54}$$

Note, different to conventional MIMO systems, that the channel matrix does not have to be full rank for SD to work, i.e., SMT-SD works for $N_t < N_r$ as well as $N_t \geq N_r$.

3.10.2.5 Computational Complexity

The detailed number of multiplication operations needed by SS–SMT–SD is shown in Table 3.9, where $\mathcal{P}_{(3.47)}$ and $\mathcal{P}_{(3.48)}$ are the number of points in the bounds (3.47) and (3.48), respectively.

Hence, the total complexity of SS–SMT–SD is

$$\mathcal{C}_{\text{SS-SMTs}} = 3N_t(2N_r + 3) + 5\mathcal{P}_{(3.47)} + 2\mathcal{P}_{(3.48)}. \tag{3.55}$$

The detailed number of multiplication operations needed by SS-SMT-SD is shown in Table 3.10, where $\mathcal{P}_{(3.53)}$ and $\mathcal{P}_{(3.54)}$ are the number of points in the bounds (3.53) and (3.54), respectively.

Hence, the total complexity of DS-SMT-SD is,

$$\mathcal{C}_{\text{DS-SMTs}} = 2N_t^2(9N_r + 26) + 10\mathcal{P}_{(3.53)} + 2\mathcal{P}_{(3.54)}. \tag{3.56}$$

Table 3.9 Detailed complexity analysis of SS-SMTs-SD.

Operation	Number of multiplications
$\lvert\bar{\mathcal{H}}_\ell\rvert_F^2$	$N_t \times 2N_r$
$\bar{\rho}_\ell$	$N_t \times (2 \times 2N_r + 2)$
$\bar{\mathbf{z}}_\ell$	$N_t \times 2$
$\bar{R}_\ell^2$	$N_t \times 2$
(3.47)	$N_t \times 3$
(3.48)	$\mathcal{P}_{(3.47)} \times 5$
(3.49)	$\mathcal{P}_{(3.48)} \times 2$

Table 3.10 Detailed complexity analysis of DS-SMTs-SD.

Operation	Number of multiplications
$\bar{\mathbf{G}}_\ell$	$N_t^2 \times 5 \times 2N_r$
$\bar{\mathbf{D}}_\ell$	$N_t^2 \times 16$
$\bar{\rho}_\ell$	$N_t^2 \times 4(2N_r + 5)$
$\bar{\mathbf{z}}_\ell$	$N_t^2 \times 9$
$\bar{R}_\ell^2$	$N_t^2 \times 4$
(3.53)	$N_t^2 \times 3$
(3.54)	$\mathcal{P}_{(3.53)} \times 10$
(3.49)	$\mathcal{P}_{(3.54)} \times 2$

3.10.2.6 Error Probability Analysis and Initial Radius

The pairwise error probability (PEP) of deciding on the point $(\tilde{\ell}_{\text{SMT-SD}}, \tilde{\imath}_{\text{SMT-SD}})$ given that the point $(\ell_t, \imath_t)$ is transmitted can be written as

$$\underset{\text{error}}{\overset{\text{SMT-SD}}{\Pr}} = \Pr\left(\left(\tilde{\ell}_{\text{SMT-SD}}, \tilde{\imath}_{\text{SMT-SD}}\right) \neq (\ell_t, \imath_t)\right). \tag{3.57}$$

The probability of error in (3.57) can be thought of as two mutually exclusive events depending on whether the transmitted point $(\ell_t, \imath_t)$ is inside the sphere or not. In other words, the probability of error for SMT–SD can be separated in two parts as [180],

$$\underset{\text{error}}{\overset{\text{SMT-SD}}{\Pr}} \leq \left(\Pr\left((\tilde{\ell}_{\text{SMT-ML}}, \tilde{\imath}_{\text{SMT-ML}}) \neq (\ell_t, \imath_t)\right) + \Pr\left((\ell_t, \imath_t) \notin \Theta_R\right)\right). \tag{3.58}$$

(1) $\Pr((\tilde{\ell}_{\text{SMT-ML}}, \tilde{\imath}_{\text{SMT-ML}}) \neq (\ell_t, \imath_t))$: The probability of deciding on the incorrect transmitted symbol and/or used antenna combination, given that the transmitted point $(\ell_t, \imath_t)$ is inside the sphere.
(2) $\Pr((\ell_t, \imath_t) \notin \Theta_R)$: The probability that the transmitted point $(\ell_t, \imath_t)$ is outside the set of points Θ_R considered by the SMT–SD.

From (3.58), SMT–SD will have a near optimum performance when,

$$\Pr((\ell_t, \imath_t) \notin \Theta_R) << \Pr((\tilde{\ell}_{\text{SMT-ML}}, \tilde{\imath}_{\text{SMT-ML}}) \neq (\ell_t, \imath_t)). \tag{3.59}$$

The probability of *not* having the transmitted point $(\ell_t, \imath_t)$ inside Θ_R can be written as

$$\Pr((\ell_t, \imath_t) \notin \Theta_R) = \Pr\left(\sum_{i=1}^{2N_r} \left|\bar{y}_r - \bar{\mathcal{H}}_{\ell_t,i}\bar{\mathcal{S}}_t\right|^2 > R^2\right)$$

$$= \Pr\left(\kappa > \left(\frac{R}{\sigma_n/\sqrt{2}}\right)^2\right)$$

$$= 1 - \frac{\gamma\left(N_r, \left(\frac{R}{\sigma_n}\right)^2\right)}{\Gamma(N_r)}, \tag{3.60}$$

where $\bar{\mathcal{H}}_{\ell_t,r}$ is the ith row of $\bar{\mathcal{H}}_{\ell_t}$, and

$$\kappa = \sum_{i=1}^{2N_r} \left|\frac{\bar{n}_i}{\sigma_n/2}\right|^2, \tag{3.61}$$

is a central chi-squared random variable (RV) with $2N_r$ degrees of freedom, $\bar{n}_i$ is the ith element of $\bar{\mathbf{n}} = [\mathrm{Re}\{\mathbf{n}\}; \mathrm{Im}\{\mathbf{n}\}]$, and the cumulative distribution function (CDF) of a chi-squared RV is given by [120],

$$F_\kappa(a, b) = \frac{\gamma(b/2, a/2)}{\Gamma(b/2)}, \tag{3.62}$$

where $\gamma(c, d)$ is the lower incomplete gamma function given by

$$\gamma(c, d) = \int_0^d t^{c-1}\, e^{-t}\, dt, \tag{3.63}$$

and $\Gamma(c)$ is the gamma function given by

$$\Gamma(c) = \int_0^\infty t^{c-1}\, e^{-t}\, dt. \tag{3.64}$$

The initial radius considered in SMT–SD is a function of the noise variance as given in [181],

$$R^2 = 2\alpha N_r \sigma_n^2, \tag{3.65}$$

where α is a constant chosen to satisfy (3.59). This can be done by setting $\Pr((\ell_t, \iota_t) \notin \Theta_R) = 10^{-6}$ and back solving (3.60). For $N_r = 1, 2, 4$, $\alpha = 13.8, 8.3$, and 5.3, respectively.

3.11 Transmitter Power Consumption Analysis

The approximate transmitter power consumption for the different SMTs and GSMTs is calculated in what follows. In particular, the transmitter designs for SSK, SM, QSSK, QSM, GSSK, GSM, GQSSK, and GQSM are considered in the analysis. The results are compared to SMX system with the previously presented transmitter design.

For power consumption analysis, the EARTH power model is considered, which describes the relation between the total power supplied or consumed by a transceiver system and the RF transmit power under the assumptions of full load and sleep mode [182, 183].

Therefore and through the EARTH model, the power consumptions for SMX, SSK, SM, QSM, and QSSK systems are calculated as follows [40, 182, Eq. (1.2), p. 7]:

$$P_t^{\{SMX\}} = (P_o N_t) + (\alpha P_{max}), \tag{3.66}$$
$$P_t^{\{SSK\}} = (\alpha P_{max}) + P_t^{\{P_{sw} N_{sw}\}}, \tag{3.67}$$
$$P_t^{\{QSSK\}} = (\alpha P_{max}) + P_t^{\{P_{sw} N_{sw}\}}, \tag{3.68}$$
$$P_t^{\{SM\}} = (P_o) + (\alpha P_{max}) + P_t^{\{P_{sw} N_{sw}\}}, \tag{3.69}$$
$$P_t^{\{QSM\}} = (P_o) + (\alpha P_{max}) + P_t^{\{P_{sw} N_{sw}\}}, \tag{3.70}$$

where P_o denotes the minimum consumed power per RF-chain, α is the slope of the load dependent power consumption, P_{max} is the total RF transmit power [182], P_{sw} is the consumed power by a single RF switch, and N_{sw} denotes the number of needed single pole double through (SPDT) RF switches to implement the transmitter of the corresponding scheme. It should be noted here that different RF switches with variable number of output terminals can be considered and will lead to different results. However, SPDT switches are widely available and achieve the least switching time, which in turn means maximum possible data rate for SMTs. Each SPDT switch is connected to two transmit antennas. Hence, the number of needed RF switches to achieve a target spectral efficiency, η, for each SMT scheme is calculated as

$$N_{sw}^{\{SSK\}} = 2^{\eta-1}, \tag{3.71}$$
$$N_{sw}^{\{QSSK\}} = 2^{\frac{\eta}{2}-1}, \tag{3.72}$$
$$N_{sw}^{\{SM\}} = 2^{((\eta-\log_2 M)-1)}, \tag{3.73}$$
$$N_{sw}^{\{QSM\}} = 2^{((\eta-\log_2 M)/2-1)}. \tag{3.74}$$

In [182], the relation between the power consumption for various base station types as a function of the RF output power is reported. Four types of base stations are considered in the conducted study in [182] including Macro, Micro, Pico, and Femto cells base stations. Any of these models can be considered in the presented comparative study between different systems. Here Macro-type base station is assumed, and the reported numbers in [182, Table 1.2, p. 8] are adopted, which are $P_o = 53$ W, $\alpha = 3.1$, and $P_{max} = 6.3$ W. In addition, the consumed power by a single SPDT RF switch is assumed to be $P_{sw} = 5$ mW [184]. It is important to note that even $P_t^{\{SM\}}$ and $P_t^{\{QSM\}}$ have the same formula, the total consumed power is not equal since the number of required RF switches to achieve a target spectral efficiency is not the same. Similarly, $P_t^{\{SSK\}}$ and $P_t^{\{QSSK\}}$ are not equal.

For the generalized version of SMTs, GSMTs, the anticipated power consumption depends on the value of M and n_u. For a specific spectral efficiency, the needed number of transmit antennas by each GSMT system can be calculated and used to compute the required number of RF switches. Assume, $n_u = 2$, the number of antennas for specific η is

$$N_t^{\{\mathrm{GSSK}\}} = \mathcal{R}\left(\left(N_t^{\{\mathrm{GSSK}\}}\right)^2 - N_t^{\{\mathrm{GSSK}\}} - 2^{(\eta+1)}\right), \tag{3.75}$$

$$N_t^{\{\mathrm{GSM}\}} = \mathcal{R}\left(\left(N_t^{\{\mathrm{GSM}\}}\right)^2 - N_t^{\{\mathrm{GSM}\}} - 2^{(\eta-\log_2(M)+1)}\right), \tag{3.76}$$

$$N_t^{\{\mathrm{GQSSK}\}} = \mathcal{R}\left(\left(N_t^{\{\mathrm{GQSSK}\}}\right)^2 - N_t^{\{\mathrm{GQSSK}\}} - 2^{(\eta+2)/2}\right), \tag{3.77}$$

$$N_t^{\{\mathrm{GQSM}\}} = \mathcal{R}\left(\left(N_t^{\{\mathrm{GQSM}\}}\right)^2 - N_t^{\{\mathrm{GQSM}\}} - 2^{(\eta-\log_2(M)+2)/2}\right), \tag{3.78}$$

where $\mathcal{R}(\cdot)$ denotes a positive root greater than one of the polynomial function.

As such, the power consumption for GSSK, GSM, GQSM, and GQSSK systems is calculated as

$$P_t^{\{\mathrm{GSSK}\}} = (\alpha P_{\max}) + P_t^{\{\mathrm{RF\text{-}sw}\}}, \tag{3.79}$$

$$P_t^{\{\mathrm{GSM}\}} = (P_o) + (\alpha P_{\max}) + P_t^{\{\mathrm{RF\text{-}sw}\}}, \tag{3.80}$$

$$P_t^{\{\mathrm{GQSSK}\}} = (\alpha P_{\max}) + P_t^{\{\mathrm{RF\text{-}sw}\}}, \tag{3.81}$$

$$P_t^{\{\mathrm{GQSM}\}} = (P_o) + (\alpha P_{\max}) + P_t^{\{\mathrm{RF\text{-}sw}\}}. \tag{3.82}$$

Again, $P_t^{\{\mathrm{GSM}\}}$ and $P_t^{\{\mathrm{GQSM}\}}$ have the same formula, but the total consumed power is not equal since the number of required RF switches to achieve a target spectral efficiency in not the same. Similarly, $P_t^{\{\mathrm{GSSK}\}}$ and $P_t^{\{\mathrm{GQSSK}\}}$ are not equal. Assuming that SPDT RF switches, where each switch can serve two transmit antennas, the number of needed RF switches for the different GSMTs is given by

$$N_{\mathrm{sw}}^{\{\mathrm{GSSK}\}} = \frac{1}{2} N_t^{\{\mathrm{GSSK}\}}, \tag{3.83}$$

$$N_{\mathrm{sw}}^{\{\mathrm{GQSSK}\}} = N_t^{\{\mathrm{GQSSK}\}}, \tag{3.84}$$

$$N_{\mathrm{sw}}^{\{\mathrm{GSM}\}} = \frac{1}{2} N_t^{\{\mathrm{GSM}\}}, \tag{3.85}$$

$$N_{\mathrm{sw}}^{\{\mathrm{GQSM}\}} = N_t^{\{\mathrm{GQSM}\}}. \tag{3.86}$$

3.11.1 Power Consumption Comparison

The transmitter power consumptions for all SMTs and GSMTs are discussed in what follows. Results for SMTs are illustrated in Figure 3.18 and for GSMTs are shown in Figure 3.19. The consumed power by each system is depicted

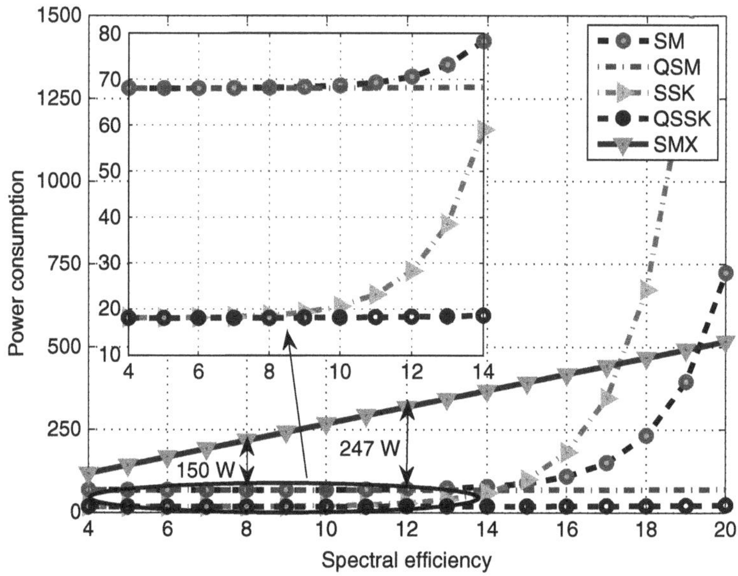

Figure 3.18 Transmitter power consumption for SM, QSM, SSK, and QSSK MIMO systems. For SM and QSM, $M = 4$ is assumed. Also, for all SMTs, SPDT RF switches are considered in all systems.

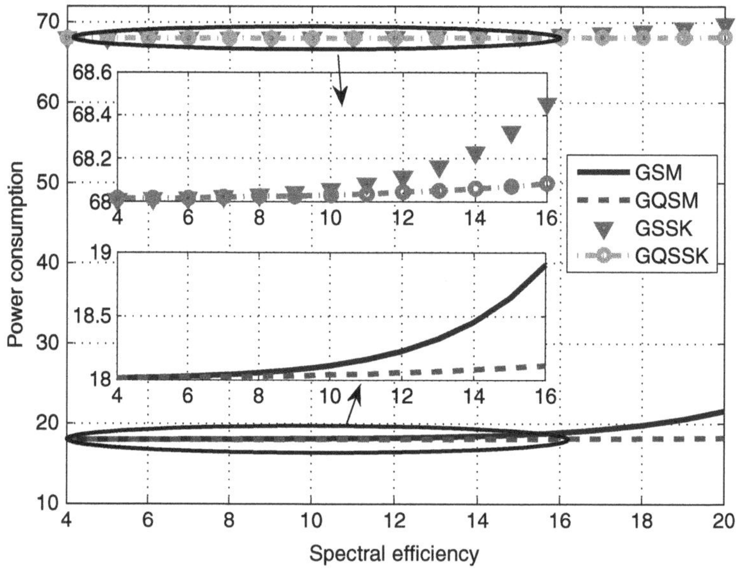

Figure 3.19 Transmitter power consumption for GSM, GQSM, GSSK, and GQSSK MIMO systems. For GSM and GQSM, $M = 4$ is assumed.

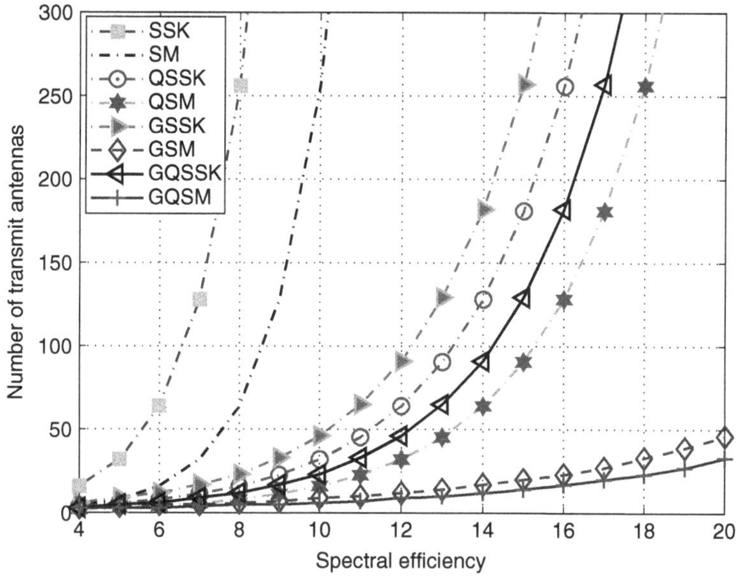

Figure 3.20 Needed number of transmit antennas to achieve a target spectral efficiency for SSK, SM, QSSK, QSM, GSSK, GSM, GQSSK, and GQSM MIMO systems. For SM, QSM, GSM, and GQSM, $M = 4$ is assumed and for all GSMTs $n_u = 2$ is considered.

versus the target spectral efficiency, which is varied from $\eta = 4$ to 20 bits. For SM, QSM, GSM, and GQSM, $M = 4$-QAM is assumed, and for GSMTs, $n_u = 2$ is considered. Besides the required number of transmit antennas to achieve a target spectral efficiency for all schemes is computed and illustrated in Figure 3.20. The number of needed RF switches can be calculated from the computed number of transmit antennas and used in the power analysis. For all systems, SPDT RF switches are considered, where each switch is connected to two transmit antennas. Interesting results are noticed in the figures, where SSK and GSSK demonstrate very low power consumption for relatively low spectral efficiencies. However, both schemes demonstrate a significant increase in power consumption at higher spectral efficiencies, where SSK demonstrates the maximum power consumption among all SMTs and SMX at $\eta = 20$ bits. This is because SSK scheme needs 524,288 RF switches at 20 bits spectral efficiency, while GSSK needs 725 RF switches at that particular spectral efficiency. However, QSSK scheme requires only 512 RF switches, and GQSSK needs 23 switches at this spectral efficiency. For the same reason, the exponential growth of the SM and GSM power consumptions at higher spectral efficiencies can be explained, where SM needs 131,072 switches and GSM needs 363 switches at $\eta = 20$ bits, while QSM and GQSM systems can be, respectively, implemented with 256 and 17 RF switches at this spectral efficiency. However, all SMTs and GSMTs except SSK and SM are shown to consume much less power than SMX system for all depicted spectral efficiencies. SM and SSK are shown to be power efficient schemes at moderate spectral efficiencies, where comparing SM to SMX at spectral efficiencies of 8 and 12 bits is shown to, respectively, provide 150 W and 247 W gains.

3.12 Hardware Cost

A rough estimate of the implementation cost for each SMTs and GSMTs system is calculated based on the available off-the-shelf components. The different implementation elements can be categorized as follows:

(1) C_{RF}: The cost of one RF-chain (includes signal modulation, pulse shaping, and I/Q modulator blocks);
(2) C_{memory}: The cost of a memory module such as microcontroller with a DAC chip;
(3) $C_{\text{S/P}}$: The cost of a serial to parallel converter;
(4) C_{sw}: The cost of one SPDT RF switch.

The cost of the required hardware items to implement the transmitter for each of SSK, QSSK, SM, QSM, and SMX is given by [40],

$$C^{\{\text{SSK}\}} = C_{\text{memory}} + C_{\text{S/P}} + \left(C_{\text{sw}} N_{\text{sw}}^{\{\text{SSK}\}}\right), \tag{3.87}$$

$$C^{\{\text{QSSK}\}} = C_{\text{memory}} + C_{\text{S/P}} + 2\left(C_{\text{sw}} N_{\text{sw}}^{\{\text{QSSK}\}}\right), \tag{3.88}$$

$$C^{\{\text{SM}\}} = C_{\text{RF}} + C_{\text{S/P}} + \left(C_{\text{sw}} N_{\text{sw}}^{\{\text{SM}\}}\right), \tag{3.89}$$

$$C^{\{\text{QSM}\}} = C_{\text{RF}} + C_{\text{S/P}} + 2\left(C_{\text{sw}} N_{\text{sw}}^{\{\text{QSM}\}}\right), \tag{3.90}$$

$$C^{\{\text{SMX}\}} = (C_{\text{RF}})N_t. \tag{3.91}$$

Similarly, the cost of the required hardware items to implement the transmitter for each of the GSMTs systems is calculated as

$$C^{\{\text{GSSK}\}} = C_{\text{memory}} + C_{\text{S/P}} + \left(C_{\text{sw}} N_{\text{sw}}^{\{\text{GSSK}\}}\right), \tag{3.92}$$

$$C^{\{\text{GQSSK}\}} = C_{\text{memory}} + C_{\text{S/P}} + 2\left(C_{\text{sw}} N_{\text{sw}}^{\{\text{GQSSK}\}}\right), \tag{3.93}$$

$$C^{\{\text{GSM}\}} = C_{\text{RF}} + C_{\text{S/P}} + \left(C_{\text{sw}} N_{\text{sw}}^{\{\text{GSM}\}}\right), \tag{3.94}$$

$$C^{\{\text{GQSM}\}} = C_{\text{RF}} + C_{\text{S/P}} + 2\left(C_{\text{sw}} N_{\text{sw}}^{\{\text{GQSM}\}}\right). \tag{3.95}$$

For evaluation and comparison purposes, the prices listed in Table 3.11 for the different elements are considered.

Table 3.11 Hardware items cost in US$.

Item	Cost ($)
C_{RF}	180 [185]
C_{Memory}	4 [186]
$C_{\text{S/P}}$	2 [187]
C_{sw}	2 [184]

3.12.1 Hardware Cost Comparison

A rough estimate for the cost of deploying the transmitter of different SMTs and GSMTs is illustrated in Figures 3.21 and 3.22, respectively. $M = 4$ is assumed for SM, QSM, GSM, and GQSM, and SPDT RF switch is assumed for all systems. Also, $n_u = 2$ is considered for generalized systems.

Similar trend as noticed for the power consumption is seen here as well. Hardware implementation costs for SSK and GSSK schemes increase exponentially with the increase of spectral efficiency due the exponential growth of the needed number of RF switches. However, it can be seen that implementing GSSK scheme costs much less than all other SMTs. This is because GSSK

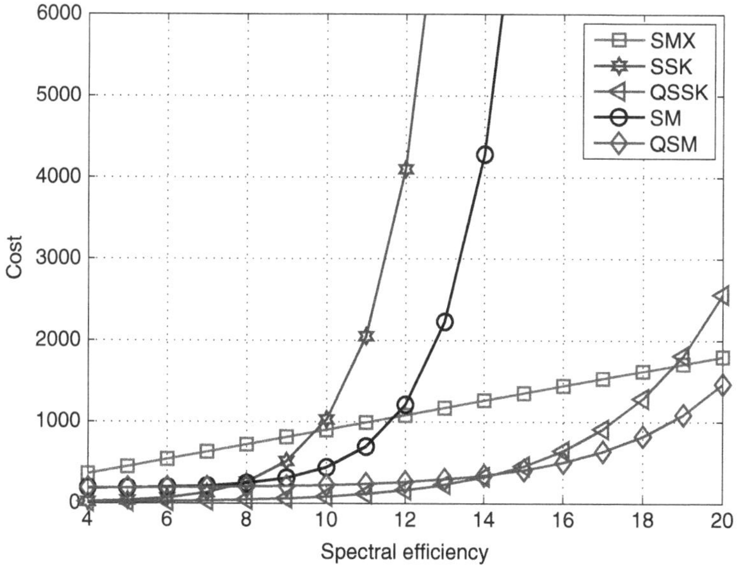

Figure 3.21 Transmitter implementation cost for SM, QSM, SSK, and QSSK assuming $M = 4$ and SPDT RF switches.

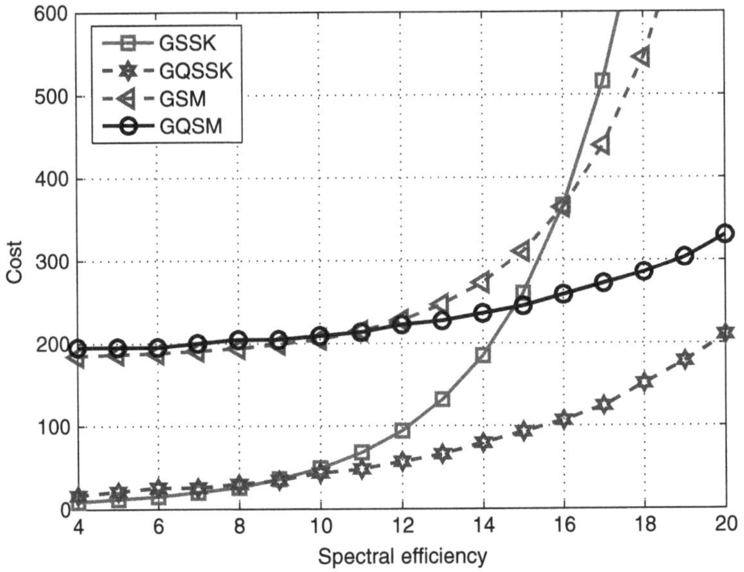

Figure 3.22 Transmitter implementation cost for GSM, GQSM, GSSK, and GQSSK assuming $M = 4$ and SPDT RF switches.

requires less number of transmit antennas for the same spectral efficiency and can be implemented without any RF-chains, which in turn means less number of RF switches and reduced cost. For $\eta > 10$ bits, the cost for implementing SSK system is shown to be very high and exceeds all other system costs. Similar trend can be seen as well for SM system and can be referred to the same reason as for SSK scheme. SMX implementation demonstrates the maximum cost for $\eta \leq 10$ bits. However, the cost of implementing SSK and SM, respectively, exceed the cost of SMX at $\eta = 10$ and $\eta = 12$ bits. The implementation cost of QSM and QSSK is shown to be moderate and much lower than all other system's costs. Similar trends can be seen for GSMT's results in Figure 3.22. Again, implementing GSM and GSSK schemes is shown to cost much more than other schemes at high spectral efficiencies.

3.13 SMTs Coherent and Noncoherent Spectral Efficiencies

The realizable spectral efficiencies for DQSM, QSM, DSM, SM, DSSK, and SSK systems with different number of transmit antennas, $N_t \rightarrow 2 : 9$, and with $M = 4$-QAM/PSK modulation are compared in Figure 3.23. It can be seen from the figure that quadrature space modulation techniques (QSMTs), such as

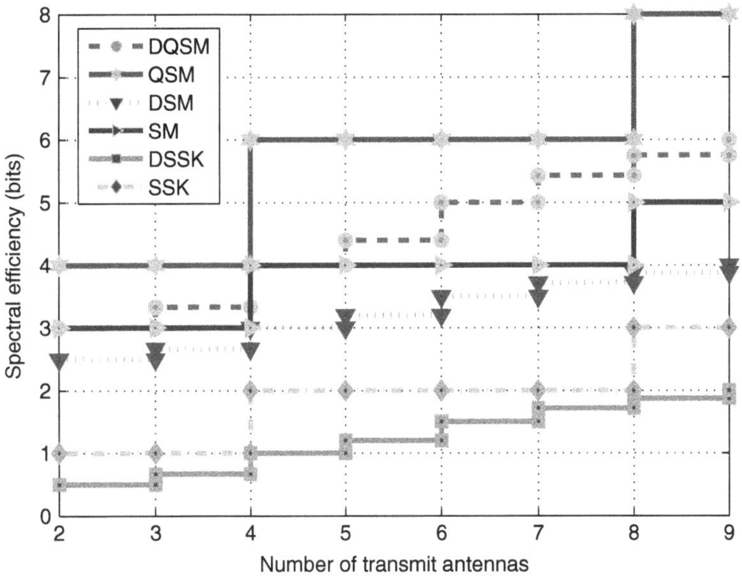

Figure 3.23 A comparison of achievable spectral efficiencies with variable number of transmit antennas, N_t, and with $M = 4$-QAM modulation for different techniques including DQSM, QSM, DSM, SM, DSSK, and SSK systems.

QSM and DQSM, offer an increase in spectral efficiency in comparison to their SMTs counterpart. Comparing QSM to SM, 1 bit more data can be attained with $N_t = 2$, which increases to 2 and 3 bits for $N_t = 4$ and 8, respectively. As discussed earlier, this enhancement equals $\log_2 N_t$, which increases logarithmically with the number of transmit antennas. SSK scheme is shown in Figure 3.23 to achieve the least spectral efficiency among all SMTs as it depends only on N_t. SM, on the other hand, increases SSK's spectral efficiency by the number of transmitted bits in the signal domain, $\log_2 M$. Similarly, QSM enhances SSK's spectral efficiency by $(\log_2 N_t + \log_2 M)$.

Finally, the spectral efficiency of differential space modulation techniques (DSMTs) is shown in Figure 3.23 to increase with an arbitrary number of transmit antennas, which is not necessarily being a power of 2. Furthermore, SMTs are shown to always achieve higher spectral efficiencies than DSMTs in Figure 3.23 for the same N_t. Also, DQSM always achieve higher spectral efficiency than that of DSM system and achieving similar spectral efficiency for both schemes with 4-QAM modulation is only possible with $N_t = 4$ for DQSM and $N_t = 9$ for DSM.

4

Average Bit Error Probability Analysis for SMTs

In this chapter, average bit error ratio (ABER) performance analysis for the different space modulation techniques (SMTs) discussed in previous chapter is presented. First, the ABER is computed for space shift keying (SSK), spatial modulation (SM), quadrature space shift keying (QSSK), and quadrature spatial modulation (QSM) over Rayleigh flat fading channels. Detailed derivation of the error probability is presented. As well, asymptotic analysis at high, but pragmatic, signal-to-noise-ratio (SNR) values are also given for all schemes. Second, the average and asymptotic error probabilities for these schemes in the presence of Gaussian imperfect channel estimation and over Rayleigh fading channels are discussed and derived [52, 65, 68, 188]. Finally, a general framework for the performance analysis of the different SMTs over arbitrary fading channels, in the presence of spatial correlation and imperfect channel estimation, is derived and thoroughly discussed [46, 48, 72–75, 189].

4.1 Average Error Probability over Rayleigh Fading Channels

The derivation of the average error probability for SMTs can be computed by deriving the pairwise error probability (PEP). The PEP is defined as the error probability that a transmitted spatial and signal symbols, $\mathcal{H}_\ell$ and $\mathcal{S}_\iota$, are received as another spatial and signal symbols, $\mathcal{H}_{\tilde{\ell}}$ and $\mathcal{S}_{\tilde{\iota}}$, respectively, and is given by $\text{PEP}((\mathcal{H}_\ell, \mathcal{S}_\iota) \rightarrow (\mathcal{H}_{\tilde{\ell}}, \mathcal{S}_{\tilde{\iota}}))$ [120].

4.1.1 SM and SSK with Perfect Channel Knowledge at the Receiver

The derivation of the error probability for SM and SSK is almost the same. SSK scheme can be treated as a special case from SM, and the probability of error for SSK scheme can be obtained as such.

Space Modulation Techniques, First Edition. Raed Mesleh and Abdelhamid Alhassi.

4.1.1.1 Single Receive Antenna ($N_r = 1$)

In what follows, the derivation is conducted for the special case where the receiver has single receive antenna. Later on, this is generalized to an arbitrary number of receive antennas.

For SM over multiple-input single-output (MISO) channel, the spatial and signal symbols are $\mathcal{H}_\ell = h_\ell$ and $\mathcal{S}_\iota = s_\iota$, respectively, with h_ℓ being the ℓth element of the N_t-length channel vector $\mathbf{h}$ and s_ι is the ιth modulation symbol. Note, $\mathrm{SNR} = 1/\sigma_n^2$, where for simplicity, $E_s = \mathrm{E}[\|\mathcal{H}_\ell \mathcal{S}_\iota\|_\mathrm{F}^2] = N_r$ is assumed.

Assuming the maximum-likelihood (ML)-optimum receiver in (3.2) is used, the PEP is calculated as

$$\begin{aligned}
\Pr_{\text{error}} &= \Pr((h_\ell, s_\iota) \rightarrow (h_{\tilde{\ell}}, s_{\tilde{\iota}})|\mathbf{h}) \\
&= \Pr(|y - h_\ell s_\iota|^2 > |y - h_{\tilde{\ell}} s_{\tilde{\iota}}|^2|\mathbf{h}) \\
&= \Pr(|(h_\ell s_\iota + n) - h_\ell s_\iota|^2 > |(h_\ell s_\iota + n) - h_{\tilde{\ell}} s_{\tilde{\iota}}|^2|\mathbf{h}) \\
&= \Pr(|n|^2 > |(h_\ell s_\iota - h_{\tilde{\ell}} s_{\tilde{\iota}}) + n|^2|\mathbf{h}) \\
&= \Pr(|n|^2 > (|h_\ell s_\iota - h_{\tilde{\ell}} s_{\tilde{\iota}}|^2 - 2\mathrm{Re}\{n^*(h_\ell s_\iota - h_{\tilde{\ell}} s_{\tilde{\iota}})\} + |n|^2)|\mathbf{h}) \\
&= \Pr(2\mathrm{Re}\{n^*(h_\ell s_\iota - h_{\tilde{\ell}} s_{\tilde{\iota}})\} > |h_\ell s_\iota - h_{\tilde{\ell}} s_{\tilde{\iota}}|^2|\mathbf{h}) \\
&= Q\left(\frac{|h_\ell s_\iota - h_{\tilde{\ell}} s_{\tilde{\iota}}|^2}{\sqrt{2\sigma_n^2 |h_\ell s_\iota - h_{\tilde{\ell}} s_{\tilde{\iota}}|^2}}\right) = Q\left(\sqrt{\frac{|h_\ell s_\iota - h_{\tilde{\ell}} s_{\tilde{\iota}}|^2}{2\sigma_n^2}}\right) = Q\left(\sqrt{\gamma^{\mathrm{SM}}}\right),
\end{aligned} \tag{4.1}$$

where $(\cdot)^*$ denotes the conjugate, $Q(x) = 1/(2\pi) \int_x^\infty \exp(t^2/2)\, dt$ is the Q-function, and

$$\gamma^{\mathrm{SM}} = \frac{|h_\ell s_\iota - h_{\tilde{\ell}} s_{\tilde{\iota}}|^2}{2\sigma_n^2} \tag{4.2}$$

is an exponential random variable (RV) with a probability distribution function (PDF) given by

$$p_{\gamma^{\mathrm{SM}}}(\gamma^{\mathrm{SM}}) = \frac{1}{\bar{\gamma}^{\mathrm{SM}}} \exp\left(-\frac{\gamma^{\mathrm{SM}}}{\bar{\gamma}^{\mathrm{SM}}}\right) \tag{4.3}$$

with $\bar{\gamma}^{\mathrm{SM}}$ denoting the mean value and is given by

$$\bar{\gamma}^{\mathrm{SM}} = \frac{1}{2\sigma_n^2} \times \begin{cases} |s_\iota - s_{\tilde{\iota}}|^2, & \text{if } \tilde{\ell} = \ell, \\ |s_\iota|^2 + |s_{\tilde{\iota}}|^2, & \text{if } \tilde{\ell} \neq \ell. \end{cases} \tag{4.4}$$

Therefore, the average PEP is

$$\Pr((h_\ell, s_\iota) \rightarrow (h_{\tilde{\ell}}, s_{\tilde{\iota}})) = \mathrm{E}_{\gamma^{\mathrm{SM}}}\left\{\Pr_{\text{error}}\right\} = \frac{1}{2}\left(1 - \sqrt{\frac{\bar{\gamma}^{\mathrm{SM}}}{1 + \bar{\gamma}^{\mathrm{SM}}}}\right). \tag{4.5}$$

The formula in (4.5) is obtained for an exponential RV from [123, 190] as

$$\mathrm{E}_{\gamma}\{Q(\sqrt{\gamma})\} = \int_{\gamma} p_{\gamma}(\gamma)Q(\sqrt{\gamma})\, d\gamma = \int_{\gamma} \frac{1}{\bar{\gamma}} \exp\left(-\frac{\gamma}{\bar{\gamma}}\right) Q(\sqrt{\gamma})\, d\gamma$$
$$= \frac{1}{2}\left(1 - \sqrt{\frac{\bar{\gamma}/2}{1+\bar{\gamma}/2}}\right). \tag{4.6}$$

Thus, the ABER of SM over MISO Rayleigh fading channels can be computed using the union bounding technique as [120]

$$\mathrm{ABER} \leq \frac{1}{2^{\eta}} \sum_{\tilde{\ell},\tilde{\imath}} \sum_{\ell,\imath} \frac{e_{\tilde{\ell},\tilde{\imath}}^{\ell,\imath}}{\eta} \frac{1}{2}\left(1 - \sqrt{\frac{\bar{\gamma}^{\mathrm{SM}}/2}{1+\bar{\gamma}^{\mathrm{SM}}/2}}\right) \tag{4.7}$$

with $e_{\tilde{\ell},\tilde{\imath}}^{\ell,\imath}$ being the number of bits in error associated with the PEP event. The derived analytical ABER in (4.7) is compared to the simulated ABER for $\eta = 4, N_t = 4$, and $\eta = 6, N_t = 8$ in Figure 4.1. The results in the figure validate the derived bound in (4.7).

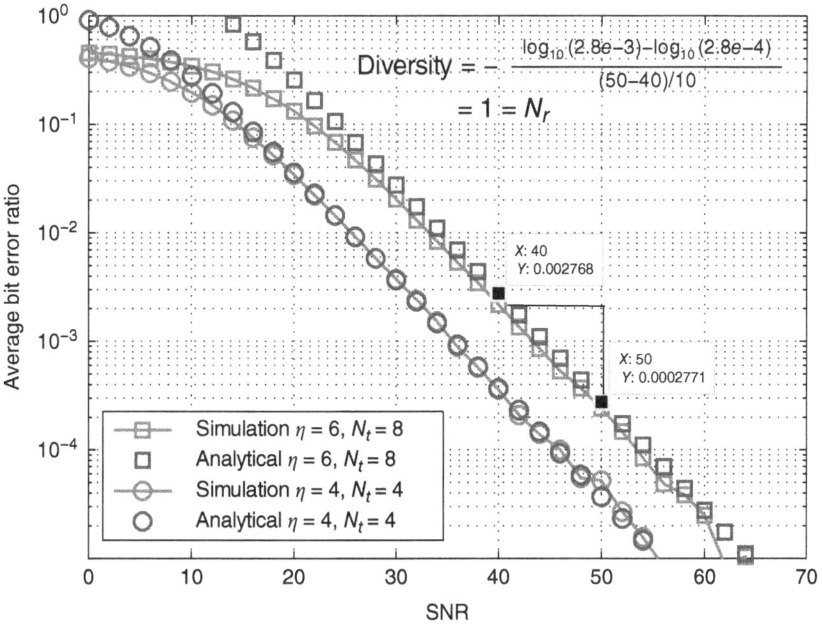

Figure 4.1 The derived analytical ABER of SM for MISO Rayleigh fading channels in (4.7) compared simulated ABER for $\eta = 4, N_t = 4$ and $\eta = 6, N_t = 8$.

4.1.1.2 Arbitrary Number of Receive Antennas (N_r)

The previous derivations consider the special case of $N_r = 1$. For an arbitrary number of receive antennas, N_r, the PEP is given by

$$\Pr_{\text{error}} = Q\left(\sqrt{\frac{\|\mathbf{h}_\ell s_\iota - \mathbf{h}_{\tilde{\ell}} s_{\tilde{\iota}}\|_F^2}{2\sigma_n^2}}\right) = Q\left(\sqrt{\sum_{r=1}^{N_r} \gamma_r^{\text{SM}}}\right) = Q(\sqrt{\upsilon^{\text{SM}}}), \tag{4.8}$$

where $\mathbf{h}_\ell$ is the ℓth column of the $N_r \times N_t$ channel matrix $\mathbf{H}$

$$\gamma_r^{\text{SM}} = \frac{|h_{\ell,r} s_\iota - h_{\tilde{\ell},r} s_{\tilde{\iota}}|^2}{2\sigma_n^2}, \tag{4.9}$$

and $\upsilon^{\text{SM}} = \sum_{r=1}^{N_r} \gamma_r^{\text{SM}}$ is a chi-squared RV with a PDF given by

$$p_{\upsilon^{\text{SM}}}(\upsilon^{\text{SM}}) = \frac{1}{\Gamma(N_r)(\bar{\gamma}^{\text{SM}})^{N_r}} \left(\upsilon^{\text{SM}}\right)^{(N_r-1)} \exp\left(-\frac{\upsilon^{\text{SM}}}{\bar{\gamma}^{\text{SM}}}\right), \tag{4.10}$$

where $h_{\ell,r}$ is the rth element of the N_r-length channel vector $\mathbf{h}_\ell$. Note, $\bar{\gamma}^{\text{SM}}$ is given in (4.4).

Thus, the average PEP is given by [123, 191],

$$\begin{aligned}\Pr((\mathbf{h}_\ell, s_\iota) \to (\mathbf{h}_{\tilde{\ell}}, s_{\tilde{\iota}})) &= \mathrm{E}_{\upsilon^{\text{SM}}}\left\{\Pr_{\text{error}}\right\} \\ &= (\alpha_a^{\text{SM}})^{N_r} \sum_{i=0}^{N_r-1} \binom{N_r - 1 + i}{i} [1 - \alpha_a^{\text{SM}}]^i,\end{aligned} \tag{4.11}$$

where $\alpha_a^{\text{SM}} = \frac{1}{2}\left(1 - \sqrt{\frac{\bar{\gamma}^{\text{SM}}/2}{1+\bar{\gamma}^{\text{SM}}/2}}\right)$. Note, from [123, 190] if υ is a chi-squared RV then,

$$\begin{aligned}\mathrm{E}_\upsilon\{Q(\sqrt{\upsilon})\} &= \int_\upsilon p_\upsilon(\upsilon) Q(\sqrt{\upsilon})\, d\upsilon \\ &= \int_\upsilon \frac{1}{\Gamma(N_r)\bar{\gamma}^{N_r}} \upsilon^{(N_r-1)} \exp\left(-\frac{\upsilon}{\bar{\gamma}}\right) Q(\sqrt{\upsilon})\, d\upsilon \\ &= \alpha_a^{N_r} \sum_{i=0}^{N_r-1} \binom{N_r - 1 + i}{i} [1 - \alpha_a]^i,\end{aligned} \tag{4.12}$$

where $\alpha_a = \frac{1}{2}\left(1 - \sqrt{\frac{\bar{\gamma}/2}{1+\bar{\gamma}/2}}\right)$.

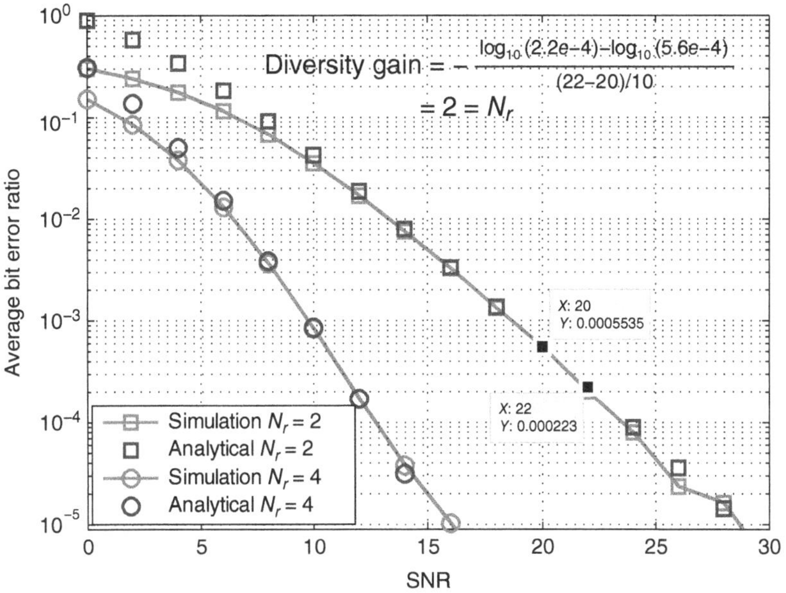

Figure 4.2 The derived analytical ABER of SM over MISO Rayleigh fading channels in (4.13) compared to the simulated ABER for $N_r = 2$ and 4, $\eta = 4$, and $N_t = 4$.

Finally, the ABER for SM over Rayleigh fading multiple-input multiple-output (MIMO) channel is derived as

$$\text{ABER} = \frac{1}{2^{\eta}} \sum_{\tilde{\ell},\tilde{\imath}} \sum_{\ell,\imath} \frac{e_{\tilde{\ell},\tilde{\imath}}^{\ell,\imath}}{\eta} (\alpha_a^{\text{SM}})^{N_r} \sum_{i=0}^{N_r-1} \binom{N_r - 1 + i}{i} [1 - \alpha_a^{\text{SM}}]^i. \quad (4.13)$$

Figure 4.2 compares the derived analytical ABER bound in (4.13) to the simulated ABER for $N_r = 2$ and 4, $\eta = 4$, and $N_t = 4$. The accuracy of the bound is obvious in the figure where a close-match between analytical and simulation curves can be clearly seen for all depicted SNR values greater than 5 dB.

4.1.1.3 Asymptotic Analysis

Another very useful formula to obtain for such systems is the approximate average error probability, generally called asymptotic error probability, at high SNR values. Such formula clearly shows the diversity gain for such systems and can be utilized for optimization studies. The asymptotic error probability is derived based on the behavior of the PDF of υ^{SM} around the origin. This can be obtained by taking Taylor series of $p_{\upsilon^{\text{SM}}}(\upsilon^{\text{SM}})$ leading to

$$p_{\upsilon^{\text{SM}}}(\upsilon^{\text{SM}}) = \frac{1}{\Gamma(N_r)(\bar{\gamma}^{\text{SM}})^{N_r}} \left(\upsilon^{\text{SM}}\right)^{(N_r-1)} + \mathcal{O}, \tag{4.14}$$

where $\mathcal{O}$ denotes higher-order terms that are ignored. The average PEP is then computed as [123, 191]

$$\begin{aligned} \mathrm{E}_{\upsilon^{\text{SM}}}\{\Pr_{\text{error}}\} &= \int_0^\infty Q(\sqrt{\upsilon}) \frac{1}{\Gamma(N_r)(\bar{\gamma}^{\text{SM}})^{N_r}} \left(\upsilon^{\text{SM}}\right)^{(N_r-1)} d\upsilon \\ &\approx \frac{2^{(N_r-1)}\Gamma(N_r+0.5)}{\sqrt{\pi}\Gamma(N_r+1)} \left(\frac{1}{\bar{\gamma}^{\text{SM}}}\right)^{N_r}. \end{aligned} \tag{4.15}$$

It is evident that SM achieves a diversity gain of N_r. This can be seen as well in Figures 4.1 and 4.2, where for MISO system in Figure 4.1, the diversity gain is 1 as there is only one receive antenna. For the MIMO system in Figure 4.2, the diversity gain is 2, which is equal to the number of receive antennas, $N_r = 2$.

For SSK, the ABER and asymptotic PEP can be computed by (4.13) and (4.15), respectively, by letting $s_\iota = s_{\tilde{\iota}} = 1$.

4.1.2 SM and SSK in the Presence of Imperfect Channel Estimation

The assumption of perfect channel knowledge at the receiver is impractical. In practical wireless systems, pilots are transmitted to estimate the channel at the receiver input. However and similar to transmitted data, transmitted pilots are corrupted by the additive white Gaussian noise (AWGN) at the receiver input, which leads to a mismatch between the exact channel and the estimated channel. The difference between both is generally called channel estimation errors (CSE). Generally, it is assumed that the true channel and the estimated channel are jointly ergodic and stationary processes. Also, the estimated channel and the estimation error are assumed to be orthogonal. Hence [192],

$$\mathbf{H} = \tilde{\mathbf{H}} + \mathbf{e}, \tag{4.16}$$

where $\tilde{\mathbf{H}}$ denotes the $N_r \times N_t$ estimated channel matrix, and $\mathbf{e}$ denotes the CSE, which is a complex Gaussian random variable with zero mean and $\sigma_{\mathbf{e}}^2$ variance, $\mathbf{e} \rightarrow \mathcal{CN}(0, \sigma_{\mathbf{e}}^2 \mathbf{I}_{N_r})$. Note that $\sigma_{\mathbf{e}}^2$ is a parameter that captures the quality of the channel estimation and can be appropriately chosen depending on the channel dynamics and estimation schemes.

The derivation of the instantaneous PEP in the presence of imperfect channel knowledge at the receiver is similar to the derivation of perfect channel knowledge and is computed in what follows.

4.1.2.1 Single Receive Antenna ($N_r = 1$)

In the presence of CSE, the ML-optimum receiver for SM is given by

$$\left[\hat{\tilde{h}}_{\ell}, \hat{s}_{\iota}\right] = \underset{\substack{\tilde{h}_{\tilde{\ell}} \in \tilde{\mathbf{h}} \\ s_{\tilde{\iota}} \in \mathcal{S}}}{\arg\min} \left\| y - \tilde{h}_{\tilde{\ell}} s_{\tilde{\iota}} \right\|_{\mathrm{F}}^{2}, \tag{4.17}$$

where $\tilde{h}_{\ell}$ is the ℓth element of the N_t-length estimated MISO channel vector $\tilde{\mathbf{h}}$.

Thus, the PEP for SM in the presence of CSE is given by

$$\begin{aligned}
\Pr_{\text{error}} &= \Pr((\tilde{h}_{\ell}, s_{\iota}) \rightarrow (\tilde{h}_{\tilde{\ell}}, s_{\tilde{\iota}})|\tilde{\mathbf{h}}) \\
&= \Pr(|y - \tilde{h}_{\ell} s_{\iota}|^2 > |y - \tilde{h}_{\tilde{\ell}} s_{\tilde{\iota}}|^2 |\tilde{\mathbf{h}}) \\
&= \Pr(|(h_{\ell} s_{\iota} + n) - \tilde{h}_{\ell} s_{\iota}|^2 > |(h_{\ell} s_{\iota} + n) - \tilde{h}_{\tilde{\ell}} s_{\tilde{\iota}}|^2 |\tilde{\mathbf{h}}) \\
&= \Pr(|((\tilde{h}_{\ell} + e) - \tilde{h}_{\ell}) s_{\iota} + n|^2 > |((\tilde{h}_{\ell} + e) s_{\iota} - \tilde{h}_{\tilde{\ell}} s_{\tilde{\iota}}) + n|^2 |\tilde{\mathbf{h}}) \\
&= \Pr(|e s_{\iota} + n|^2 > |(\tilde{h}_{\ell} s_{\iota} - \tilde{h}_{\tilde{\ell}} s_{\tilde{\iota}}) + (e s_{\iota} + n)|^2 |\tilde{\mathbf{h}}) \\
&= \Pr(|\bar{n}|^2 > (|\tilde{h}_{\ell} s_{\iota} - \tilde{h}_{\tilde{\ell}} s_{\tilde{\iota}}|^2 - 2\mathrm{Re}\{\bar{n}^* (\tilde{h}_{\ell} s_{\iota} - \tilde{h}_{\tilde{\ell}} s_{\tilde{\iota}})\} + |\bar{n}|^2) |\tilde{\mathbf{h}}) \\
&= \Pr(2\mathrm{Re}\{\bar{n}^* (\tilde{h}_{\ell} s_{\iota} - \tilde{h}_{\tilde{\ell}} s_{\tilde{\iota}})\} > |\tilde{h}_{\ell} s_{\iota} - \tilde{h}_{\tilde{\ell}} s_{\tilde{\iota}}|^2 |\tilde{\mathbf{h}}) \\
&= Q\left(\sqrt{\frac{|\tilde{h}_{\ell} s_{\iota} - \tilde{h}_{\tilde{\ell}} s_{\tilde{\iota}}|^2}{2(\sigma_{\mathbf{e}}^2 |s_{\iota}|^2 + \sigma_n^2)}}\right) = Q\left(\sqrt{\gamma_{\mathbf{e}}^{\mathrm{SM}}}\right),
\end{aligned} \tag{4.18}$$

where $\bar{n} = e s_{\iota} + n$ is a complex Gaussian RV with zero mean and variance $(\sigma_{\mathbf{e}}^2 |s_{\iota}|^2 + \sigma_n^2)$. Moreover,

$$\gamma_{\mathbf{e}}^{\mathrm{SM}} = \frac{|\tilde{h}_{\ell} s_{\iota} - \tilde{h}_{\tilde{\ell}} s_{\tilde{\iota}}|^2}{2(\sigma_{\mathbf{e}}^2 |s_{\iota}|^2 + \sigma_n^2)} \tag{4.19}$$

is an exponential RV with a mean value

$$\bar{\gamma}_{\mathbf{e}}^{\mathrm{SM}} = \varphi_{\mathbf{e}} \times \begin{cases} |s_{\iota} - s_{\tilde{\iota}}|^2 & \text{if } \tilde{\ell} = \ell, \\ (|s_{\iota}|^2 + |s_{\tilde{\iota}}|^2) & \text{if } \tilde{\ell} \neq \ell, \end{cases} \tag{4.20}$$

where

$$\varphi_{\mathbf{e}} = \frac{1 + \sigma_{\mathbf{e}}^2}{2(\sigma_{\mathbf{e}}^2 |s_{\iota}|^2 + \sigma_n^2)}. \tag{4.21}$$

Using the same method as discussed in Section 4.1.1.1 for perfect channel knowledge at the receiver, the ABER for SM over MISO Rayleigh fading channel with imperfect channel knowledge is given by [120, 123]

$$\mathrm{ABER} = \frac{1}{2^{\eta}} \sum_{\tilde{\ell}, \tilde{\iota}} \sum_{\ell, \iota} \frac{e_{\tilde{\ell}, \tilde{\iota}}^{\ell, \iota}}{\eta} \frac{1}{2} \left(1 - \sqrt{\frac{\bar{\gamma}_{\mathbf{e}}^{\mathrm{SM}}/2}{1 + \bar{\gamma}_{\mathbf{e}}^{\mathrm{SM}}/2}}\right). \tag{4.22}$$

4.1.2.2 Arbitrary Number of Receive Antennas (N_r)

For an arbitrary number of receive antennas N_r, the PEP is given by

$$\Pr_{\text{error}} = Q\left(\sqrt{\sum_{r=1}^{N_r} \gamma_{e,r}^{\text{SM}}}\right) = Q\left(\sqrt{\upsilon_{\mathbf{e}}^{\text{SM}}}\right), \tag{4.23}$$

where $\upsilon_{\mathbf{e}}^{\text{SM}} = \sum_{r=1}^{N_r} \gamma_{e,r}^{\text{SM}}$ is a chi-squared RV with a PDF

$$p_{\upsilon_{\mathbf{e}}^{\text{SM}}}(\upsilon_{\mathbf{e}}^{\text{SM}}) = \frac{1}{\Gamma(N_r)(\bar{\gamma}_{\mathbf{e}}^{\text{SM}})^{N_r}} (\upsilon_{\mathbf{e}}^{\text{SM}})^{(N_r-1)} \exp\left(-\frac{\upsilon_{\mathbf{e}}^{\text{SM}}}{\bar{\gamma}_{\mathbf{e}}^{\text{SM}}}\right) \tag{4.24}$$

and

$$\gamma_{e,r}^{\text{SM}} = \frac{|\tilde{h}_{\ell,r} s_\iota - \tilde{h}_{\tilde{\ell},r} s_{\tilde{\iota}}|^2}{2(\sigma_{\mathbf{e}_{\tilde{\mathbf{H}}}}^2 |s_\iota|^2 + \sigma_n^2)}, \tag{4.25}$$

where $\tilde{h}_{\ell,r}$ is the rth element of the N_r-length channel vector $\tilde{\mathbf{h}}_\ell$, and $\tilde{\mathbf{h}}_\ell$ is the ℓth vector of the $N_r \times N_t$ estimated channel matrix $\tilde{\mathbf{H}}$. Note, $\bar{\gamma}_{\mathbf{e}}^{\text{SM}}$ is given in (4.20).

Thus, and using similar procedure as in Section 4.1.1.2, the average PEP is given by [123, 191]

$$\mathrm{E}_{\tilde{\mathbf{H}}}\left\{\Pr_{\text{error}}\right\} = (\alpha_{a,e}^{\text{SM}})^{N_r} \sum_{i=0}^{N_r-1} \binom{N_r - 1 + i}{i} \left[1 - \alpha_{a,e}^{\text{SM}}\right]^i, \tag{4.26}$$

where $\alpha_{a,e}^{\text{SM}} = \frac{1}{2}\left(1 - \sqrt{\frac{\bar{\gamma}_{\mathbf{e}}^{\text{SM}}/2}{1+\bar{\gamma}_{\mathbf{e}}^{\text{SM}}/2}}\right)$.

Finally, the ABER for SM over MIMO Rayleigh fading channels in the presence of CSE is given by

$$\text{ABER} = \frac{1}{2^\eta} \sum_{\tilde{\ell},\tilde{\iota}} \sum_{\ell,\iota} \frac{e_{\tilde{\ell},\tilde{\iota}}^{\ell,\iota}}{\eta} (\alpha_{a,e}^{\text{SM}})^{N_r} \sum_{i=0}^{N_r-1} \binom{N_r - 1 + i}{i} \left[1 - \alpha_{a,e}^{\text{SM}}\right]^i. \tag{4.27}$$

The derived ABER analytical bounds in (4.22) for MISO and in (4.27) for MIMO are depicted in Figure 4.3 and compared with the simulated ABER, for $\sigma_{\mathbf{e}}^2 = \sigma_n^2$, $N_r = 1$ and 2, $\eta = 4$, and $N_t = 2$. The depicted results validate the accuracy of the derived formulas for the ABER.

4.1.2.3 Asymptotic Analysis

The asymptotic PEP in the presence of CSE can be computed as

$$\mathrm{E}_{\tilde{\mathbf{H}}}\left\{\Pr_{\text{error}}\right\} \approx \frac{2^{(N_r-1)}\Gamma(N_r + 0.5)}{\sqrt{\pi}\Gamma(N_r + 1)} \left(\frac{1}{\bar{\gamma}_{\mathbf{e}}^{\text{SM}}}\right)^{N_r}. \tag{4.28}$$

The obtained asymptotic formula allows deep investigations of several cases.

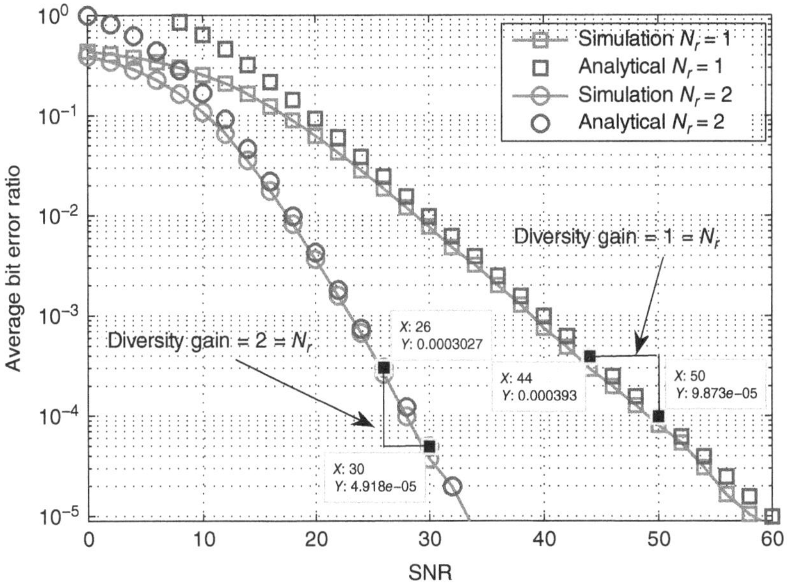

Figure 4.3 The derived analytical ABER of SM over Rayleigh fading channels in the presence of CSE in (4.22) for MISO and in (4.27) for MIMO compared to the simulated ABER for $\sigma_{\mathbf{e}}^2 = \sigma_n^2$, $N_r = 1$and 2, $\eta = 4$, and $N_t = 2$.

(1) *Case I – Fixed* $\sigma_{\mathbf{e}}^2$*:*
Assuming fixed CSE, i.e. the CSE is not a function of the SNR, the asymptotic PEP is given by

$$\mathrm{E}_{\tilde{\mathbf{H}}}\left\{\Pr_{\text{error}}\right\} = \frac{2^{(N_r-1)}\Gamma(N_r+0.5)}{\sqrt{\pi}\Gamma(N_r+1)} \times \Xi_{\text{SM}}^{N_r} \times \left(\frac{1+\sigma_{\mathbf{e}}^2}{2\sigma_{\mathbf{e}}^2|s_{\tilde{l}}|^2}\right)^{-N_r}, \tag{4.29}$$

where $\Xi_{\text{SM}} = \varphi_{\mathbf{e}}/\bar{\gamma}_{\mathbf{e}}^{\text{SM}}$. It is clear from (4.29) that increasing the SNR has no impact on the average PEP, which leads to zero diversity order and an error floor will occur.

(2) *Case II – CSE is a function of SNR:*
In this case, the pilot symbols are assumed to be transmitted with the same energy as the symbols and the channel estimation error decreases as the SNR increases. Hence, the average asymptotic PEP is given by

$$\mathrm{E}_{\tilde{\mathbf{H}}}\left\{\Pr_{\text{error}}\right\} \approx \frac{2^{(N_r-1)}\Gamma(N_r+0.5)}{\sqrt{\pi}\Gamma(N_r+1)} \times \Xi_{\text{SM}}^{N_r} \times (2(1+|s_{\tilde{l}}|^2))^{N_r}\left(\frac{1}{\sigma_n^2}\right)^{-N_r}. \tag{4.30}$$

The diversity order in this case is N_r, and increasing the SNR enhances the performance. This gain can be seen in Figure 4.3 where the ABER results at high SNR have diversity gains of 1 and 2 for $N_r = 1$ and $N_r = 2$, respectively.

In the case of SSK scheme, the ABER and the asymptotic PEP can be calculated using (4.27) and (4.28), respectively, by letting $s_{\tilde{\iota}} = s_{\iota} = 1$.

4.1.3 QSM with Perfect Channel Knowledge at the Receiver

The average error probability of QSM and QSSK can be derived following similar steps as discussed before for SM and SSK systems [65, 70].

Assuming the ML-optimum receiver in (3.2) is used, the PEP is given by

$$\begin{aligned}
\Pr_{\text{error}} &= \Pr((h_{\ell^{\text{Re}}}, h_{\ell^{\text{Im}}}, s_{\iota}) \rightarrow (h_{\tilde{\ell}^{\text{Re}}}, h_{\tilde{\ell}^{\text{Im}}}, s_{\tilde{\iota}})|\mathbf{h}) \\
&= \Pr(|y - (h_{\ell^{\text{Re}}} s_{\iota}^{\text{Re}} + h_{\ell^{\text{Im}}} s_{\iota}^{\text{Im}})|^2 > |y - (h_{\tilde{\ell}^{\text{Re}}} s_{\tilde{\iota}}^{\text{Re}} + h_{\tilde{\ell}^{\text{Im}}} s_{\tilde{\iota}}^{\text{Im}})|^2 |\mathbf{h}) \\
&= \Pr(|n|^2 > |((h_{\ell^{\text{Re}}} s_{\iota}^{\text{Re}} - h_{\tilde{\ell}^{\text{Re}}} s_{\tilde{\iota}}^{\text{Re}}) + (h_{\ell^{\text{Im}}} s_{\iota}^{\text{Im}} - h_{\tilde{\ell}^{\text{Im}}} s_{\tilde{\iota}}^{\text{Im}})) + n|^2 |\mathbf{h}) \\
&= \Pr(|n|^2 > (|\Psi|^2 - 2\text{Re}\{n^* \Psi\} + |n|^2)|\mathbf{h}) \\
&= \Pr(2\text{Re}\{n^* \Psi\} > |\Psi|^2 |\mathbf{h}) = Q\left(\sqrt{\frac{|\Psi|^2}{2\sigma_n^2}}\right) = Q(\sqrt{\gamma^{\text{QSM}}}),
\end{aligned} \tag{4.31}$$

where a^{Re} and a^{Im} denote the real and imaginary parts of the complex number a, respectively; $\Psi = ((h_{\ell^{\text{Re}}} s_{\iota}^{\text{Re}} - h_{\tilde{\ell}^{\text{Re}}} s_{\tilde{\iota}}^{\text{Re}}) + (h_{\ell^{\text{Im}}} s_{\iota}^{\text{Im}} - h_{\tilde{\ell}^{\text{Im}}} s_{\tilde{\iota}}^{\text{Im}}))$, and $\gamma^{\text{QSM}} = |\Psi|^2 / 2\sigma_n^2$ is an exponential RV with a mean value given by

$$\bar{\gamma}^{\text{QSM}} = \frac{1}{2\sigma_n^2} \begin{cases} (|s_{\iota}|^2 + |s_{\tilde{\iota}}|^2) & \text{if } \ell^{\text{Re}} \neq \tilde{\ell}^{\text{Re}}, \ell^{\text{Im}} \neq \tilde{\ell}^{\text{Im}} \\ (|s_{\iota}^{\text{Re}} - s_{\tilde{\iota}}^{\text{Re}}|^2 + |s_{\iota}^{\text{Im}}|^2 + |s_{\tilde{\iota}}^{\text{Im}}|^2) & \text{if } \ell^{\text{Re}} = \tilde{\ell}^{\text{Re}}, \ell^{\text{Im}} \neq \tilde{\ell}^{\text{Im}} \\ (|s_{\iota}^{\text{Re}}|^2 + |s_{\tilde{\iota}}^{\text{Re}}|^2 + |s_{\iota}^{\text{Im}} - s_{\tilde{\iota}}^{\text{Im}}|^2) & \text{if } \ell^{\text{Re}} \neq \tilde{\ell}^{\text{Re}}, \ell^{\text{Im}} = \tilde{\ell}^{\text{Im}} \\ (|s_{\iota}^{\text{Re}} - s_{\tilde{\iota}}^{\text{Re}}|^2 + |s_{\iota}^{\text{Im}} - s_{\tilde{\iota}}^{\text{Im}}|^2) & \text{if } \ell^{\text{Re}} = \tilde{\ell}^{\text{Re}}, \ell^{\text{Im}} = \tilde{\ell}^{\text{Im}} \end{cases}. \tag{4.32}$$

Therefore, from [123], and as discussed in Section 4.1.1.1, the average PEP is

$$\begin{aligned}
\Pr((h_{\ell^{\text{Re}}}, h_{\ell^{\text{Im}}}, s_{\iota}) \rightarrow (h_{\tilde{\ell}^{\text{Re}}}, h_{\tilde{\ell}^{\text{Im}}}, s_{\tilde{\iota}})) &= \text{E}\left\{\Pr_{\text{error}}\right\} \\
&= \frac{1}{2}\left(1 - \sqrt{\frac{\bar{\gamma}^{\text{QSM}}/2}{1 + \bar{\gamma}^{\text{QSM}}/2}}\right).
\end{aligned} \tag{4.33}$$

Thus, the ABER of QSM is

$$\text{ABER} = \frac{1}{2^\eta} \sum_{\ell,\iota} \sum_{\tilde{\ell},\tilde{\iota}} \frac{e_{\tilde{\ell},\tilde{\iota}}^{\ell,\iota}}{\eta} \frac{1}{2}\left(1 - \sqrt{\frac{\bar{\gamma}^{\text{QSM}}/2}{1 + \bar{\gamma}^{\text{QSM}}/2}}\right), \tag{4.34}$$

where $\ell = \{\ell^{\text{Re}}, \ell^{\text{Im}}\}$.

For MIMO system with N_r receive antennas, using the same methodology as discussed earlier, the ABER for QSM is

$$\text{ABER} = \frac{1}{2^{\eta}} \sum_{\ell,\iota} \sum_{\tilde{\ell},\tilde{\iota}} \frac{e_{\tilde{\ell},\tilde{\iota}}^{\ell,\iota}}{\eta} \left(\alpha_a^{\text{QSM}}\right)^{N_r} \sum_{i=0}^{N_r-1} \binom{N_r - 1 + i}{i} [1 - \alpha_a^{\text{QSM}}]^i, \quad (4.35)$$

where $\alpha_a^{\text{QSM}} = \frac{1}{2}\left(1 - \sqrt{\frac{\bar{\gamma}^{\text{QSM}}/2}{1+\bar{\gamma}^{\text{QSM}}/2}}\right)$.

Taking the Taylor series and ignoring higher-order terms give the asymptotic average PEP of QSM as

$$\Pr((h_{\ell\text{Re}}, h_{\ell\text{Im}}, s_{\iota}) \rightarrow (h_{\tilde{\ell}\text{Re}}, h_{\tilde{\ell}\text{Im}}, s_{\tilde{\iota}})) \approx \frac{2^{N_r-1}\Gamma(N_r + 0.5)}{\sqrt{\pi}\Gamma(N_r + 1)} \left(\frac{1}{\bar{\gamma}^{\text{QSM}}}\right)^{N_r}, \quad (4.36)$$

where a diversity gain of N_r is obtained. Figure 4.4 depicts the derived ABER analytical bounds (4.34) for MISO and in (4.35) for MIMO systems

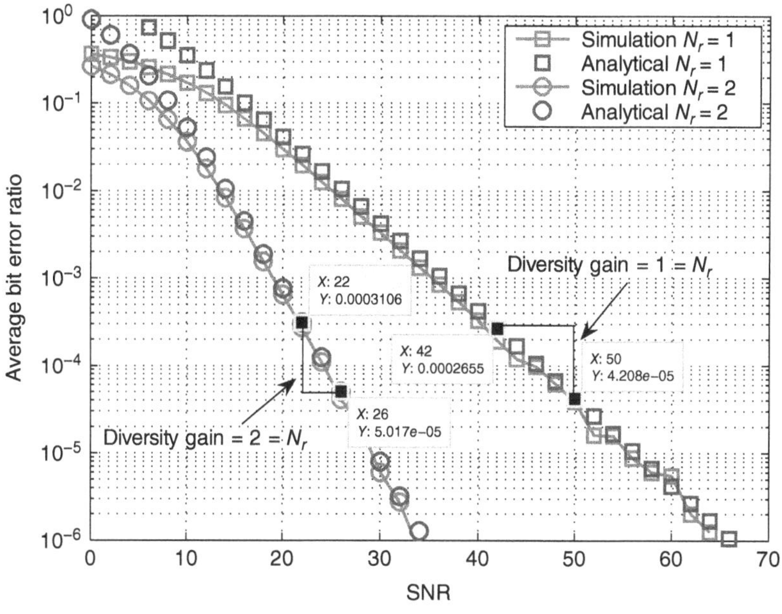

Figure 4.4 The derived analytical ABER of QSM over Rayleigh fading channels in (4.34) for MISO and in (4.35) for MIMO compared to the simulated ABER for $N_r = 1$ and 2, $\eta = 4$, and $N_t = 2$.

and compare it to the simulated ABER for QSM with $\eta = 4$, $N_t = 2$, and $N_r = 1$ and 2. From the figure, simulation and analytical results demonstrate close-match, and the diversity gains equal the number of receive antennas.

For QSSK, the ABER and PEP can be calculated using (4.34) and (4.36), respectively, by letting $s_\iota = s_{\tilde{\iota}} = (1+j)/\sqrt{2}$.

4.1.4 QSM in the Presence of Imperfect Channel Estimation

Following similar steps as discussed in Section 4.1.2.2 for SM, the ABER of QSM over MIMO Rayleigh fading channel in the presence of CSE is

$$\text{ABER} = \frac{1}{2^\eta} \sum_{\ell,\iota} \sum_{\tilde{\ell},\tilde{\iota}} \frac{e^{\ell,\iota}_{\tilde{\ell},\tilde{\iota}}}{\eta} \left(\alpha_{a,e}^{\text{QSM}}\right)^{N_r} \sum_{i=0}^{N_r-1} \binom{N_r - 1 + i}{i} [1 - \alpha_{a,e}^{\text{QSM}}]^i, \quad (4.37)$$

where $\alpha_{a,e}^{\text{QSM}} = \frac{1}{2}\left(1 - \sqrt{\frac{\bar{\gamma}_{\mathbf{e}}^{\text{QSM}}/2}{1+\bar{\gamma}_{\mathbf{e}}^{\text{QSM}}/2}}\right)$ and

$$\bar{\gamma}_{\mathbf{e}}^{\text{QSM}} = \varphi_{\mathbf{e}} \times \begin{cases} (|s_\iota|^2 + |s_{\tilde{\iota}}|^2) & \text{if } \ell^{\text{Re}} \neq \tilde{\ell}^{\text{Re}}, \ell^{\text{Im}} \neq \tilde{\ell}^{\text{Im}} \\ (|s_\iota^{\text{Re}} - s_{\tilde{\iota}}^{\text{Re}}|^2 + |s_\iota^{\text{Im}}|^2 + |s_{\tilde{\iota}}^{\text{Im}}|^2) & \text{if } \ell^{\text{Re}} = \tilde{\ell}^{\text{Re}}, \ell^{\text{Im}} \neq \tilde{\ell}^{\text{Im}} \\ (|s_\iota^{\text{Re}}|^2 + |s_{\tilde{\iota}}^{\text{Re}}|^2 + |s_\iota^{\text{Im}} - s_{\tilde{\iota}}^{\text{Im}}|^2) & \text{if } \ell^{\text{Re}} \neq \tilde{\ell}^{\text{Re}}, \ell^{\text{Im}} = \tilde{\ell}^{\text{Im}} \\ (|s_\iota^{\text{Re}} - s_{\tilde{\iota}}^{\text{Re}}|^2 + |s_\iota^{\text{Im}} - s_{\tilde{\iota}}^{\text{Im}}|^2) & \text{if } \ell^{\text{Re}} = \tilde{\ell}^{\text{Re}}, \ell^{\text{Im}} = \tilde{\ell}^{\text{Im}} \end{cases} . \quad (4.38)$$

Note, $\varphi_{\mathbf{e}}$ is given in (4.21). Figure 4.5 shows the derived bound and compares it to the simulated ABER for $\eta = 4$, $N_t = 2$, and $N_r = 1$ and 2. The figure shows that the simulated ABER is bounded by (4.37). Furthermore and from the figure, QSM offers a diversity gain equal to the number of receive antennas.

The asymptotic error probability for QSM in the presence of CSE can be computed using similar steps as discussed earlier and is given by

$$\Pr((\tilde{h}_{\ell^{\text{Re}}}, \tilde{h}_{\ell^{\text{Im}}}, s_\iota) \to (\tilde{h}_{\tilde{\ell}^{\text{Re}}}, \tilde{h}_{\tilde{\ell}^{\text{Im}}}, s_{\tilde{\iota}})) \approx \frac{2^{N_r-1}\Gamma(N_r + 0.5)}{\sqrt{\pi}\Gamma(N_r + 1)} \left(\frac{1}{\bar{\gamma}_{\mathbf{e}}^{\text{QSM}}}\right)^{N_r} . \quad (4.39)$$

The asymptotic error probability can be analyzed for different cases based on the value of $\sigma_{\mathbf{e}}^2$:

Case I – $\sigma_{\mathbf{e}}^2 \neq (\text{SNR}^{-1} = \sigma_n^2)$: Assuming a fixed channel estimation error that remains constant even if the SNR changes, the average asymptotic PEP

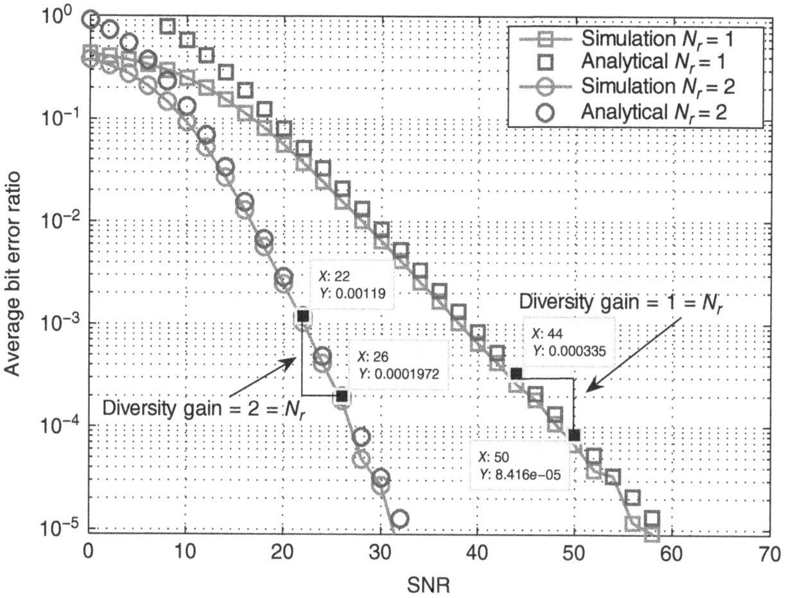

Figure 4.5 The derived analytical ABER of QSM over Rayleigh fading channels in the presence of CSE in (4.37) compared to the simulated ABER for $N_r = 1$and 2, $\eta = 4$, and $N_t = 2$.

is given by

$$E\left\{\Pr_{\text{error}}\right\} \approx \frac{2^{(N_r-1)}\Gamma(N_r+0.5)}{\sqrt{\pi}\Gamma(N_r+1)} \times \Xi_{\text{QSM}}^{N_r} \times \left(\frac{2\sigma_{\mathbf{e}}^2|s_{\tilde{l}}|^2}{1+\sigma_{\mathbf{e}}^2}\right)^{N_r}, \tag{4.40}$$

where $\Xi_{\text{QSM}} = \varphi_{\mathbf{e}}/\bar{\gamma}_{\mathbf{e}}^{\text{QSM}}$. From (4.40), it can be seen that the diversity gain is zero, where changing the SNR would not change the PEP performance.

Case II – $\sigma_{\mathbf{e}}^2 = (\text{SNR}^{-1} = \sigma_n^2)$: this is the general case, where pilot symbols are transmitted with the same energy as the data symbols. The asymptotic PEP is given by

$$E\{\Pr_{\text{error}}\} \approx \frac{2^{(N_r-1)}\Gamma(N_r+0.5)}{\sqrt{\pi}\Gamma(N_r+1)} \times \Xi_{\text{QSM}}^{N_r} \times (2(1+|s_{\tilde{l}}|^2))^{N_r}\left(\frac{1}{\sigma_n^2}\right)^{-N_r}. \tag{4.41}$$

Similar to the assumption of perfect channel knowledge, a diversity gain of N_r is achieved here as well.

4.2 A General Framework for SMTs Average Error Probability over Generalized Fading Channels and in the Presence of Spatial Correlation and Imperfect Channel Estimation

The previous analysis for SM, SSK, QSSK, and QSM systems is valid only over Rayleigh flat fading channels where the channel phase distribution is uniform. The performance of SMTs over generalized fading channels, such as Nakagami-m, Rician, $\eta-\mu$, and others, attracted significant attention in literature [45, 84, 170, 193–197]. However, in all these analyses, the phase distribution of the generalized channel fading distribution is assumed to be uniform, which is needed for mathematical tractability and simplified analysis. Though, this assumption leads to inaccurate conclusions. It was shown in [195, 196] that increasing the m value of the Nakagami-m distribution enhances the performance of SMTs. This conclusion means that SMTs performance will be significantly enhanced for very large values of m. However, for $m \to \infty$, the Nakagami-m channel becomes Gaussian, and MIMO communication over Gaussian channels is impossible since it is not possible to resolve the different channel paths. This conclusion was reported recently, and a general framework for SMTs performance analysis over generalized fading channels is presented in [48, 73, 75]. In this section, the general framework for SMTs performance analysis over generalized fading channels and in the presence of spatial correlation and imperfect channel estimation is discussed in detail.

Consider a general MIMO system where the transmitter is equipped with N_t transmit antennas and the receiver has N_r receive antennas. The received signal at any particular time is given by

$$\mathbf{y} = \widehat{\mathbf{H}}\mathbf{x}_t + \mathbf{n}, \tag{4.42}$$

where $\widehat{\mathbf{H}}$ denotes a correlated MIMO channel matrix given by

$$\widehat{\mathbf{H}} = \mathbf{R}_{\text{Rx}}^{\frac{1}{2}}\mathbf{H}\mathbf{R}_{\text{Tx}}^{\frac{1}{2}} \tag{4.43}$$

where $\mathbf{R}_{\text{Rx}}^{\frac{1}{2}}$ and $\mathbf{R}_{\text{Tx}}^{\frac{1}{2}}$ being, respectively, the receiver and transmitter spatial correlation matrices as defined in Chapter 2, and $\mathbf{H}$ is the MIMO channel matrix. Assuming CSE at the receiver, the ML-optimum receiver is written as

$$\hat{\mathbf{x}} = \arg\min_{\mathbf{x}\in\mathbf{X}} \left\|\mathbf{y} - \widehat{\mathbf{H}}_{\mathbf{e}}\mathbf{x}\right\|_{\text{F}}^{2}. \tag{4.44}$$

where $\mathbf{X}$ is 2^{η} size space containing all possible transmitted vectors $\mathbf{x}$, and $\widehat{\mathbf{H}}_{\mathbf{e}}$ is the estimated MIMO channel matrix

$$\widehat{\mathbf{H}} = \widehat{\mathbf{H}}_{\mathbf{e}} + \mathbf{e}. \tag{4.45}$$

Thus, the PEP of SMTs over generalized and correlated MIMO channel matrix in the presence of CSE is given by

$$
\begin{aligned}
\Pr_{\text{error}} &= \Pr(\mathbf{x}_t \rightarrow \mathbf{x}|\widehat{\mathbf{H}}_\mathbf{e}) \\
&= \Pr(\|\mathbf{y} - \widehat{\mathbf{H}}_\mathbf{e}\mathbf{x}_t\|_F^2 > \|\mathbf{y} - \widehat{\mathbf{H}}_\mathbf{e}\mathbf{x}\|_F^2|\widehat{\mathbf{H}}_\mathbf{e}) \\
&= \Pr(\|(\widehat{\mathbf{H}}\mathbf{x}_t + \mathbf{n}) - \widehat{\mathbf{H}}_\mathbf{e}\mathbf{x}_t\|_F^2 > \|(\widehat{\mathbf{H}}\mathbf{x}_t + \mathbf{n}) - \widehat{\mathbf{H}}_\mathbf{e}\mathbf{x}\|_F^2|\widehat{\mathbf{H}}_\mathbf{e}) \\
&= \Pr(\|(\widehat{\mathbf{H}} - \widehat{\mathbf{H}}_\mathbf{e})\mathbf{x}_t + \mathbf{n}\|_F^2 > \|\widehat{\mathbf{H}}\mathbf{x}_t - \widehat{\mathbf{H}}_\mathbf{e}\mathbf{x} + \mathbf{n}\|_F^2|\widehat{\mathbf{H}}_\mathbf{e}) \\
&= \Pr(\|((\widehat{\mathbf{H}}_\mathbf{e} + \mathbf{e}) - \widehat{\mathbf{H}}_\mathbf{e})\mathbf{x}_t + \mathbf{n}\|_F^2 > \|(\widehat{\mathbf{H}}_\mathbf{e} + \mathbf{e})\mathbf{x}_t - \widehat{\mathbf{H}}_\mathbf{e}\mathbf{x} + \mathbf{n}\|_F^2|\widehat{\mathbf{H}}_\mathbf{e}) \\
&= \Pr(\|\mathbf{e}\mathbf{x}_t + \mathbf{n}\|_F^2 > \|\widehat{\mathbf{H}}_\mathbf{e}(\mathbf{x}_t - \mathbf{x}) + (\mathbf{e}\mathbf{x}_t + \mathbf{n})\|_F^2|\widehat{\mathbf{H}}_\mathbf{e}) \\
&= \Pr(\|\bar{\mathbf{n}}\|_F^2 > \|\widehat{\mathbf{H}}_\mathbf{e}\boldsymbol{\Psi}\|_F^2 - 2\mathrm{Re}\{\bar{\mathbf{n}}^H\widehat{\mathbf{H}}_\mathbf{e}\boldsymbol{\Psi}\} + \|\bar{\mathbf{n}}\|_F^2|\widehat{\mathbf{H}}_\mathbf{e}) \\
&= \Pr(2\mathrm{Re}\{\bar{\mathbf{n}}^H\widehat{\mathbf{H}}_\mathbf{e}\boldsymbol{\Psi}\} > \|\widehat{\mathbf{H}}_\mathbf{e}\boldsymbol{\Psi}\|_F^2|\widehat{\mathbf{H}}_\mathbf{e}) \\
&= Q\left(\sqrt{\frac{\|\widehat{\mathbf{H}}_\mathbf{e}\boldsymbol{\Psi}\|_F^2}{2\sigma_{\bar{\mathbf{n}}}^2}}\right) = \frac{1}{\pi}\int_0^{\pi/2} \exp\left(-\frac{\bar{\gamma}\|\widehat{\mathbf{H}}_\mathbf{e}\boldsymbol{\Psi}\|_F^2}{2\sin^2\theta}\right) d\theta,
\end{aligned}
\tag{4.46}
$$

where $\boldsymbol{\Psi} = (\mathbf{x}_t - \mathbf{x})$, $\bar{\mathbf{n}} = \mathbf{e}\mathbf{x}_t + \mathbf{n} \sim \mathcal{CN}(\mathbf{0}_{N_r}, \sigma_{\bar{\mathbf{n}}}^2\mathbf{I}_{N_r})$, $\sigma_{\bar{\mathbf{n}}}^2 = \sigma_\mathbf{e}^2\|\mathbf{x}_t\|_F^2 + \sigma_n^2$, $\bar{\gamma} = 1/(2\sigma_{\bar{\mathbf{n}}}^2)$, and from [198], the alternative integral expression of the Q-function is $Q(x) = \frac{1}{\pi}\int_0^{\pi/2} \exp(-\frac{x^2}{2\sin^2\theta})d\theta$.

The average PEP is computed by taking the expectation of (4.46) as

$$
\mathrm{E}\left\{\Pr_{\text{error}}\right\} = \Pr(\mathbf{x}_t \rightarrow \mathbf{x}) = \frac{1}{\pi}\int_0^{\pi/2} \mathcal{M}_\vartheta\left(-\frac{\bar{\gamma}}{2\sin^2\theta}\right) d\theta, \tag{4.47}
$$

where $\mathcal{M}_\vartheta(\cdot)$ denotes the moment-generation function (MGF) of $\vartheta = \|\widehat{\mathbf{H}}_\mathbf{e}\boldsymbol{\Psi}\|_F^2$.

From [199], the Frobenius norm of $\|\widehat{\mathbf{H}}_\mathbf{e}\boldsymbol{\Psi}\|_F^2$ can be expanded as

$$
\|\widehat{\mathbf{H}}_\mathbf{e}\boldsymbol{\Psi}\|_F^2 = \mathrm{tr}(\widehat{\mathbf{H}}_\mathbf{e}\boldsymbol{\Psi}\boldsymbol{\Psi}^H\widehat{\mathbf{H}}_\mathbf{e}^H) = \widehat{\mathbf{H}}_{\mathbf{e},\mathbf{v}}^H(\mathbf{I}_{N_r} \otimes \boldsymbol{\Psi}\boldsymbol{\Psi}^H)\widehat{\mathbf{H}}_{\mathbf{e},\mathbf{v}}, \tag{4.48}
$$

where $\widehat{\mathbf{H}}_{\mathbf{e},\mathbf{v}} = \mathrm{vec}(\widehat{\mathbf{H}}_\mathbf{e}^H)$, $\mathrm{vec}(\mathbf{B})$ is the vectorization operator, where the columns of the matrix $\mathbf{B}$ are stacked in column vector, and $\mathrm{tr}(\cdot)$ is the trace function.

Let $\mathbf{v}$ be an identical and independently distributed (i.i.d.) complex random vector with mean $\mathbf{u}_\mathbf{v}$, covariance matrix $\mathbf{L}_\mathbf{v}$, and real and imaginary components with equal mean and variance. From [200] and for any Hermitian matrix $\mathbf{Q}$, the MGF of $f = \mathbf{v}^H\mathbf{Q}\mathbf{v}$ is

$$
\mathcal{M}_f(s) = \frac{\exp(s \times \mathbf{u}_\mathbf{v}{}^H\mathbf{Q}(\mathbf{I} - s\mathbf{L}_\mathbf{v}\mathbf{Q})^{-1}\mathbf{u}_\mathbf{v})}{|\mathbf{I}_{N_rN_t} - s\mathbf{L}_\mathbf{v}\mathbf{Q}|}. \tag{4.49}
$$

Hence, from (4.49) and (4.48), the MGF in (4.47) can be written as

$$\mathcal{M}_{\vartheta}\left(-\frac{\bar{\gamma}}{2\sin^2\theta}\right) = \frac{\exp\left(-\frac{\bar{\gamma}}{2\sin^2\theta} \times \mathbf{u}_{\tilde{\mathbf{H}}}^H \Lambda \left(\mathbf{I}_{N_rN_t} + \frac{\bar{\gamma}}{2\sin^2\theta}\mathbf{L}_{\tilde{\mathbf{H}}}\Lambda\right)^{-1} \mathbf{u}_{\tilde{\mathbf{H}}}\right)}{\left|\mathbf{I} + \frac{\bar{\gamma}}{2\sin^2\theta}\mathbf{L}_{\tilde{\mathbf{H}}}\Lambda\right|}, \tag{4.50}$$

where $\boldsymbol{\Lambda} = \mathbf{I}_{N_r} \otimes \boldsymbol{\Psi}\boldsymbol{\Psi}^H$ and

$$\mathbf{u}_{\tilde{\mathbf{H}}} = u_{\mathbf{H}}\mathbf{R}_{\mathbf{s}}^{\frac{1}{2}} \times \mathbf{1}_{N_rN_t}, \tag{4.51}$$

$$\mathbf{L}_{\tilde{\mathbf{H}}} = \sigma_{\mathbf{H}}^2\mathbf{R}_{\mathbf{s}} + \sigma_{\mathbf{e}}^2\mathbf{I}_{N_rN_t}, \tag{4.52}$$

where $u_{\mathbf{H}}$ and $\sigma_{\mathbf{H}}^2$ are the mean and variance of the channel $\mathbf{H}$, respectively, $\mathbf{R}_s = \mathbf{R}_{\mathrm{Rx}} \otimes \mathbf{R}_{\mathrm{Tx}}$, with $\otimes$ being the Kronecker product, and $\mathbf{1}_N$ is an N-length all ones vector.

Substituting (4.50) in (4.47) gives

$$\begin{aligned} \mathrm{E}\left\{\Pr_{\text{error}}\right\} &= \frac{1}{\pi}\int_0^{\pi/2} \frac{\exp\left(-\frac{\bar{\gamma}}{2\sin^2\theta} \times \mathbf{u}_{\tilde{\mathbf{H}}}^H \Lambda \left(\mathbf{I}_{N_rN_t} + \frac{\bar{\gamma}}{2\sin^2\theta}\mathbf{L}_{\tilde{\mathbf{H}}}\Lambda\right)^{-1} \mathbf{u}_{\tilde{\mathbf{H}}}\right)}{\left|\mathbf{I}_{N_rN_t} + \frac{\bar{\gamma}}{2\sin^2\theta}\mathbf{L}_{\tilde{\mathbf{H}}}\Lambda\right|} d\theta \\ &\leq \frac{1}{2}\frac{\exp\left(-\frac{\bar{\gamma}}{2} \times \mathbf{u}_{\tilde{\mathbf{H}}}^H \Lambda \left(\mathbf{I}_{N_rN_t} + \frac{\bar{\gamma}}{\sqrt{2}}\mathbf{L}_{\tilde{\mathbf{H}}}\Lambda\right)^{-1} \mathbf{u}_{\tilde{\mathbf{H}}}\right)}{\left|\mathbf{I} - \frac{\bar{\gamma}}{\sqrt{2}}\mathbf{L}_{\tilde{\mathbf{H}}}\Lambda\right|}. \end{aligned} \tag{4.53}$$

Finally, using (4.53), the ABER performance of SMTs over generalized correlated channels in the presence of CSE can be bounded by

$$\mathrm{ABER} \leq \frac{1}{2^\eta}\sum_{\ell,\imath}\sum_{\tilde{\ell},\tilde{\imath}} \frac{e_{\tilde{\ell},\tilde{\imath}}^{\ell,\imath}}{\eta}\frac{1}{2}\frac{\exp\left(-\frac{\bar{\gamma}}{2} \times \mathbf{u}_{\tilde{\mathbf{H}}}^H \Lambda \left(\mathbf{I}_{N_rN_t} + \frac{\bar{\gamma}}{\sqrt{2}}\mathbf{L}_{\tilde{\mathbf{H}}}\Lambda\right)^{-1} \mathbf{u}_{\tilde{\mathbf{H}}}\right)}{\left|\mathbf{I}_{N_rN_t} + \frac{\bar{\gamma}}{\sqrt{2}}\mathbf{L}_{\tilde{\mathbf{H}}}\Lambda\right|}. \tag{4.54}$$

Note, the ABER bound in (4.54) even though derived for SMTs is also valid for classical spatial multiplexing (SMX) MIMO systems as shown in the case of Rician fading channels in [199] and generalized fading channels in [54, 73, 74, 201]. Figure 4.6 validates the derived bound in (4.54), where it compares it to the simulated ABER of two SMT systems, SM and QSM, over correlated Rayleigh and Nakagami-$m = 4$ fading channels, and in the presence of CSE, where $\sigma_{\mathbf{e}}^2 = \sigma_n^2$, $\eta = 6$, and $N_t = N_r = 4$. As can be seen from the figure, analytical and simulation results demonstrate close match for a wide and pragmatic range of SNR values and for the different channel conditions.

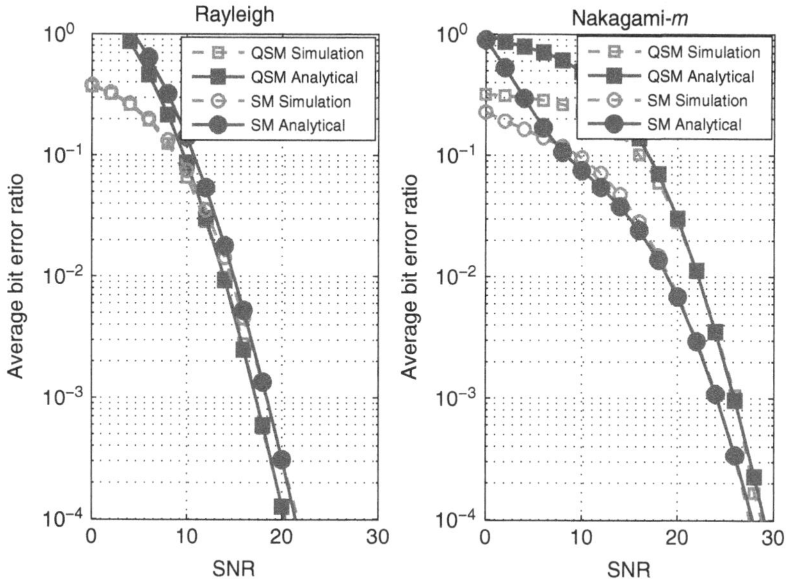

Figure 4.6 The derived analytical ABER of SMTs in (4.54) compared with the simulated ABER of SM and QSM over correlated Rayleigh and Nakagami-$m = 4$ fading channels in the presence of CSE, where $\sigma_{\mathbf{e}}^2 = \sigma_n^2$, $\eta = 6$, and $N_t = N_r = 4$.

4.3 Average Error Probability Analysis of Differential SMTs

Performance analysis of differential space modulation techniques (DSMT) is presented hereinafter. The PEP for a DSMT can be formulated as follows:

$$\begin{aligned}\Pr_{\text{error}} &= \Pr(\mathbf{X}_t \rightarrow \mathbf{X}_i|\mathbf{H})\\ &= \Pr(\|\mathbf{Y}_t - \mathbf{Y}_{t-1}\mathbf{X}_t\|_{\mathrm{F}}^2 > \|\mathbf{Y}_t - \mathbf{Y}_{t-1}\mathbf{X}_i\|_{\mathrm{F}}^2|\mathbf{H})\\ &= \Pr(\|\mathbf{Z}_t - \mathbf{Z}_{t-1}\mathbf{X}_t\|_{\mathrm{F}}^2 > \|\mathbf{Y}_{t-1}\dot{\mathbf{\Psi}} + \mathbf{Z}_t - \mathbf{Z}_{t-1}\mathbf{X}_t\|_{\mathrm{F}}^2|\mathbf{H})\\ &= \Pr(\|\mathbf{Z}_t - \mathbf{Z}_{t-1}\mathbf{X}_t\|_{\mathrm{F}}^2 > \|\dot{\mathbf{G}} + \mathbf{Z}_t - \mathbf{Z}_{t-1}\mathbf{X}\|_{\mathrm{F}}^2|\mathbf{H}),\end{aligned} \tag{4.55}$$

where $\dot{\mathbf{\Psi}} = \mathbf{X}_t - \mathbf{X}_i$, and $\dot{\mathbf{G}} = \mathbf{H}\mathbf{X}_{t-1}\dot{\mathbf{\Psi}}$.

Now, the left-hand side of the inequality in (4.55) can be written as

$$\|\mathbf{Z}_t - \mathbf{Z}_{t-1}\mathbf{X}_t\|_{\mathrm{F}}^2 = \|\mathbf{Z}_t\|_{\mathrm{F}}^2 - 2\mathrm{Re}\{\mathrm{tr}(\mathbf{Z}_t^H\mathbf{Z}_{t-1}\mathbf{X}_t)\} + \|\mathbf{Z}_{t-1}\mathbf{X}_t\|_{\mathrm{F}}^2, \tag{4.56}$$

Moreover, the right-hand side can be written as

$$\begin{aligned}\|\dot{\mathbf{G}} + \mathbf{Z}_t - \mathbf{Z}_{t-1}\mathbf{X}\|_{\mathrm{F}}^2 &= \|\dot{\mathbf{G}}\|_{\mathrm{F}}^2 - 2\mathrm{Re}\{\mathrm{tr}(\dot{\mathbf{G}}^H(\mathbf{Z}_t - \mathbf{Z}_{t-1}\mathbf{X}))\}\\ &\quad + \|\mathbf{Z}_t\|_{\mathrm{F}}^2 - 2\mathrm{Re}\{\mathrm{tr}(\mathbf{Z}_t^H\mathbf{Z}_{t-1}\mathbf{X})\} + \|\mathbf{Z}_{t-1}\mathbf{X}\|_{\mathrm{F}}^2.\end{aligned} \tag{4.57}$$

From (4.56) and (4.57), the PEP in (4.55) can be simplified to

$$\Pr_{\text{error}} = \Pr(2\text{Re}\{\text{tr}(\dot{\mathbf{G}}^H(\mathbf{Z}_t - \mathbf{Z}_{t-1}\mathbf{X}) - \mathbf{Z}_t^H\mathbf{Z}_{t-1}\dot{\mathbf{\Psi}})\} > \|\dot{\mathbf{G}}\|_\text{F}^2). \tag{4.58}$$

Note that $\|\mathbf{Z}_{t-1}\mathbf{X}_t\|_\text{F}^2 = \|\mathbf{Z}_{t-1}\mathbf{X}\|_\text{F}^2$ because $\mathbf{X}_t\mathbf{X}_t{}^H = \mathbf{X}_i\mathbf{X}_i{}^H = \mathbf{I}_{N_t}$.

Unfortunately, no closed-form expression is available for the PEP in (4.58). Therefore, an approximate expression will be targeted in what follows. The approximation is based on assuming that $\mathbf{Z}_t^H\mathbf{Z}_{t-1} \approx 0$ for high SNR values [[120], p. 274]. Notice that both $\mathbf{Z}_t$ and $\mathbf{Z}_{t-1}$ are Gaussian random variables with zero mean and σ_n^2 variance. Therefore, the mean of their product is also zero, and the variance is σ_n^4. Consequently, as SNR increases, σ_n^4 approaches zero and so the variance of $\mathbf{Z}_t^H\mathbf{Z}_{t-1}$.

Based on the previous approximation, (4.58) can be rewritten as

$$\begin{aligned}\Pr_{\text{error}} &\approx \Pr(2\text{Re}\{\text{tr}(\dot{\mathbf{G}}^H(\mathbf{Z}_t - \mathbf{Z}_{t-1}\mathbf{X}))\} > \|\dot{\mathbf{G}}\|_\text{F}^2)\\ &\approx Q(\dot{\gamma}\|\dot{\mathbf{G}}\|_\text{F}^2) \approx \frac{1}{\pi}\int_0^{\frac{\pi}{2}} \exp\left(-\frac{\dot{\gamma}\|\dot{\mathbf{G}}\|_\text{F}^2}{2\sin^2\theta}\right) d\theta.\end{aligned} \tag{4.59}$$

The average PEP is computed by taking the expectation of the PEP in (4.59)

$$\text{E}\left\{\Pr_{\text{error}}\right\} \approx \frac{1}{\pi}\int_0^{\pi/2} \mathcal{M}_{\dot{\vartheta}}\left(-\frac{\dot{\gamma}}{2\sin^2\theta}\right) d\theta, \tag{4.60}$$

where $\mathcal{M}_{\dot{\vartheta}}$ denotes the MGF of $\dot{\vartheta} = \|\dot{\mathbf{G}}\|_\text{F}^2$.

Based on [123], the variable $\|\dot{\mathbf{G}}\|_\text{F}^2$ can be expanded as

$$\|\dot{\mathbf{G}}\|_\text{F}^2 = \text{tr}\{\dot{\mathbf{D}}\dot{\mathbf{\Psi}}\dot{\mathbf{\Psi}}^H\dot{\mathbf{D}}^H\} = \text{vec}(\dot{\mathbf{D}}^H)^H(\mathbf{I}_{N_r} \otimes \dot{\mathbf{\Psi}}\dot{\mathbf{\Psi}}^H)\text{vec}(\dot{\mathbf{D}}^H). \tag{4.61}$$

As in the previous section, the MGF of the variable $\dot{\vartheta}$ can be expressed using the quadratic form expression as [48, 73, 200]

$$\mathcal{M}_{\dot{\vartheta}}(s) = \frac{\exp(s \times \mathbf{u}_{\dot{\mathbf{D}}}{}^H\dot{\Lambda}(\mathbf{I}_{N_rN_t} - s\mathbf{L}_{\dot{\mathbf{D}}}\dot{\Lambda})^{-1}\mathbf{u}_{\dot{\mathbf{D}}})}{|\mathbf{I}_{N_tN_r} - s\mathbf{L}_{\dot{\mathbf{D}}}\dot{\Lambda}|}, \tag{4.62}$$

where $\mathbf{u}_{\dot{\mathbf{D}}}$ and $\mathbf{L}_{\dot{\mathbf{D}}}$ are the mean vector and the covariance matrix of the vector $\text{vec}(\dot{\mathbf{D}}^H)$, respectively, and $\dot{\mathbf{\Lambda}} = \mathbf{I}_{N_r} \otimes \dot{\mathbf{\Psi}}\dot{\mathbf{\Psi}}^H$.

Using (4.62), the average PEP can be upper bounded by

$$\text{E}\{\Pr_{\text{error}}\} \leq \frac{1}{2}\frac{\exp\left(\frac{\dot{\gamma}}{2} \times \mathbf{u}_{\dot{\mathbf{D}}}{}^H\dot{\Lambda}(\mathbf{I}_{N_rN_t} + \frac{\dot{\gamma}}{\sqrt{2}}\mathbf{L}_{\dot{\mathbf{D}}}\dot{\Lambda})^{-1}\mathbf{u}_{\dot{\mathbf{D}}}\right)}{\left|\mathbf{I}_{N_tN_r} + \frac{\dot{\gamma}}{\sqrt{2}}\mathbf{L}_{\dot{\mathbf{D}}}\dot{\Lambda}\right|}. \tag{4.63}$$

Finally, the ABER performance of DSMT can be bounded as

$$\text{ABER} \leq \frac{1}{2^\eta}\sum_{\mathbf{X}_t\in\dot{\mathcal{X}}}\sum_{\mathbf{X}_i\in\dot{\mathcal{X}}} \frac{e_{\mathbf{X}_i}^{\mathbf{X}_t}}{\eta}\frac{1}{2}\frac{\exp\left(\frac{\dot{\gamma}}{2} \times \mathbf{u}_{\dot{\mathbf{D}}}{}^H\dot{\Lambda}(\mathbf{I}_{N_rN_t} + \frac{\dot{\gamma}}{\sqrt{2}}\mathbf{L}_{\dot{\mathbf{D}}}\dot{\Lambda})^{-1}\mathbf{u}_{\dot{\mathbf{D}}}\right)}{\left|\mathbf{I}_{N_tN_r} + \frac{\dot{\gamma}}{\sqrt{2}}\mathbf{L}_{\dot{\mathbf{D}}}\dot{\Lambda}\right|}, \tag{4.64}$$

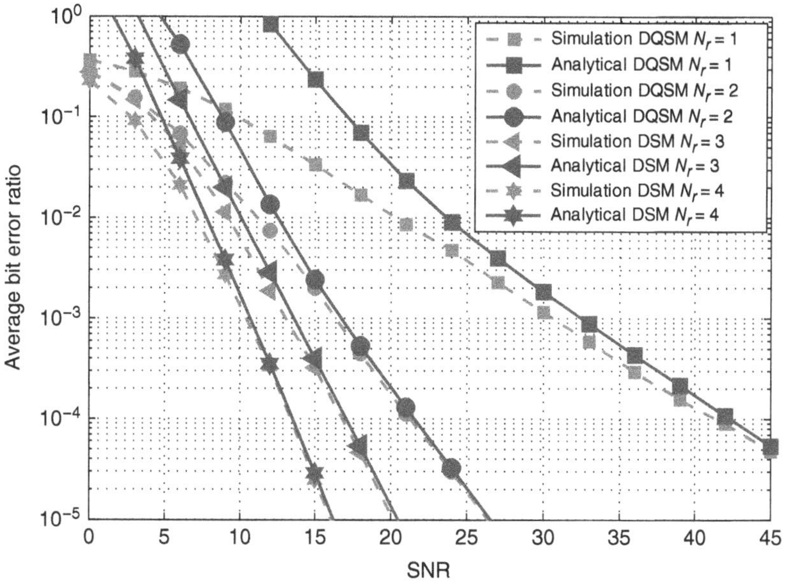

Figure 4.7 The derived analytical ABER of DSMTs in (4.64) compared with the simulated ABER of DSM and DQSM for $M = 4$-QAM, $N_t = 3$, $N_r = 1, 2, 3$, and 4.

where $e_{\mathbf{x}_i}^{\mathbf{x}_t}$ is the number of bits error associated with the corresponding PEP event. In Figure 4.7, the derived bound in (4.64) is compared to the simulated ABER of differential spatial modulation (DSM) with $N_r = 3$ and 4, and differential quadrature spatial modulation (DQSM) with $N_r = 1$ and 2, where $N_t = 3$, and $M = 4$-quadrature amplitude modulation (QAM) is used. From Figure 4.7, the simulated ABER validates the bound where it closely follows the bound for a wide range of SNR values.

4.4 Comparative Average Bit Error Rate Results

In this section, Monte Carlo simulation results are presented to study the ABER performance of SMTs, generalized space modulation techniques (GSMTs), quadrature space modulation techniques (QSMTs), and DSMTs system with different configurations.

4.4.1 SMTs, GSMTs, and QSMTs ABER Comparisons

In Figure 4.8, the ABER of SMTs is presented and compared with SMX for $\eta = 8$ bits. From the figure, it can be seen that QSSK offers the best performance

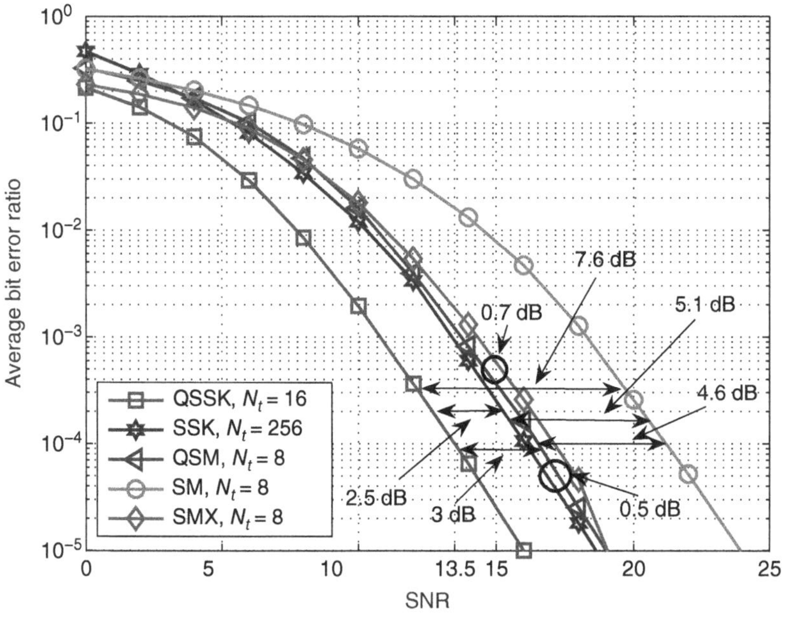

Figure 4.8 ABER performance comparison between the SMTs, QSSK, SSK, QSM and SM, and SMX systems over Rayleigh fading channels for different number of transmit antennas and modulation orders achieving $\eta = 8$ bits with $N_r = 4$ antennas.

compared to the rest of the SMTs, 2.5, 3, 3.7, and 7.6 dB better than SSK, QSM, SMX, and SM, respectively. This is because in QSSK all information bits are modulated in the spatial domain which is in Rayleigh fading channels more robust that the signal domain. This can also be seen in SSK where it performs better than QSM, SMX, and SM, 0.5, 1.2, and 5.1 dB, respectively. QSM and SM modulate part of the information bits in the signal domain as will as the spatial domain. However and even though QSSK and SSK modulate all information bits in the spatial domain, SSK performs 2.5 dB less than QSSK. This can be attributed to the need of much more transmit antennas in SSK than QSSK. In Figure 4.8, QSSK needs only $N_t = 16$, while SSK requires $N_t = 256$. Comparing QSM to SM, QSM modulates more bits in the spatial domain than SM, and it offers 4.6 dB better performance than SM. With $N_t = 8$, QSM uses 4-QAM to achieve $\eta = 8$ bits. However, SM uses 32-QAM to achieve the same spectral efficiency assuming similar number of transmit antennas. Finally, it can be seen that SMTs that modulate large bits in the spatial domain performs better than SMX. Yet, SMTs like SM that uses large signal constellation diagram perform worse than SMX.

The ABER results of SMTs and SMX over Nakagami-$m = 4$ fading channels for $\eta = 8$ and $N_r = 4$ are presented in Figure 4.9, where $N_t = 16$ for QSSK,

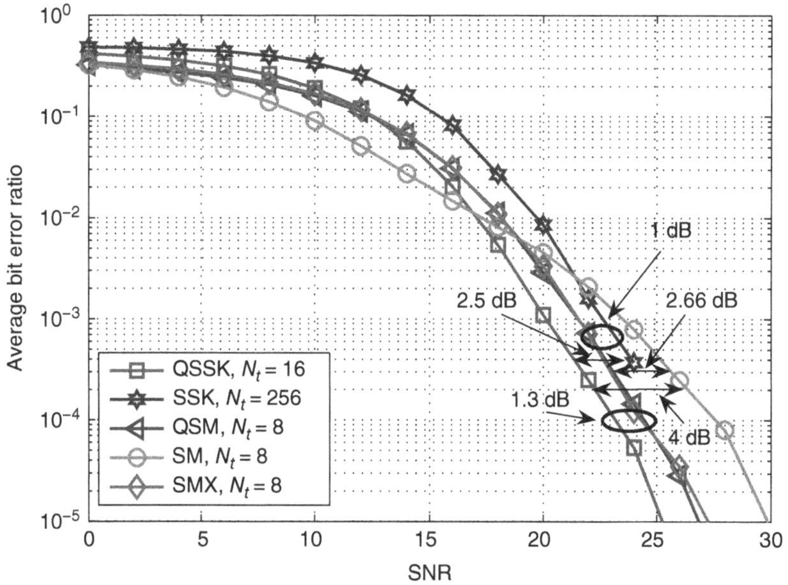

Figure 4.9 ABER performance comparison between the SMTs, QSSK, SSK, QSM and SM, and SMX systems over Nakagami-m = 4 fading channels for different number of transmit antennas and modulation orders achieving η = 8 bits with N_r = 4 antennas.

$N_t = 256$ for SSK, and $N_t = 8$ for SM, QSM, and SMX. As in the case for Rayleigh fading channels in Figure 4.8, QSSK over Nakagami-m fading channels offers the best performance with 1.3, 2.5, and 4 dB better performance than SMX and QSM, SSK, and SM, respectively. This can be attributed to the same reason as before where there exist no signal constellations and all the information bits are modulated in the spatial domain. Compared to Rayleigh, the performance gap between QSSK and the other systems is smaller, because it is harder to distinguish the different transmit antennas in Nakagami-m fading channel compared to Rayleigh fading channels. This can also be seen in SSK, where it performs 1 dB worse than QSM even though it outperforms QSM performance over Rayleigh fading channels as shown in Figure 4.8. Furthermore, SM still demonstrates the worst performance as it uses large signal constellation diagram compared to all other systems. Finally, QSM is shown to offer nearly the same performance as SMX.

Figure 4.10 depicts the ABER performance of generalized quadrature space shift keying (GQSSK) with $N_t = 8$, generalized space shift keying (GSSK) with $N_t = 24$, generalized quadrature spatial modulation (GQSM) with $N_t = 5$, and generalized spatial modulation (GSM) with $N_t = 8$, and compares their performance to SMX with $N_t = 8$ over Rayleigh fading channel. All systems achieve $\eta = 8$, and $N_r = 4$ is considered in all systems. From the figure, GQSSK offers

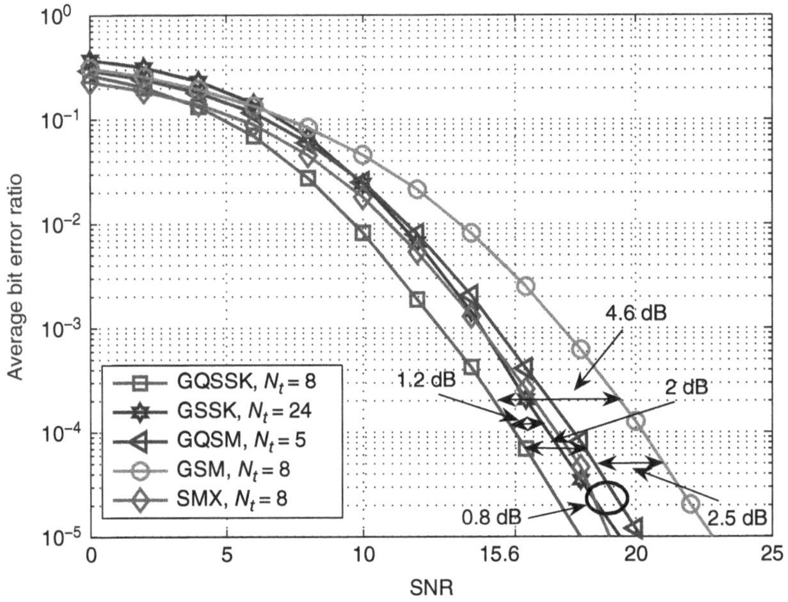

Figure 4.10 ABER performance comparison between GSMTs, GQSSK, GSSK, GQSM and GSM, and SMX systems over Rayleigh fading channels for different number of transmit antennas and modulation orders achieving $\eta = 8$ bits with $N_r = 4$ antennas.

the best performance with 1.2, 2, and 4.6 dB better performance than GSSK and SMX, GQSM, and GSM, respectively. As in QSSK, GQSSK offers this performance because all information bits are modulated in the spatial domain, while using small number of transmit antennas. This also can be seen for GSSK in Figure 4.10, where it offers the same performance as SMX and better performance than both GQSM and GSM. However, GQSSK performs better than GSSK as it uses less number of transmit antennas. Comparing the performance of QSSK in Figure 4.8 to GQSSK in Figure 4.10, it can be observed that QSSK achieves an ABER of 10^{-4} at an SNR = 13.5 dB while GQSSK achieves the same ABER at an SNR = 15.6 dB, i.e. GQSSK even though it uses less number of transmit antennas it performs 2.1 dB worse than QSSK. Even though GSMTs in general reduce the number of transmit antennas, they create correlation between the different spatial symbols, which increases the ABER. That is unlike QSMTs, where the reduction in the number of transmit antennas and the increase in the number of bits modulated in the spatial domain are attained while keeping the spatial symbols orthogonal.

In summary, the more bits modulated in the spatial domain, the better is the performance of SMTs. Increasing the number of bits modulated in the spatial domain is best achieved through QSMTs. To increase the data rate, it is better to increase the number of transmit antennas, which as discussed in previous chapter comes at no extra hardware or computational complexity, nor energy consumption. Another way is to use GSMTs and generalized quadrature space modulation techniques (GQSMTs), which need less number of transmit antennas. However, they offer slightly worse performance than SMTs and QSMTs. Finally, SMTs offer the same performance as SMX or better, if the number of bits modulated in the signal domain is relatively similar to SMX. Though, as the number of bits modulated in the signal domain for SMTs increases, as compared to SMX, the worse the performance will be.

4.4.2 Differential SMTs Results

A comparison between DQSM with $N_t = 2$ and 4-QAM and DSM with $N_t = 4$ and 8-phase shift keying (PSK) is illustrated in Figure 4.11. Results for QSM with $N_t = 4$ and 4-QAM, SM with $N_t = 4$ and 4-QAM, and SMX over Rayleigh fading channels and assuming $\eta = 4$ are depicted as well. Note, DSM and DQSM do not need channel state information (CSI) at the receiver. However, SM, QSM, and SMX assume full CSI knowledge at the receiver.

Interesting conclusions can be obtained from the depicted results in Figure 4.11. First, the ABER curves of DSM and DQSM intersect at a

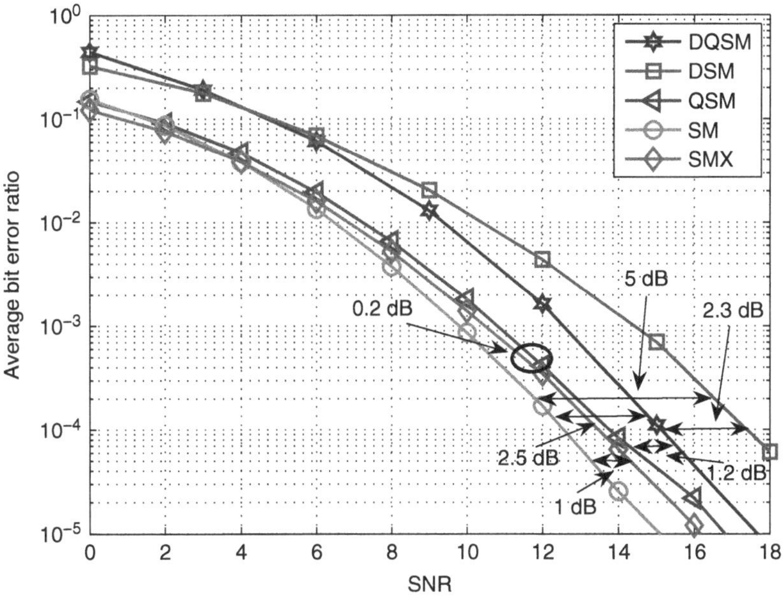

Figure 4.11 ABER performance comparison between DQSM with $N_t = 2$ and 4-QAM, DSM with $N_t = 4$ and 8-PSK, QSM with $N_t = 4$ and 4-QAM, SM with $N_t = 4$ and 4-QAM, and SMX with $N_t = 4$ and BPSK over Rayleigh fading channels for $\eta = 4$ bits and $N_r = 4$ antennas.

SNR ~ 4 dB, where DSM demonstrates slightly better performance for SNR < 4 dB, whereas DQSM outperforms DSM for SNR > 4 dB. At pragmatic ABER values of about 10^{-4}, DQSM outperforms DSM by around 2.3 dB. In addition, around the same ABER, SM is shown in Figure 4.11 to outperform DQSM and DSM by about 2.5 dB and 5 dB, respectively. Furthermore, QSM outperforms DQSM and DSM by about 1.2 and 2.5 dB, respectively.

Finally, it can be seen from Figure 4.11 that SMX offers better performance than both DSMTs, while offering nearly the same performance as QSM, and 2.5 dB worse performance than SM. It should be noted, though, that different configurations might lead to different performances, but the trend is likely to remain the same.

In summary, DSMTs offer a relatively good performance without the need for a full CSI at the receiver, while retaining all advantages of SMTs since one antenna is active at a time and transmitter deployment through single radio frequency (RF)-chain is anticipated.

5

Information Theoretic Treatment for SMTs

Previously discussed space modulation techniques (SMTs) are novel wireless communication systems that deploy multiple transmit antennas at the transmitter and uses spatial symbols to convey additional information bits. They propose a new way to convey information between a source and destination nodes that is not trivial. One of the major elements to fully understand the capabilities of SMTs is the derivation of the capacity for such techniques. Several attempts were made in literature to derive the capacity of SMTs, and different assumptions were made to facilitate such analysis [202–206]. Most existing studies derive SMTs capacity by following conventional multiple–input multiple–output (MIMO) capacity derivation. A common assumption in literature is that SMTs have two information symbols: spatial and signal symbols. Spatial information symbols are represented by the indexes of the different transmit antennas, while signal symbols are drawn from arbitrary signal constellation diagram. However, representing the spatial symbols by the indexes of transmit antennas is not accurate. Assuming for instance that there exist multiple transmit antennas and are located at the same spatial location in space, the size of the spatial constellation diagram is one and no data can be conveyed through spatial symbols. Hence, the indexes of the transmit antennas are not the source of spatial information, and the spatial bits are encoded in the Euclidean difference between the different channel paths associated with the different transmit antennas. As such, spatial symbols should be represented by the channel vectors associated with the transmit antennas. The assumption that the indexes of the transmit antennas are the source of the spatial information led to the conclusion that SMT capacity is different for different channel statistics and can be achieved if the signal constellation follows a complex Gaussian distribution, similar to conventional spatial multiplexing (SMX) systems [25]. However, in SMT, data are transmitted by an ordinary symbol drawn from arbitrary constellation diagram and by activating a single antenna among the set of available transmit antennas. Hence, the capacity analysis for SMTs is not trivial and requires investigation beyond existing theory.

Space Modulation Techniques, First Edition. Raed Mesleh and Abdelhamid Alhassi.

Considering the working mechanism of SMTs where two information symbols are transmitted and jointly decoded, a joint consideration of spatial and signal symbols is needed when deriving the theoretical capacity. In this chapter, information theoretic treatment of SMTs is presented and discussed. It is shown that the mutual information of SMTs approaches the capacity limit if the distribution of the signal constellation symbols multiplied by the spatial constellation symbols follows a complex Gaussian distribution. Therefore and for each channel statistics, the distribution of the signal symbols must be shaped such that the product of the channel and the symbol is complex Gaussian distribution. First attempts in this direction were foreseen in [51, 55], where capacity analysis for quadrature spatial modulation (QSM) over line-of-sight (LOS) and 3D-millimeter-wave (mmWave) channels was reported.

5.1 Evaluating the Mutual Information

5.1.1 Classical Spatial Multiplexing MIMO

To fully understand the differences between both classical MIMO systems and SMTs, the mutual information of SMX-MIMO system is derived. In SMX-MIMO system, there exist no spatial symbols and only signal symbols are present. Incoming data bits modulate complex signal symbols, and these symbols are the only means for conveying information. In principle, N_t signal symbols are transmitted simultaneously from N_t transmit antennas [25, 207–209].

By definition, the mutual information, $I(\mathbf{X};\mathbf{Y})$, is the amount of information gained about the transmitted vector space $\mathbf{X}$ when knowing the received vector space $\mathbf{Y}$, and is given by

$$I(\mathbf{X};\mathbf{Y}) = \mathrm{E}_{\mathbf{H}}\{I(\mathbf{X};\mathbf{Y}|\mathbf{H})\} = \mathrm{E}_{\mathbf{H}}\{H(\mathbf{Y}|\mathbf{H}) - H(\mathbf{Y}|\mathbf{X},\mathbf{H})\}, \tag{5.1}$$

where $H(\cdot)$ denotes the entropy function.

The entropy of the received vector $\mathbf{Y}$ knowing $\mathbf{H}$ is

$$\begin{aligned} H(\mathbf{Y}|\mathbf{H}) &= -\int_{\mathbf{Y}} p_{\mathbf{Y}|\mathbf{H}}(\mathbf{y}|\mathbf{H})\log_2 p_{\mathbf{Y}|\mathbf{H}}(\mathbf{y}|\mathbf{H})\, d\mathbf{y} \\ &= -\mathrm{E}_{\mathbf{Y}|\mathbf{H}}\{\log_2 p_{\mathbf{Y}|\mathbf{H}}(\mathbf{y}|\mathbf{H})\}, \end{aligned} \tag{5.2}$$

where $\|\cdot\|_{\mathrm{F}}$ is the Frobenius norm, and $p_{\mathbf{Y}|\mathbf{H}}(\cdot)$ is the probability distribution function (PDF) of $\mathbf{Y}$ given $\mathbf{H}$, and is given by

$$\begin{aligned} p_{\mathbf{Y}|\mathbf{H}}(\mathbf{y}|\mathbf{H}) &= \int_{\mathbf{X}} p_{\mathbf{X}}(\mathbf{x}_t) p_{(\mathbf{Y}|\mathbf{X},\mathbf{H})}(\mathbf{y}|\mathbf{x}_t,\mathbf{H})\, d\mathbf{x}_t \\ &= \frac{1}{(\pi\sigma_n^2)^{N_r}} \mathrm{E}_{\mathbf{X}}\left\{\exp\left(\frac{-\|\mathbf{y}-\mathbf{H}\mathbf{x}_t\|_{\mathrm{F}}^2}{\sigma_n^2}\right)\right\}, \end{aligned} \tag{5.3}$$

where $p_{(\mathbf{Y}|\mathbf{X},\mathbf{H})}$ is the PDF of $\mathbf{Y}$ given $\mathbf{X}$ and $\mathbf{H}$, and is given by

$$p_{(\mathbf{Y}|\mathbf{X},\mathbf{H})}(\mathbf{y}|\mathbf{x}_t,\mathbf{H}) = \frac{1}{(\pi\sigma_n^2)^{N_r}}\exp\left(\frac{-\|\mathbf{y}-\mathbf{H}\mathbf{x}_t\|_{\mathrm{F}}^2}{\sigma_n^2}\right). \tag{5.4}$$

The received vector $\mathbf{y}$ was defined previously in Chapter 2 and is given by

$$\mathbf{y} = \mathbf{H}\mathbf{x}_t + \mathbf{n}, \tag{5.5}$$

where $\mathbf{n}$ is the noise vector with each element $n \sim \mathcal{CN}(0,\sigma_n^2)$. Hence, assuming deterministic $\mathbf{H}$ and $\mathbf{x}_t$, $\mathbf{y} \sim \mathcal{CN}(\mathbf{H}\mathbf{x}_t, \sigma_n^2\mathbf{I}_{N_r})$.

From (5.2) and (5.3), the entropy of $\mathbf{y}$ given $\mathbf{H}$ is

$$\begin{aligned} H(\mathbf{Y}|\mathbf{H}) &= -\mathrm{E}_{\mathbf{Y}|\mathbf{H}}\left\{\log_2\left(\left(\pi\sigma_n^2\right)^{-N_r}\mathrm{E}_{\mathbf{X}}\left\{\exp\left(\frac{-\|\mathbf{y}-\mathbf{H}\mathbf{x}_t\|_{\mathrm{F}}^2}{\sigma_n^2}\right)\right\}\right)\right\} \\ &= N_r\log_2\left(\pi\sigma_n^2\right) \\ &\quad - \mathrm{E}_{\mathbf{Y}|\mathbf{H}}\left\{\log_2\left(\mathrm{E}_{\mathbf{X}}\left\{\exp\left(\frac{-\|\mathbf{y}-\mathbf{H}\mathbf{x}_t\|_{\mathrm{F}}^2}{\sigma_n^2}\right)\right\}\right)\right\}. \end{aligned} \tag{5.6}$$

The entropy of $\mathbf{Y}$ given $\mathbf{X}$ and $\mathbf{H}$ is [25]

$$\begin{aligned} H(\mathbf{Y}|\mathbf{X},\mathbf{H}) &= \mathrm{E}_{\mathbf{X}}\{H(\mathbf{Y}|\mathbf{X}=\mathbf{x}_t,\mathbf{H})\} \\ &= \mathrm{E}_{\mathbf{X}}\{-\mathrm{E}_{\mathbf{Y}|\mathbf{X},\mathbf{H}}\{\log_2 p_{\mathbf{Y}|\mathbf{X},\mathbf{H}}(\mathbf{y}|\mathbf{x}_t,\mathbf{H})\}\} \\ &= \mathrm{E}_{\mathbf{X}}\left\{-\mathrm{E}_{\mathbf{Y}|\mathbf{X},\mathbf{H}}\left\{\log_2\left(\frac{1}{\left(\pi\sigma_n^2\right)^{N_r}}\exp\left(\frac{-\|\mathbf{y}-\mathbf{H}\mathbf{x}_t\|_{\mathrm{F}}^2}{\sigma_n^2}\right)\right)\right\}\right\} \\ &= N_r\log_2\left(\pi\sigma_n^2\right) + \frac{\mathrm{E}_{\mathbf{X}}\left\{\overbrace{\mathrm{E}_{\mathbf{Y}|\mathbf{X},\mathbf{H}}\left\{\|\mathbf{y}-\mathbf{H}\mathbf{x}_t\|_{\mathrm{F}}^2\right\}}^{N_r\sigma_n^2}\right\}}{\sigma_n^2}\log_2(\exp(1)) \\ &= N_r\log_2\left(\pi\sigma_n^2\exp(1)\right). \end{aligned} \tag{5.7}$$

Finally, substituting (5.6) and (5.7) in (5.1) gives

$$\begin{aligned} I(\mathbf{X};\mathbf{Y}) &= -N_r\log_2(\exp(1)) \\ &\quad - \mathrm{E}_{\mathbf{H}}\left\{\mathrm{E}_{\mathbf{Y}|\mathbf{H}}\left\{\log_2\left(\mathrm{E}_{\mathbf{X}}\left\{\exp\left(\frac{-\|\mathbf{y}-\mathbf{H}\mathbf{x}_t\|_{\mathrm{F}}^2}{\sigma_n^2}\right)\right\}\right)\right\}\right\}. \end{aligned} \tag{5.8}$$

5.1.2 SMTs

When deriving the mutual information for SMTs, the way information bits are modulated and transmitted needs to be considered. In SMX, as discussed above, all transmitted information bits are modulated in the N_t-length vector $\mathbf{x}_t$, which is transmitted simultaneously over the MIMO channel matrix $\mathbf{H}$. However, the communication protocol is totally different in SMTs.

For simplicity, let us first consider a space shift keying (SSK) transmitted signal over a multiple-input single-output (MISO) channel. In an SSK system, incoming information bits activate a transmit antenna index, ℓ, to transmit a constant symbol, say +1. Hence, the received signal y^{SSK} is

$$y^{\text{SSK}} = h_\ell + n, \tag{5.9}$$

where h_ℓ is the ℓth channel element. In (5.9), the information bits are not modulated in ℓ. Rather, h_ℓ is the spatial symbol that conveys information bits. To explain this further, consider a single-input single-output (SISO)-additive white Gaussian noise (AWGN) channel transmitting the symbol $x_\iota \in \mathbf{X}$. Hence, the received signal is

$$y^{\text{SISO}} = x_\iota + n. \tag{5.10}$$

In (5.10), the information bits are modulated in x_ι, and ι is just an index representing which symbol out of the available symbols is transmitted and contains no information. Now, comparing (5.10) with (5.9) clearly highlight that the incoming data bits modulate a spatial symbol h_ℓ from $\mathcal{H}$, which is done by activating only one antenna at a time. Therefore, ℓ is just an index that contains no information, and all information bits are modulated in the different $\mathbf{h}_\ell$ vectors.

To elaborate further, let us now compare the received signals in spatial modulation (SM) and SMX for a MISO system,

$$y^{\text{SM}} = h_\ell s_\iota + n, \tag{5.11}$$

$$y^{\text{SMX}} = \mathbf{hx} + n, \tag{5.12}$$

where $\mathbf{h}$ is an N_t-dimensional channel vector. From (5.11) and (5.12), it can be seen that, different to SMX, SM modulates information bits in the channel and in the transmitted signal, where the information bits are transmitted in the spatial and signal symbols h_ℓ and s_ι, respectively. Furthermore, as in SSK, ℓ is used as an index to differentiate between the different N_t channel elements of $\mathbf{h}$ and carries no information. The different N_t elements of $\mathbf{h}$ are the spatial symbols, h_ℓ, that carry information bits.

In summary, the information bits in SMTs are modulated in the spatial symbol, $\mathcal{H}_\ell$, and the signal symbol, $\mathcal{S}_\iota$. Therefore, the mutual information is the amount of information gained about both the spatial and signal constellation spaces $\mathcal{H}$ and $\mathcal{S}$ by knowing the received vector space $\mathbf{Y}$, and is given by

$$I(\mathcal{H}, \mathcal{S}; \mathbf{Y}) = H(\mathbf{Y}) - H(\mathbf{Y}|\mathcal{H}, \mathcal{S}). \tag{5.13}$$

It is important to note that there is no averaging over the channel $\mathbf{H}$ in (5.13) since $\mathbf{H}$ is used to convey information, where the spatial constellation space, $\mathcal{H}$, is generated from $\mathbf{H}$.

The entropy of $\mathbf{Y}$ is

$$H(\mathbf{Y}) = -\mathrm{E}_{\mathbf{Y}}\{\log_2 p_{\mathbf{Y}}(\mathbf{y})\}, \tag{5.14}$$

where

$$\begin{aligned} p_{\mathbf{Y}}(\mathbf{y}) &= \int_{\mathcal{H}}\int_{\mathcal{S}} p_{\mathcal{H}}(\mathcal{H}_\ell) p_{\mathcal{S}}(\mathcal{S}_\iota) p_{(\mathbf{Y}|\mathcal{H},\mathcal{S})}(\mathbf{y}|\mathcal{H}_\ell, \mathcal{S}_\iota)\, d\mathcal{H}_\ell\, d\mathcal{S}_\iota \\ &= \frac{1}{\left(\pi\sigma_n^2\right)^{N_r}} \mathrm{E}_{\mathcal{H},\mathcal{S}}\left\{\exp\left(\frac{-\|\mathbf{y} - \mathcal{H}_\ell \mathcal{S}_\iota\|_{\mathrm{F}}^2}{\sigma_n^2}\right)\right\}, \end{aligned} \tag{5.15}$$

where $p_{\mathbf{Y}|\mathcal{H},\mathcal{S}}$ is the PDF of receive vector space $\mathbf{Y}$ given spatial and signal constellation diagrams $\mathcal{H}$ and $\mathcal{S}$, respectively, and is given by

$$p_{\mathbf{Y}|\mathcal{H},\mathcal{S}}(\mathbf{y}|\mathcal{H}_\ell, \mathcal{S}_\iota) = \frac{1}{\left(\pi\sigma_n^2\right)^{N_r}} \exp\left(\frac{-\|\mathbf{y} - \mathcal{H}_\ell \mathcal{S}_\iota\|_{\mathrm{F}}^2}{\sigma_n^2}\right), \tag{5.16}$$

where the received vector $\mathbf{y}$, as defined in Chapter 3, is given by

$$\mathbf{y} = \mathcal{H}_\ell \mathcal{S}_\iota + \mathbf{n}. \tag{5.17}$$

Therefore, assuming deterministic $\mathcal{H}_\ell$ and $\mathcal{S}_\iota$, $\mathbf{y} \sim \mathcal{CN}(\mathcal{H}_\ell \mathcal{S}_\iota, \sigma_n^2 \mathbf{I}_{N_r})$.

From (5.14) and (5.15), the entropy of $\mathbf{Y}$ is

$$\begin{aligned} H(\mathbf{Y}) &= -\mathrm{E}_{\mathbf{Y}}\left\{\log_2\left(\frac{1}{\left(\pi\sigma_n^2\right)^{N_r}} \mathrm{E}_{\mathcal{H},\mathcal{S}}\left\{\exp\left(\frac{-\|\mathbf{y} - \mathcal{H}_\ell \mathcal{S}_\iota\|_{\mathrm{F}}^2}{\sigma_n^2}\right)\right\}\right)\right\} \\ &= N_r \log_2\left(\pi\sigma_n^2\right) \\ &\quad - \mathrm{E}_{\mathbf{Y}}\left\{\log_2\left(\mathrm{E}_{\mathcal{H},\mathcal{S}}\left\{\exp\left(\frac{-\|\mathbf{y} - \mathcal{H}_\ell \mathcal{S}_\iota\|_{\mathrm{F}}^2}{\sigma_n^2}\right)\right\}\right)\right\}. \end{aligned} \tag{5.18}$$

The entropy of $\mathbf{Y}$ knowing $\mathcal{H}$ and $\mathcal{S}$ is

$$\begin{aligned} &H(\mathbf{Y}|\mathcal{H}, \mathcal{S}) \\ &= \mathrm{E}_{\mathcal{H},\mathcal{S}}\{H(\mathbf{Y}|\mathcal{H} = \mathcal{H}_\ell, \mathcal{S} = \mathcal{S}_\iota)\} \\ &= \mathrm{E}_{\mathcal{H},\mathcal{S}}\{-\mathrm{E}_{\mathbf{Y}|\mathcal{H},\mathcal{S}}\{\log_2 p_{\mathbf{Y}|\mathcal{H},\mathcal{S}}(\mathbf{y}|\mathcal{H}_\ell, \mathcal{S}_\iota)\}\} \\ &= \mathrm{E}_{\mathcal{H},\mathcal{S}}\left\{-\mathrm{E}_{\mathbf{Y}|\mathcal{H},\mathcal{S}}\left\{\log_2\left(\frac{1}{\left(\pi\sigma_n^2\right)^{N_r}} \exp\left(\frac{-\|\mathbf{y} - \mathcal{H}_\ell \mathcal{S}_\iota\|_{\mathrm{F}}^2}{\sigma_n^2}\right)\right)\right\}\right\} \\ &= N_r \log_2\left(\pi\sigma_n^2\right) + \frac{\mathrm{E}_{\mathcal{H},\mathcal{S}}\left\{\overbrace{\mathrm{E}_{\mathbf{Y}|\mathcal{H}_\ell,\mathcal{S}_\iota}\left\{\|\mathbf{y} - \mathcal{H}_\ell \mathcal{S}_\iota\|_{\mathrm{F}}^2\right\}}^{N_r\sigma_n^2}\right\}}{\sigma_n^2} \log_2(\exp(1)) \\ &= N_r \log_2\left(\pi\sigma_n^2 \exp(1)\right). \end{aligned} \tag{5.19}$$

Finally, substituting (5.18) and (5.19) in (5.13), the mutual information for SMTs is

$$I(\mathcal{H}, \mathcal{S}; \mathbf{Y}) = -N_r \log_2(\exp(1)) - \mathrm{E}_{\mathbf{Y}} \left\{ \log_2 \left(\mathrm{E}_{\mathcal{H},\mathcal{S}} \left\{ \exp \left(\frac{-\|\mathbf{y} - \mathcal{H}_\ell \mathcal{S}_\iota\|_\mathrm{F}^2}{\sigma_n^2} \right) \right\} \right) \right\}. \tag{5.20}$$

Unfortunately, no closed-form expression is available for (5.8) and (5.20), and numerical methods should be used.

5.2 Capacity Analysis

5.2.1 SMX

By definition, the capacity is the maximum number of bits that can be transmitted without any errors. Hence, the capacity for SMX is given by [116]

$$\begin{aligned} C &= \mathrm{E}_{\mathbf{H}} \left\{ \max_{p_\mathbf{X}} \{ I(\mathbf{X}; \mathbf{Y}|\mathbf{H}) \} \right\} \\ &= \mathrm{E}_{\mathbf{H}} \left\{ \max_{p_\mathbf{X}} \{ H(\mathbf{Y}|\mathbf{H}) \} - H(\mathbf{Y}|\mathbf{X}, \mathbf{H}) \right\}, \end{aligned} \tag{5.21}$$

where

(1) the maximization is done over the choices of the PDF of possible transmitted vector space $\mathbf{X}$, $p_\mathbf{X}$,
(2) the mutual information $I(\mathbf{X}, \mathbf{Y})$ is given in (5.1),
(3) and $H(\mathbf{Y}|\mathbf{X}, \mathbf{H})$ does not depend on the distribution of $\mathbf{X}$. Therefore, the maximization is reduced to the maximization of $H(\mathbf{Y}|\mathbf{H})$.

From [3], the distribution that maximizes the entropy is the zero mean complex Gaussian distribution ($\mathcal{CN}(0, \sigma^2)$). As such, $H(\mathbf{Y}|\mathbf{H})$ is maximized if $\mathbf{Y} \sim \mathcal{CN}(\mathbf{0}_{N_r}, \sigma_\mathbf{y}^2 \mathbf{I}_{N_r})$ with $\sigma_\mathbf{y}^2$ denoting the variance of $\mathbf{Y}$, and $\mathbf{0}_N$ is an N-length all zeros vector. From (5.5), the received vector is complex Gaussian distributed if the transmitted vector space is also a complex Gaussian distributed, $\mathbf{X} \sim \mathcal{CN}(\mathbf{0}_{N_t}, \frac{1}{N_t} \mathbf{I}_{N_t})$.

Assuming complex Gaussian-distributed transmitted vector, the PDF of $\mathbf{Y}$ given $\mathbf{H}$ is given by

$$p_{\mathbf{Y}|\mathbf{H}}(\mathbf{y}|\mathbf{H}) = \frac{\exp \left(-\mathbf{y}^H \left| \frac{1}{N_t} \mathbf{H}\mathbf{H}^H + \sigma_n^2 \mathbf{I}_{N_r} \right|^{-1} \mathbf{y} \right)}{\pi^{N_r} \left| \frac{1}{N_t} \mathbf{H}\mathbf{H}^H + \sigma_n^2 \mathbf{I}_{N_r} \right|}. \tag{5.22}$$

where $|\cdot|$ denotes the determinant.

From (5.22) and by following similar steps as discussed for (5.7), the maximum entropy of $\mathbf{Y}$ is

$$\max_{p_{\mathbf{X}}}\{H(\mathbf{Y}|\mathbf{H})\} = N_r \log_2\left(\pi\sigma_n^2 \exp(1)\right) + \log_2\left|\mathbf{I}_{N_r} + \mathrm{SNR} \times \frac{1}{N_t}\mathbf{H}\mathbf{H}^H\right|. \tag{5.23}$$

Thus, by substituting (5.7) and (5.23) in (5.21), the capacity of SMX is derived as [25]

$$\begin{aligned} C &= \mathrm{E}_{\mathbf{H}}\left\{ N_r \log_2\left(\pi\sigma_n^2 \exp(1)\right) + \log_2\left|\mathbf{I}_{N_r} + \mathrm{SNR} \times \frac{1}{N_t}\mathbf{H}\mathbf{H}^H\right| \right. \\ &\quad \left. -N_r \log_2\left(\pi\sigma_n^2 \exp(1)\right)\right\} \\ &= \mathrm{E}_{\mathbf{H}}\left\{ \log_2\left|\mathbf{I}_{N_r} + \mathrm{SNR} \times \frac{1}{N_t}\mathbf{H}\mathbf{H}^H\right| \right\}. \end{aligned} \tag{5.24}$$

Note, $\mathrm{E}\left\{\|\mathbf{Hx}\|_{\mathrm{F}}^2\right\} = N_r$ is assumed that leads to $\mathrm{SNR} = 1/\sigma_n^2$.

5.2.2 SMTs

5.2.2.1 Classical SMTs Capacity Analysis

In most studies, when calculating the capacity of SMTs, the spatial information bits are assumed to be conveyed through the index of the activated transmit antennas ℓ and not the different $\mathcal{H}_\ell$ spatial symbols. Hence, the mutual information is written as [205, 210]

$$I(\ell, \mathcal{S}; \mathbf{Y}|\mathbf{H}) = \mathrm{E}_\ell\{I(\mathcal{S}; \mathbf{Y}|\ell, \mathbf{H})\} + I(\ell; \mathbf{Y}|\mathbf{H}), \tag{5.25}$$

where the chain rule for information is used [211], since both the spatial and signal constellation symbols are assumed to be independent. The capacity is then calculated by maximizing the mutual information in (5.25) over the choice of $p_{\mathcal{S}}$,

$$\begin{aligned} C &= \mathrm{E}_{\mathbf{H}}\left\{ \max_{p_{\mathcal{S}}}\{I(\ell, \mathcal{S}; \mathbf{Y}|\mathbf{H})\} \right\} = \mathrm{E}_{\mathbf{H}}\left\{ \max_{p_{\mathcal{S}}}\left\{\mathrm{E}_\ell\{I(\mathcal{S}; \mathbf{Y}|\ell, \mathbf{H})\}\right\} \right. \\ &\quad \left. + I(\ell; \mathbf{Y}|\mathbf{H}) \right\} = \mathrm{E}_{\mathbf{H}}\left\{ \underbrace{\mathrm{E}_\ell\left\{\max_{p_{\mathcal{S}}}\{I(\mathcal{S}; \mathbf{Y}|\ell, \mathbf{H})\}\right\}}_{C_1} + \underbrace{I(\ell; \mathbf{Y}|\mathbf{H})}_{C_2} \right\}, \end{aligned} \tag{5.26}$$

where it is assumed that the PDF of $\mathcal{S}$ that maximizes C_1 would also maximize C_2.

The right-hand side of (5.26) is assumed to be the maximum mutual information between signal constellation symbols $\mathcal{S}$ and the received vector $\mathbf{Y}$,

$$C_1 = \mathrm{E}_\ell \left\{ \max_{p_S} \{ I(\mathcal{S}; \mathbf{Y}|\ell, \mathbf{H}) \} \right\}$$
$$= \mathrm{E}_\ell \left\{ \max_{p_S} \{ H(\mathbf{Y}|\ell, \mathbf{H}) \} - H(\mathbf{Y}|\mathcal{S}, \ell, \mathbf{H}) \right\}, \tag{5.27}$$

where $H(\mathbf{Y}|\mathcal{S}, \ell, \mathbf{H})$ does not depend on the distribution of $\mathcal{S}$, and therefore, the maximization is reduced to the maximization of $H(\mathbf{Y}|\ell, \mathbf{H})$.

As discussed earlier, and from [3], the entropy is maximized by a zero mean complex Gaussian random variable (RV). Hence, the entropy $H(\mathbf{Y}|\ell, \mathbf{H})$ is maximized when $\mathbf{Y} \sim \mathcal{CN}(\mathbf{0}_{N_r}, \sigma_\mathbf{y}^2 \mathbf{I}_{N_r})$, which is achieved when $\mathcal{S} \sim \mathcal{CN}(\mathbf{0}_{N_r}, \mathbf{I}_{N_r})$. Hence and following the same steps as in (5.19),

$$\max_{p_S} \{ H(\mathbf{Y}|\ell, \mathbf{H}) \} = -\mathrm{E}_{\mathbf{Y}|\ell,\mathbf{H}} \left\{ \log_2 (p_{\mathbf{Y}|\ell,\mathbf{H}}(\mathbf{y}|\ell, \mathbf{H})) \right\}$$
$$= N_r \log_2 \left(\pi \sigma_n^2 \exp(1) \right) + \log_2 \left| \mathbf{I}_{N_r} + \mathrm{SNR} \times \mathcal{H}_\ell \mathcal{H}_\ell^H \right|, \tag{5.28}$$

where

$$p_{\mathbf{Y}|\ell,\mathbf{H}}(\mathbf{y}|\ell, \mathbf{H}) = \frac{\exp \left(-\mathbf{y}^H | \mathcal{H}_\ell^H \mathcal{H}_\ell + \sigma_n^2 \mathbf{I}_{N_r} |^{-1} \mathbf{y} \right)}{\pi^{N_r} | \mathcal{H}_\ell^H \mathcal{H}_\ell + \sigma_n^2 \mathbf{I}_{N_r} |}. \tag{5.29}$$

The received vector $\mathbf{y}$ knowing the transmitted signal symbol $\mathcal{S}_\iota$, the indexes of the active transmit antennas ℓ and the channel matrix $\mathbf{H}$, is $\mathbf{y} \sim \mathcal{CN}(\mathcal{H}_\ell \mathcal{S}_\iota, \sigma_n^2 \mathbf{I}_{N_r})$. Hence and from (5.19)

$$H(\mathbf{Y}|\mathcal{S}, \ell, \mathbf{H}) = N_r \log_2 \left(\pi \sigma_n^2 \exp(1) \right). \tag{5.30}$$

Substituting (5.28) and (5.30) in (5.27) gives

$$C_1 = \mathrm{E}_\ell \left\{ \log_2 | \mathbf{I}_{N_r} + \mathrm{SNR} \times \mathcal{H}_\ell \mathcal{H}_\ell^H | \right\}. \tag{5.31}$$

The left-hand side of (5.26) can be written as

$$C_2 = I(\ell; \mathbf{Y}|\mathbf{H})$$
$$= \sum_\ell \int_\mathbf{Y} p_{(\ell,\mathbf{Y}|\mathbf{H})}(\ell, \mathbf{y}|\mathbf{H}) \log_2 \left(\frac{p_{(\ell,\mathbf{Y}|\mathbf{H})}(\ell, \mathbf{y}|\mathbf{H})}{p_{\ell|\mathbf{H}}(\ell|\mathbf{H}) p_{\mathbf{Y}|\mathbf{H}}(\mathbf{y}|\mathbf{H})} \right) d\mathbf{y}$$
$$= \sum_\ell \int_\mathbf{Y} p_{(\ell,\mathbf{Y}|\mathbf{H})}(\ell, \mathbf{y}|\mathbf{H}) \log_2 \left(\frac{p_{\mathbf{Y}|\ell,\mathbf{H}}(\mathbf{y}|\ell, \mathbf{H})}{p_{\mathbf{Y}|\mathbf{H}}(\mathbf{y}|\mathbf{H})} \right) d\mathbf{y}$$
$$= \sum_\ell \int_\mathbf{Y} p_\ell(\ell) p_{\mathbf{Y}|\ell,\mathbf{H}}(\mathbf{y}|\ell, \mathbf{H}) \log_2 \left(\frac{p_{\mathbf{Y}|\ell,\mathbf{H}}(\mathbf{y}|\ell, \mathbf{H})}{p_{\mathbf{Y}|\mathbf{H}}(\mathbf{y}|\mathbf{H})} \right) d\mathbf{y}$$
$$= \mathrm{E}_\ell \left\{ \int_\mathbf{Y} p_{\mathbf{Y}|\ell,\mathbf{H}}(\mathbf{y}|\ell, \mathbf{H}) \log_2 \left(\frac{p_{\mathbf{Y}|\ell,\mathbf{H}}(\mathbf{y}|\ell, \mathbf{H})}{p_{\mathbf{Y}|\mathbf{H}}(\mathbf{y}|\mathbf{H})} \right) d\mathbf{y} \right\}$$

$$= \mathrm{E}_{\ell}\left\{\overbrace{\int_{\mathbf{Y}} p_{\mathbf{Y}|\ell,\mathbf{H}}(\mathbf{y}|\ell,\mathbf{H})\log_2(p_{\mathbf{Y}|\ell,\mathbf{H}}(\mathbf{y}|\ell,\mathbf{H}))\,d\mathbf{y}}^{-H(\mathbf{Y}|\ell,\mathbf{H})}\right\}$$

$$- \mathrm{E}_{\ell}\left\{\int_{\mathbf{Y}} p_{\mathbf{Y}|\ell,\mathbf{H}}(\mathbf{y}|\ell,\mathbf{H})\log_2(p_{\mathbf{Y}|\mathbf{H}}(\mathbf{y}|\mathbf{H}))\,d\mathbf{y}\right\}$$

$$= -N_r\log_2\left(\pi\sigma_n^2\exp(1)\right) - \mathrm{E}_{\ell}\left\{\log_2|\mathbf{I}_{N_r} + \mathrm{SNR}\times\mathcal{H}_{\ell}\mathcal{H}_{\ell}^H|\right\}$$

$$- \mathrm{E}_{\ell}\left\{\mathrm{E}_{\mathbf{Y}|\ell,\mathbf{H}}\left\{\log_2\left(p_{\mathbf{Y}|\mathbf{H}}(\mathbf{y}|\mathbf{H})\right)\right\}\right\}. \tag{5.32}$$

Note, ℓ is assumed to be a discrete RV.

Substituting (5.31) and (5.32) in (5.26), the capacity is formulated as,

$$C = -N_r\log_2\left(\pi\sigma_n^2\exp(1)\right)$$
$$- \mathrm{E}_{\mathbf{H}}\left\{\mathrm{E}_{\ell}\left\{\mathrm{E}_{\mathbf{Y}|\ell,\mathbf{H}}\left\{\log_2\left(p_{\mathbf{Y}|\mathbf{H}}(\mathbf{y}|\mathbf{H})\right)\right\}\right\}\right\}. \tag{5.33}$$

From [210], and as can be seen from (5.33), the distribution of the antenna index ℓ plays a major role in the capacity. Therefore, the maximization in (5.26) should have been performed over the choices of the distribution of the transmit antenna indexes as well as the signal symbols. Thus, the capacity in (5.33) is rewritten as

$$C = \mathrm{E}_{\mathbf{H}}\left\{\max_{p_{\mathcal{S}},p_{\ell}}\{I(\ell,\mathcal{S};\mathbf{Y}|\mathbf{H})\}\right\}$$

$$= \mathrm{E}_{\mathbf{H}}\left\{\max_{p_{\mathcal{S}},p_{\ell}}\left\{\mathrm{E}_{\ell}\{I(\mathcal{S};\mathbf{Y}|\ell,\mathbf{H})\} + I(\ell;\mathbf{Y}|\mathbf{H})\right\}\right\}$$

$$= \mathrm{E}_{\mathbf{H}}\left\{\max_{p_{\ell}}\left\{\underbrace{\mathrm{E}_{\ell}\left\{\max_{p_{\mathcal{S}}}\{I(\mathcal{S};\mathbf{Y}|\ell,\mathbf{H})\}\right\}}_{C_1} + \underbrace{I(\ell;\mathbf{Y}|\mathbf{H})}_{C_2}\right\}\right\}$$

$$= -N_r\log_2\left(\pi\sigma_n^2\exp(1)\right)$$
$$- \mathrm{E}_{\mathbf{H}}\left\{\max_{p_{\ell}}\left\{\mathrm{E}_{\ell}\left\{\mathrm{E}_{\mathbf{Y}|\ell,\mathbf{H}}\left\{\log_2\left(p_{\mathbf{Y}|\mathbf{H}}(\mathbf{y}|\mathbf{H})\right)\right\}\right\}\right\}\right\}. \tag{5.34}$$

Because the signal constellation symbols are assumed to be continuous and the antenna indexes are discrete, obtaining closed-form solution of (5.34) is very sophisticated [210]. Therefore, most of the existing literature attempts to calculate the capacity by assuming the distribution of the antenna indexes to be discrete uniform (DU) [203–205, 210, 212, 213].

Assuming that ℓ follows a DU distribution, $p_{\ell} = 1/2^{\eta_{\mathcal{H}_{\ell}}}, \ell \in \{1, 2, \ldots, 2^{\eta_{\mathcal{H}}}\}$, where $\eta_{\mathcal{H}}$ is the number of bits modulated in the spatial domain. The capacity in (5.34) becomes

$$C = -N_r \log_2\left(\pi\sigma_n^2 \exp(1)\right) - \mathrm{E}_{\mathbf{H}}\left\{\frac{1}{2^{\eta_H}}\sum_{\ell}\mathrm{E}_{\mathrm{Y}|\ell,\mathbf{H}}\left\{\log_2\left(p_{\mathrm{Y}|\mathbf{H}}(\mathbf{y}|\mathbf{H})\right)\right\}\right\}. \tag{5.35}$$

The PDF of $\mathbf{y}$ given $\mathbf{H}$, and assuming DU distributed ℓ is

$$p_{\mathrm{Y}|\mathbf{H}}(\mathbf{y}|\mathbf{H}) = \sum_{\ell} p_{\ell}(\ell) p_{\mathrm{Y}|\ell,\mathbf{H}}(\mathbf{y}|\ell,\mathbf{H}) = \frac{1}{2^{\eta_{H_\ell}}}\sum_{\ell} p_{\mathrm{Y}|\ell,\mathbf{H}}(\mathbf{y}|\ell,\mathbf{H}), \tag{5.36}$$

where $p_{\mathrm{Y}|\ell,\mathbf{H}}$ is given in (5.29).

Plugging (5.36) in (5.35),

$$\begin{aligned} C &= -N_r \log_2\left(\pi\sigma_n^2 \exp(1)\right) \\ &\quad - \mathrm{E}_{\mathbf{H}}\left\{\frac{1}{2^{\eta_{H_\ell}}}\sum_{\ell}\mathrm{E}_{\mathrm{Y}|\ell,\mathbf{H}}\left\{\log_2\left(\frac{1}{2^{\eta_{H_\ell}}}\sum_{\ell'} p_{\mathrm{Y}|\ell,\mathbf{H}}(\mathbf{y}|\ell',\mathbf{H})\right)\right\}\right\} \\ &= \eta_{H_\ell} - N_r \log_2\left(\pi\sigma_n^2 \exp(1)\right) \\ &\quad - \frac{1}{2^{\eta_{H_\ell}}}\mathrm{E}_{\mathbf{H}}\left\{\sum_{\ell}\mathrm{E}_{\mathrm{Y}|\ell,\mathbf{H}}\left\{\log_2\left(\sum_{\ell'} p_{\mathrm{Y}|\ell,\mathbf{H}}(\mathbf{y}|\ell',\mathbf{H})\right)\right\}\right\} \\ &= \eta_H + \eta_S, \end{aligned} \tag{5.37}$$

where

$$\eta_S = -N_r \log_2\left(\pi\sigma_n^2 \exp(1)\right) - \frac{1}{2^{\eta_{H_\ell}}}\mathrm{E}_{\mathbf{H}}\left\{\sum_{\ell}\mathrm{E}_{\mathrm{Y}|\ell,\mathbf{H}}\left\{\log_2\left(\sum_{\ell'} p_{\mathrm{Y}|\ell,\mathbf{H}}(\mathbf{y}|\ell',\mathbf{H})\right)\right\}\right\} \tag{5.38}$$

is the number of bit modulated in the signal domain.

From (5.37), and noting that ℓ is assumed to follow a DU distribution rather than maximizing over it, the maximum number of bits that could be transmitted in the spatial domain is capped. However, in capacity, the number of bit should increase to infinity as the signal-to-noise-ratio (SNR) increases. Hence, the capacity in (5.37) is the maximum mutual information that can be transmitted assuming that DU distributed ℓ.

Finally, in (5.34) and (5.37), it is clear that complex Gaussian-distributed symbols would maximize C_1. However, it is not certain that such distribution would maximize C_2 as well. This is because C_2 depends on the joint distribution of both spatial and signal symbols [213]. Also, the distinction of which transmit antennas are activated depends on the Euclidean difference among channel paths from each transmit antenna to all receive antennas, and not on the indexes of the active antennas [213]. As discussed earlier, if two transmit antennas are located at the same spatial position, they will have identical channel paths and the cardinality of the spatial constellation diagram is one. Thereby, no information bits can be conveyed in the spatial domain even though two or more transmit antennas exist. Therefore, as will be shown in next section, the capacity for SMTs should be calculated by maximizing over the spatial and signal symbols instead of the antenna indexes and signal symbol [51, 55, 56].

5.2.2.2 SMTs Capacity Analysis by Maximing over Spatial and Constellation Symbols

As explained in the previous section and Section 5.1.2, the SMT capacity should be derived by maximizing the mutual information over the PDFs of the two different SMTs symbols, spatial and signal symbols. Therefore, the capacity for SMTs is defined as

$$\begin{aligned} C &= \max_{p_{\mathcal{H}}, p_{\mathcal{S}}} \{I(\mathcal{H}, \mathcal{S}; \mathbf{Y})\} \\ &= \max_{p_{\mathcal{H}}, p_{\mathcal{S}}} \{H(\mathbf{Y})\} - H(\mathbf{Y}|\mathcal{H}, \mathcal{S}). \end{aligned} \tag{5.39}$$

where

(1) the mutual information $I(\mathcal{H}, \mathcal{S}; \mathbf{Y})$ is given in (5.13),
(2) and $H(\mathbf{Y}|\mathcal{H}, \mathcal{S})$ does not depend on $\mathcal{H}$ nor $\mathcal{S}$. Therefore, the maximization is reduced to the maximization of $H(\mathbf{Y})$.

The received signal in (5.17) can be rewritten as

$$\mathbf{y} = \mathcal{R}_{\ell,\iota} + \mathbf{n}, \tag{5.40}$$

where $\mathcal{R}_{\ell,\iota} = \mathcal{H}_\ell \mathcal{S}_\iota$ is the SMT symbol. The received signal, $\mathbf{y}$, is complex Gaussian distributed if the SMT symbol, $\mathcal{R}_{\ell,\iota}$, also follows a complex Gaussian distribution, i.e., $\mathcal{R}_{\ell,\iota} \sim \mathcal{CN}(\mathbf{0}_{N_r}, \mathbf{I}_{N_r})$.

Assuming that the transmitted SMT symbol is complex Gaussian distributed, the PDF of the received vector space, $\mathbf{Y}$, is

$$\begin{aligned} p_{\mathbf{Y}}(\mathbf{y}) &= \frac{1}{\pi^{N_r} |\mathbf{I}_{N_r} + \sigma_n^2 \mathbf{I}_{N_r}|} \exp\left(\mathbf{y}^H (\mathbf{I}_{N_r} + \sigma_n^2 \mathbf{I}_{N_r})^{-1} \mathbf{y}\right) \\ &= \frac{1}{\pi^{N_r} \left(1+\sigma_n^2\right)^{N_r}} \exp\left(\frac{\|\mathbf{y}\|_{\mathrm{F}}^2}{1+\sigma_n^2}\right). \end{aligned} \tag{5.41}$$

From (5.41), and following the same steps as discussed for (5.19), the maximum entropy of $\mathbf{y}$ is

$$\max_{p_{\mathcal{H}}, p_{\mathcal{S}}} \{H(\mathbf{Y})\} = N_r \log_2\left(\pi\left(1+\sigma_n^2\right)\exp(1)\right). \tag{5.42}$$

Hence, and using (5.19), the capacity of SMT in (5.39) can be derived as

$$\begin{aligned} C &= N_r \log_2\left(\pi\left(1+\sigma_n^2\right)\exp(1)\right) - N_r \log_2\left(\pi\sigma_n^2 \exp(1)\right) \\ &= N_r \log_2\left(1+\frac{1}{\sigma_n^2}\right) = N_r \log_2(1+\mathrm{SNR}). \end{aligned} \tag{5.43}$$

Comparing the existing SMT capacity in (5.34), and SMX capacity in (5.24), to SMTs capacity in (5.43), the following can be observed:

(1) Similar to SMX and classical SMT analysis, the capacity equation does not depend on the constellation symbols.
(2) In the derived SMTs capacity in (5.43), the channel is a mean to convey information and the capacity does not depend on the channel. Therefore, there is no averaging over the channel.

In summary, SMTs have only one single theoretical capacity formula regardless of the considered fading channel. Such capacity is achievable if the SMT symbol $\mathcal{R}_{\ell,\iota}$ is complex Gaussian distributed. However, for any particular channel, proper shaping of the constellation symbols is needed to achieve the theoretical capacity.

To illustrate this, the derived capacity in (5.43) is compared to the MIMO capacity formula in [202–206], which is given in (5.24) for $N_t = N_r = 4$, and the results are depicted in Figure 5.1. In [202–206], SMTs capacity is different for different techniques and antithetic channel statistics. Yet, it is shown in [51, 55] and as discussed in this chapter that single capacity formula is derived for all SMTs and for any channel statistics. It is evident from the figure that the

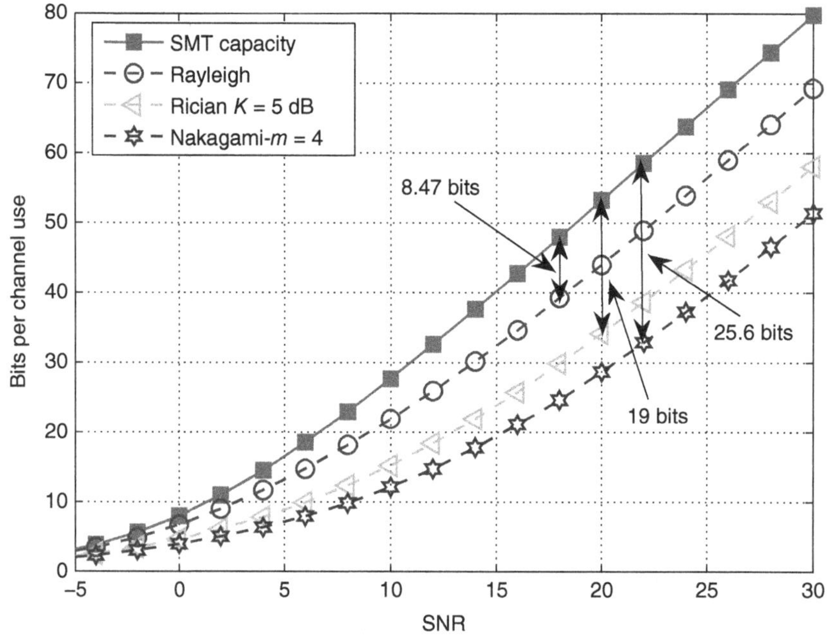

Figure 5.1 Comparison between the derived capacity in (5.43) and the MIMO capacity (5.24) over Rayleigh, Rician $K = 5$ dB, and Nakagami-$m = 4$ fading channel, for $N_t = N_r = 8$.

expected anticipated classical SMTs capacity falls far below the true capacity for such techniques. It is shown in Figure 5.1 that SMTs can actually achieve 8.47, 19, and 25.6 bits more than the classical capacity for Rayleigh, Nakagami-q, and Nakagami-m channels, respectively.

Such capacity is attainable, as discussed previously, if $\mathcal{R}_{\ell,\iota} \sim \mathcal{CN}(\mathbf{0}_{N_r}, \mathbf{I}_{N_r})$. Thus,

(1) using the product distribution theory [214];
(2) assuming $N_r = 1$ for simplicity and without losing generality;
(3) knowing that a zero mean complex normal RV, $\upsilon \sim \mathcal{CN}(0,1)$, has a Rayleigh distribution amplitude, uniformly distribution phase, and a joint PDF given by

$$p_{\upsilon}(\upsilon) = \frac{1}{2\pi} \times 2|\upsilon| \exp(-|\upsilon|^2); \tag{5.44}$$

the distribution of the used signal constellation symbols has to be shaped depending on the distribution of the channel so that it solves

$$\overbrace{\frac{1}{\pi}|\mathcal{R}_{\ell,\iota}| \exp(-|\mathcal{R}_{\ell,\iota}|^2)}^{p_{\mathcal{R}_{\ell,\iota}}} = \int_{\mathcal{H}_\ell} \frac{p_{\mathcal{H}_\ell}(\mathcal{H}_\ell)}{|\mathcal{H}_\ell|} p_{S_\iota}\left(\frac{\mathcal{R}_{\ell,\iota}}{\mathcal{H}_\ell}\right) d\mathcal{H}_\ell. \tag{5.45}$$

In the following section, illustrative examples for SSK and SM systems are presented to highlight how the theoretical channel capacity can be achieved.

5.3 Achieving SMTs Capacity

5.3.1 SSK

In SSK systems, only spatial constellation symbols exist and no signal symbols are transmitted, i.e. $\mathcal{R}_{\ell,\iota} = \mathcal{H}_\ell$. Therefore, the capacity is achieved when each element of $\mathcal{H}_\ell$ follows $\sim \mathcal{CN}(0,1)$. Hence, SSK can achieve the capacity when the channel follows a Rayleigh distribution. However, the spatial symbol diagram $\mathcal{H}$ of an SSK system over Rayleigh fading channel does not always follow a complex Gaussian distribution. In a small-scale SSK system (i.e. assuming small number of transmit antennas), $\mathcal{H}$ at each time instant is actually uniformly distributed. However and for large-scale SSK, the elements of $\mathcal{H}$ are complex Gaussian distributed. Therefore, it is anticipated that SSK system will achieve the capacity when deploying large-scale MIMO configuration over Rayleigh fading channel. To further explain this, the histograms for the real parts of the spatial constellation diagram, Re$\{\mathcal{H}\}$, for an SSK system over Rayleigh fading channel, with $N_t = 4$ and 512 are compared to the PDF of a Gaussian distribution in Figure 5.2. It is shown in the figure that $\mathcal{H}$ is uniformly distributed

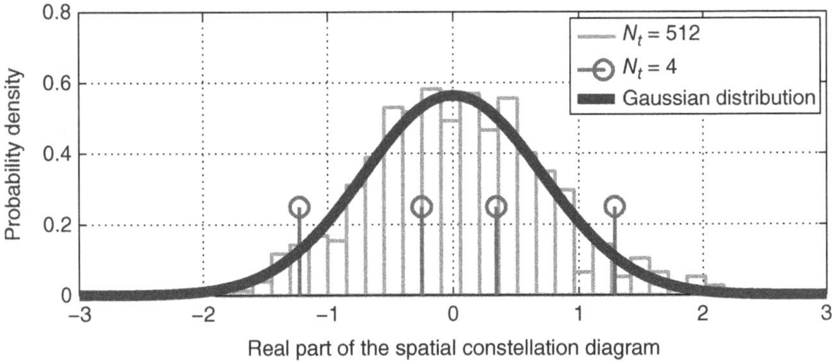

Figure 5.2 Histogram of the real part of the spatial constellation diagram, Re$\{\mathcal{H}\}$, for SSK with $N_t = 4$ and 512 compared to the PDF of the Gaussian distribution.

for a small number of transmit antennas, $N_t = 4$; whereas, it follows a Gaussian distribution for large number of transmit antennas, $N_t = 512$. This is because at each time instance, 4 unique possible symbols exist for $N_t = 4$ and they are chosen equally probable. Therefore, $\mathcal{H}$ follows a uniform distribution. However, for $N_t = 512$, there are 512 spatial symbols that are not necessarily unique, i.e. symbols would repeat and occur more than others following a Gaussian distribution. As such, for large number of transmit antennas even though the spatial symbols are chosen equally likely, the chosen symbol is not unique in the set of spatial symbols $\mathcal{H}$, and probability it occurs follows a Gaussian distribution.

The mutual information performance of SSK system over Rayleigh fading channel for variable number of transmit antennas from $N_t = 2 \sim 2048$ with multiple of 2 step size are depicted in Figure 5.3. The derived theoretical capacity in (5.43) for SMTs is depicted as well. In all results, $N_r = 2$ is assumed. It can be seen from the figure that for a small number of transmit antennas, SSK mutual information, are far below the capacity. However, as the number of antennas increases, the gap between the mutual information curves and the capacity curve dilutes. For instance, with $N_t = 1024$ and $N_t = 2048$, an SSK is shown to follow the capacity up to 6 bits per channel use, before it starts to deviate and floor at 10 bits and 11 bits, respectively, which is the maximum number of bits that can be transmitted by SSK system using 1024 and 2048 transmit antennas, respectively. However, using $N_t = 4$, the mutual information curve follows the capacity for up to 0.5 bits before it starts to deviate and floor. The reason for this is discussed before, where SSK system is shown to achieve the capacity if $\mathcal{H}_\ell \sim \mathcal{CN}(\mathbf{0}_{N_r}, \mathbf{I}_{N_r})$, which is approximately attainable for large number of transmit antennas over Rayleigh fading channel.

The mutual information results for SSK systems over different channel distributions including Rayleigh, Rician with $K = 5$ dB, and Nakagami-$m = 4$ for $N_t = 512$ and $N_r = 2$ are shown in Figure 5.4. Again and as discussed before,

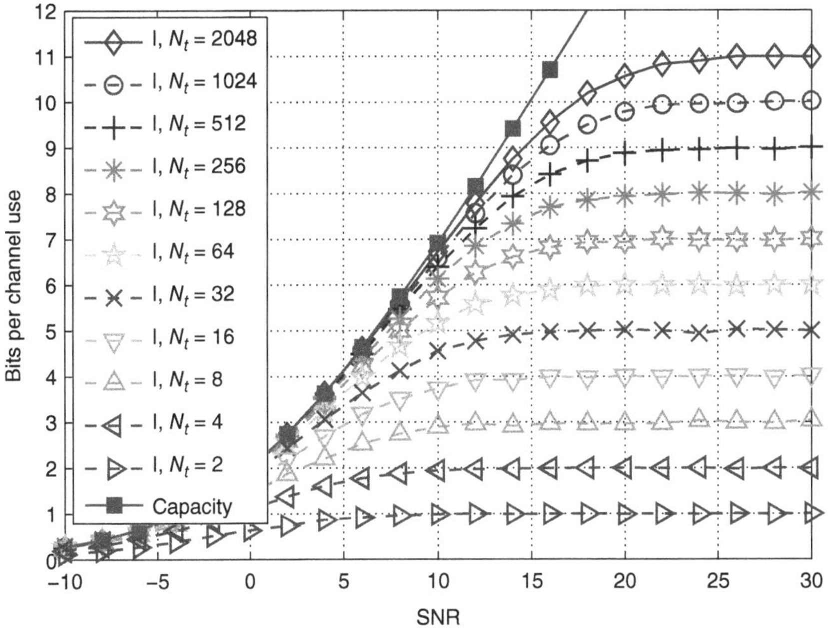

Figure 5.3 The capacity of SMTs compared to simulated mutual information of SSK over Rayleigh fading channel for $N_t = 2^{1:11}$ and $N_r = 2$.

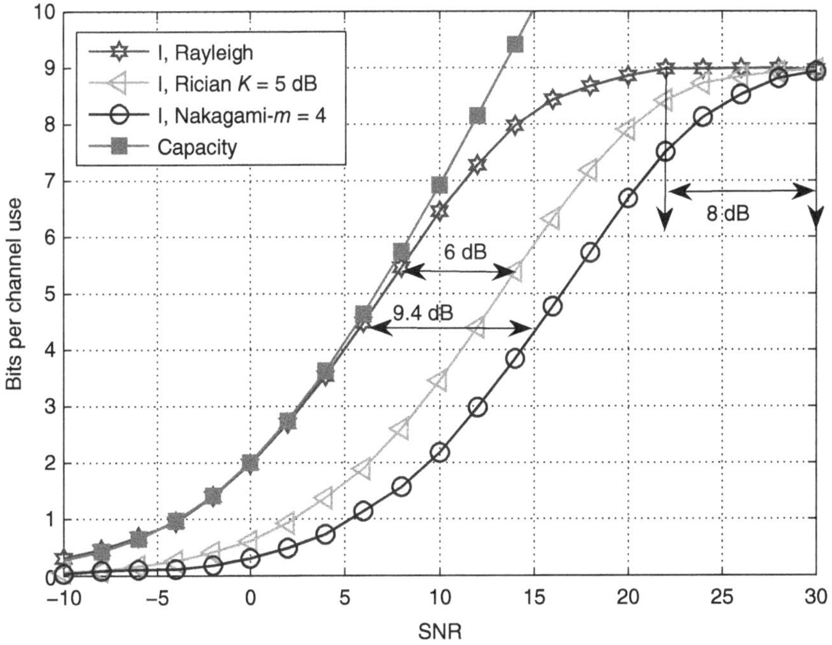

Figure 5.4 The capacity of SMT compared to simulated mutual information of SSK over Rayleigh, Rician with $K = 5$ dB, and Nakagami-$m = 4$, where $N_t = 512$ and $N_r = 2$.

$\mathcal{H}_\ell \sim \mathcal{CN}(\mathbf{0}_{N_r}, \mathbf{I}_{N_r})$ if N_t was large enough and the channel follows a Rayleigh distribution. Furthermore, in Figure 5.4, the mutual information of SSK system over Rician and Nakagami-m fading channels are, respectively, 6 and 9.4 dB worse than that of the Rayleigh fading channel. Moreover, SSK over Rayleigh fading channel reaches the maximum mutual information of $\eta_{\mathcal{H}} = \log_2 512 =$ 9 bits 8 dB earlier than Rician and Nakagami-m fading channels. Hence, for an SSK system to achieve the capacity, a large number of transmit antennas are needed along with Rayleigh fading channel.

5.3.2 SM

Considering a transmission of an SM signal over Rayleigh fading channel gives

$$\mathcal{R}_{\ell,\iota} = \overbrace{\mathbf{r}_{\mathbf{h}_\ell} e^{j\Theta_{\mathbf{h}_\ell}}}^{\mathcal{H}_\ell} \times \overbrace{s_\iota e^{j\phi_{s_\iota}}}^{\mathcal{S}_\iota} = \mathbf{r}_{\mathbf{h}_\ell} r_{s_\iota} e^{j(\Theta_{\mathbf{h}_\ell} + \phi_{s_\iota})}, \tag{5.46}$$

where $\mathbf{r}_{\mathbf{h}_\ell}$ and $\Theta_{\mathbf{h}_\ell}$ are the amplitude and the phase of $\mathcal{H}_\ell = \mathbf{h}_\ell$, respectively, and r_{s_ι} and ϕ_{s_ι} are the amplitude and the phase of $\mathcal{S}_\iota = s_\iota$, respectively. Now,

(1) A complex Gaussian distribution random variable has a Rayleigh distributed amplitude and uniform distributed phase.
(2) Rayleigh fading channel has Rayleigh distributed amplitude and uniform distributed phase.

Hence, for $\mathcal{R}_{\ell,\iota} \sim \mathcal{CN}(\mathbf{0}_{N_r}, \mathbf{I}_{N_r})$, the amplitude and phase of s_ι should be $r_{s_\iota} = 1$ and $\phi_{s_\iota} \sim \mathcal{U}(-\pi, \pi)$, i.e. the constellation symbols have to follow a circular uniform (CU) distribution. Incoming data bits are usually assumed to be uniformly distributed. Therefore, different to any other distribution, CU signal constellation symbols can be realizable. For instance, phase shift keying (PSK) are distributed according to discrete CU distribution. Hence, CU distribution can be approximated by a PSK signal constellation diagram. The larger the size of the used PSK modulation, the closer the approximation to CU. This means that SM performance over Rayleigh fading channels can be enhanced by transmitting more bits in the signal domain.

Figure 5.5 shows the histogram of the phase of a randomly generated symbols modulated using $M = 4, 16$, and 128 size PSK modulation, and the PDF of $\mathcal{U}(-\pi, \pi)$. From the figure, it can be seen that as the size of the considered PSK modulation increases, the phase distribution of the generated symbols gets closer to $\mathcal{U}(-\pi, \pi)$, where for $M = 128$ it is discrete uniform distributed $\mathcal{DU}(-\pi, \pi)$. Hence, CU distributed symbols can be simply achieved by using large size PSK constellation diagram.

Figures 5.6–5.9 show the PDF of $\sim \mathcal{CN}(0, 1/2)$ plotted against the histogram of the real part of the resultant transmitted SM symbol ($\mathrm{Re}\{\mathcal{R}_{\ell,\iota}\}$), respectively, over MISO Rayleigh and Rician ($K = 5$ dB) fading channels with

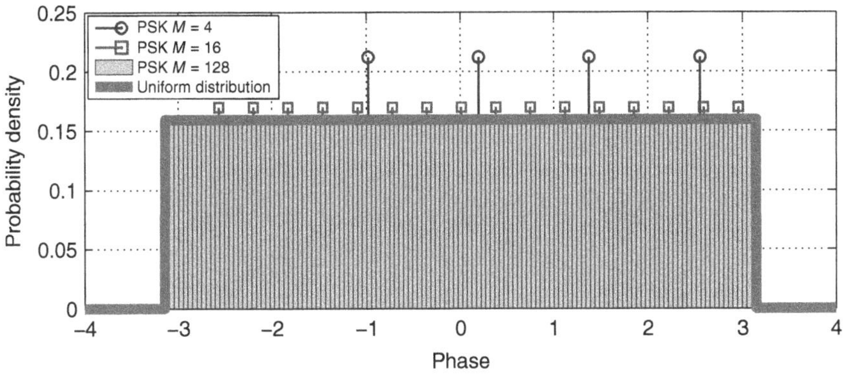

Figure 5.5 Histogram of the phase of a randomly generated symbols modulated using $M = 4$-,16-, and 128-size PSK modulation compared to the uniform distribution.

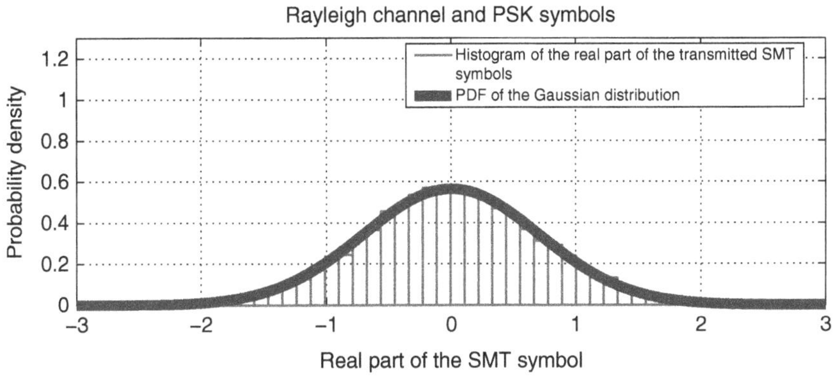

Figure 5.6 PDF of $\sim \mathcal{CN}(0, 1/2)$ plotted against histogram of Re{**Hx**} using 128-PSK modulation over MISO Rayleigh fading channel with $N_t = 128$.

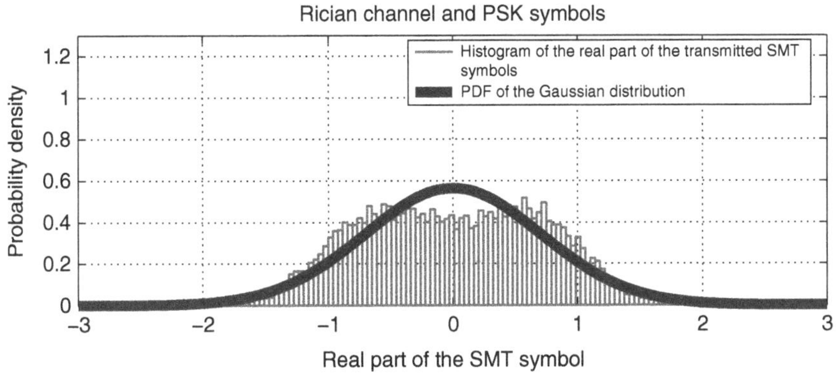

Figure 5.7 PDF of $\sim \mathcal{CN}(0, 1/2)$ plotted against histogram of Re{**Hx**} using 128-PSK modulation over MISO Rician fading channel with $K = 5$ dB and $N_t = 128$.

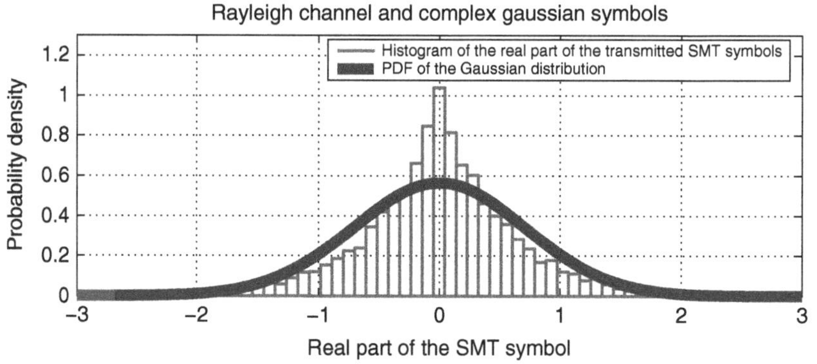

Figure 5.8 PDF of $\sim \mathcal{CN}(0, 1/2)$ plotted against histogram of Re{**Hx**} using 128 complex Gaussian-distributed symbols over MISO Rayleigh fading channel with $N_t = 128$.

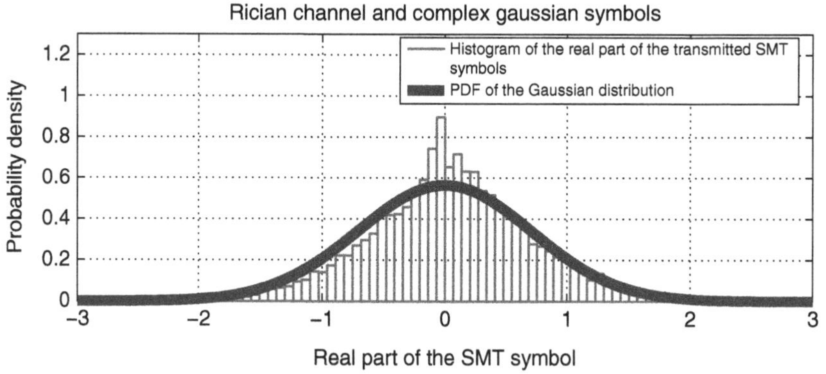

Figure 5.9 PDF of $\sim \mathcal{CN}(0, 1/2)$ plotted against histogram of Re{**Hx**} using 128 complex Gaussian-distributed symbols over MISO Rician fading channel with $K = 5$ dB and $N_t = 128$.

$N_t = 128$. In Figures 5.6 and 5.7, 128-PSK constellation diagram is considered. Whereas, in Figures 5.8 and 5.9, 128-Gaussian-distributed symbols are assumed to form the signal constellation diagram. Figures 5.8 and 5.9 show that Gaussian-distributed symbols will not lead to $\mathcal{R}_{\ell,l}$ being a complex Gaussian distribution. Though, considering PSK as illustrated in Figure 5.6, the distribution of $\mathcal{R}_{\ell,l}$ is shown to accurately follow a complex Gaussian distribution. Therefore, SM is anticipated to achieve the capacity when implemented on a large-scale MIMO system and using CU distributed symbols over Rayleigh fading channels. However and for non-Rayleigh fading channels, CU distribution will not lead to $\mathcal{R}_{\ell,l}$ being a complex Gaussian distribution as shown in Figure 5.7. Hence and for each channel type, the used constellation symbols need to be shaped such that they solve (5.45).

An example for the previous discussion is shown in Figure 5.10, where the mutual information performance for SM system over Rayleigh fading channels assuming large-scale MIMO configuration with $N_t = 1024$, $N_r = 2$ and with 1024-PSK constellation diagram and Gaussian-distributed signal constellation are depicted. The theoretical capacity in (5.43) is shown in the figure as well. The results show that CU distribution, which is obtainable through 1024-PSK constellation diagram, is required to achieve the theoretical capacity limit. This contradicts conventional theory for MIMO system where complex Gaussian-distribution symbols are needed to achieve the capacity. As illustrated in Figure 5.10, Gaussian-distributed signal constellation symbols, even though not practically possible to generate, perform 1 bit less than that of CU distribution symbols. These results highlight that SM systems can achieve the capacity with large number of transmit antennas and large PSK constellation size.

It should be also noted that the deviation from the theoretical capacity curve at high SNR is mainly due to the limited number of spatial and signal symbols. For mutual information curves to follow the capacity curve over the entire SNR

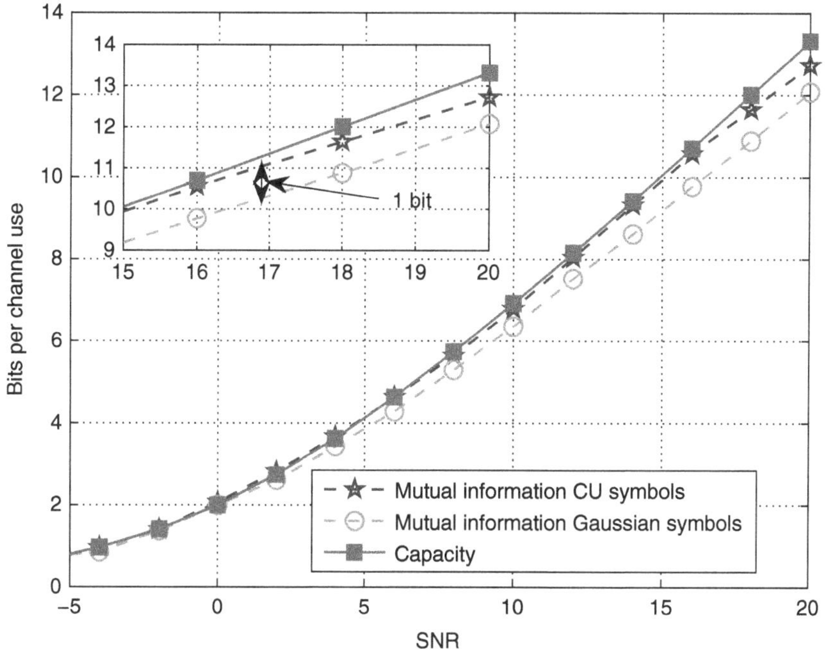

Figure 5.10 The capacity of SMT compared to the simulated mutual information of SM using PSK constellations, and Gaussian-distributed constellations, over Rayleigh fading channel, where $M = 1024$, $N_t = 1024$, and $N_r = 2$.

range, continuous distributions with infinite number of symbols are required, which is not attainable.

5.4 Information Theoretic Analysis in the Presence of Channel Estimation Errors

5.4.1 Evaluating the Mutual Information

5.4.1.1 Classical Spatial Multiplexing MIMO

The mutual information for SMX in the presence of channel estimation errors (CSE) can be written as [152]

$$I(\mathbf{X};\mathbf{Y}) = \mathrm{E}_{\tilde{\mathbf{H}}}\left\{I\left(\mathbf{X};\mathbf{Y}|\tilde{\mathbf{H}}\right)\right\} = \mathrm{E}_{\tilde{\mathbf{H}}}\left\{H\left(\mathbf{Y}|\tilde{\mathbf{H}}\right) - H\left(\mathbf{Y}|\mathbf{X},\tilde{\mathbf{H}}\right)\right\}, \tag{5.47}$$

where $\tilde{\mathbf{H}}$ is the estimated channel and is related to the perfect channel by

$$\mathbf{H} = \tilde{\mathbf{H}} + \mathbf{e}, \tag{5.48}$$

where $\mathbf{e}$ is an $N_r \times N_t$ channel estimation error matrix, assumed to follow a complex Gaussian distribution with zero mean and variance $\sigma_{\mathbf{e}}^2$, $\mathbf{e} \sim \mathcal{CN}\left(\mathbf{0}_{N_r}, \sigma_{\mathbf{e}}^2\mathbf{I}_{N_r}\right)$, with $\sigma_{\mathbf{e}}^2$ being a parameter that captures the quality of the channel estimation as discussed in previous chapters. Assuming, for instance, least square (LS) channel estimation algorithm at the receiver, $\sigma_{\mathbf{e}}^2 = \sigma_n^2$ [192].

The entropy of $\mathbf{Y}$ given the estimated channel $\tilde{\mathbf{H}}$, is written as

$$H\left(\mathbf{Y}|\tilde{\mathbf{H}}\right) = -\mathrm{E}_{\mathbf{Y}|\tilde{\mathbf{H}}}\left\{\log_2 p_{\mathbf{Y}|\tilde{\mathbf{H}}}\left(\mathbf{y}|\tilde{\mathbf{H}}\right)\right\}, \tag{5.49}$$

where $p_{\mathbf{y}|\tilde{\mathbf{H}}}(\cdot)$ is the PDF of $\mathbf{Y} = \mathbf{y}$, knowing $\tilde{\mathbf{H}}$. Plugging (5.48) in (5.5) then gives

$$\mathbf{y} = \mathbf{H}\mathbf{x}_t + \mathbf{n} = \left(\tilde{\mathbf{H}} + \mathbf{e}\right)\mathbf{x}_t + \mathbf{n} = \tilde{\mathbf{H}}\mathbf{x}_t + \mathbf{e}\mathbf{x}_t + \mathbf{n}. \tag{5.50}$$

Knowing that $\mathbf{n} \sim (\mathbf{0}_{N_r}, \sigma_n^2\mathbf{I}_{N_r})$, the PDF of $\mathbf{Y} = \mathbf{y}$ and $\mathbf{X} = \mathbf{x}_t$ knowing $\tilde{\mathbf{H}}$ is

$$p_{(\mathbf{Y}|\mathbf{X},\tilde{\mathbf{H}})}\left(\mathbf{y}|\mathbf{x}_t,\tilde{\mathbf{H}}\right) = \frac{1}{\pi^{N_r}\left(\sigma_n^2 + \sigma_{\mathbf{e}}^2\|\mathbf{x}_t\|_{\mathrm{F}}^2\right)^{N_r}} \exp\left(-\frac{\|\mathbf{y} - \tilde{\mathbf{H}}\mathbf{x}_t\|_{\mathrm{F}}^2}{\sigma_n^2 + \sigma_{\mathbf{e}}^2\|\mathbf{x}_t\|_{\mathrm{F}}^2}\right), \tag{5.51}$$

and consequently

$$\begin{aligned} p_{\mathbf{Y}|\tilde{\mathbf{H}}}\left(\mathbf{y}|\tilde{\mathbf{H}}\right) &= \int_{\mathbf{X}} p_{\mathbf{X}}(\mathbf{x}_t) \times p_{(\mathbf{Y}|\mathbf{X},\tilde{\mathbf{H}})}\left(\mathbf{y}|\mathbf{x}_t,\tilde{\mathbf{H}}\right)\, d\mathbf{x}_t \\ &= \mathrm{E}_{\mathbf{X}}\left\{\frac{\exp\left(-\frac{\|\mathbf{y}-\tilde{\mathbf{H}}\mathbf{x}_t\|_{\mathrm{F}}^2}{\sigma_n^2+\sigma_{\mathbf{e}}^2\|\mathbf{x}_t\|_{\mathrm{F}}^2}\right)}{\pi^{N_r}\left(\sigma_n^2 + \sigma_{\mathbf{e}}^2\|\mathbf{x}_t\|_{\mathrm{F}}^2\right)^{N_r}}\right\}, \end{aligned} \tag{5.52}$$

where $p_{\mathbf{X}}$ it the PDF of $\mathbf{X}$.

Plugging (5.52) in (5.49) gives

$$H\left(\mathbf{Y}|\tilde{\mathbf{H}}\right) = N_r\log_2(\pi) - \mathrm{E}_{\mathbf{Y}|\tilde{\mathbf{H}}}\left\{\log_2\left(\mathrm{E}_{\mathbf{X}}\left\{\frac{\exp\left(-\frac{\|\mathbf{y}-\tilde{\mathbf{H}}\mathbf{x}_t\|_{\mathrm{F}}^2}{\sigma_n^2+\sigma_{\mathbf{e}}^2\|\mathbf{x}_t\|_{\mathrm{F}}^2}\right)}{\left(\sigma_n^2+\sigma_{\mathbf{e}}^2\|\mathbf{x}_t\|_{\mathrm{F}}^2\right)^{N_r}}\right\}\right)\right\}. \quad (5.53)$$

The entropy of $\mathbf{Y}$, knowing $\mathbf{X}$, and the estimated channel, $\tilde{\mathbf{H}}$, is

$$\begin{aligned} H\left(\mathbf{Y}|\mathbf{X},\tilde{\mathbf{H}}\right) &= \mathrm{E}_{\mathbf{X}}\left\{-\mathrm{E}_{\mathbf{Y}|\mathbf{X},\tilde{\mathbf{H}}}\left\{\log_2 p_{(\mathbf{Y}|\mathbf{X},\tilde{\mathbf{H}})}\left(\mathbf{y}|\mathbf{x}_t,\tilde{\mathbf{H}}\right)\right\}\right\} \\ &= \mathrm{E}_{\mathbf{X}}\left\{N_r\log_2\left(\pi\left(\sigma_n^2+\sigma_{\mathbf{e}}^2\|\mathbf{x}_t\|_{\mathrm{F}}^2\right)\exp(1)\right)\right\}, \end{aligned} \quad (5.54)$$

where the same steps as in (5.7) were followed, and $p_{(\mathbf{y}|\mathbf{x}_t,\tilde{\mathbf{H}})}$ is given in (5.51).

Finally, from (5.49), (5.54), and (5.47),

$$\begin{aligned} I(\mathbf{X};\mathbf{Y}) =& -\mathrm{E}_{\mathbf{X}}\left\{N_r\log_2\left(\left(\sigma_n^2+\sigma_{\mathbf{e}}^2\|\mathbf{x}_t\|_{\mathrm{F}}^2\right)\exp(1)\right)\right\} \\ & -\mathrm{E}_{\tilde{\mathbf{H}}}\left\{\mathrm{E}_{\mathbf{Y}|\tilde{\mathbf{H}}}\left\{\log_2\left(\mathrm{E}_{\mathbf{X}}\left\{\frac{\exp\left(-\frac{\|\mathbf{y}-\tilde{\mathbf{H}}\mathbf{x}_t\|_{\mathrm{F}}^2}{\sigma_n^2+\sigma_{\mathbf{e}}^2\|\mathbf{x}_t\|_{\mathrm{F}}^2}\right)}{\left(\sigma_n^2+\sigma_{\mathbf{e}}^2\|\mathbf{x}_t|_{\mathrm{F}}^2\right)^{N_r}}\right\}\right)\right\}\right\}. \end{aligned} \quad (5.55)$$

Comparing the mutual information of SMX in the presence of CSE in (5.55) to that with no CSE in (5.8), it can be seen that CSE can be thought of as an additional noise term with a power proportional to the power of the transmitted signal. The mutual information with no CSE in (5.8) can be easily deduced from (5.55) by setting $\sigma_{\mathbf{e}}^2 = 0$.

5.4.1.2 SMTs

In the previous section, the presence of channel estimation errors were shown to degrade the performance of typical MIMO systems. It is also anticipated that CSE will have similar impact on SMTs as it impacts both transmitted symbols. However and in SMTs, CSE will cause a mismatch in the considered spatial constellation diagram at the receiver. It is like using quadrature amplitude modulation (QAM) constellation diagram at the transmitter while considering PSK at the receiver. Therefore, for SMTs in the presence of CSE at the receiver, the spatial constellation diagram, $\tilde{\mathcal{H}}$, is different than $\mathcal{H}$, where $\tilde{\mathcal{H}}$ is generated from the estimated channel $\tilde{\mathbf{H}}$ instead of $\mathbf{H}$. Now, assume that the ℓth spatial symbol, $\mathcal{H}_\ell$, is transmitted, and the receiver decodes $\tilde{\mathcal{H}}_\ell$, there will be no errors as the same spatial symbol is decoded. As such and even though the received spatial symbol is not identical to the transmitted spatial symbol, $\mathcal{H}_\ell \neq \tilde{\mathcal{H}}_\ell$, it still carries the same information about the transmitted spatial symbol. Therefore, the mutual information for SMTs in the presence of CSE is formulated as

$$I\left(\tilde{\mathcal{H}},\mathcal{S};\mathbf{Y}\right) = H(\mathbf{Y}) - H\left(\mathbf{Y}|\tilde{\mathcal{H}},\mathcal{S}\right). \quad (5.56)$$

The right-hand side of (5.56) gives

$$H(\mathbf{Y}) = -\mathrm{E}_{\mathbf{Y}}\{\log_2 p_{\mathbf{Y}}(\mathbf{y})\}, \quad (5.57)$$

where $p_{\mathbf{Y}}$ is the PDF of the received vector $\mathbf{Y}$, and is given by

$$\begin{aligned} p_{\mathbf{Y}}(\mathbf{y}) &= \int_{\tilde{\mathcal{H}},\mathcal{S}} p_{\tilde{\mathcal{H}}}\left(\tilde{\mathcal{H}}_\ell\right) \times p_{\mathcal{S}}(\mathcal{S}_\iota) \times p_{(\mathbf{Y}|\tilde{\mathcal{H}},\mathcal{S})}\left(\mathbf{y}|\tilde{\mathcal{H}}_\ell, \mathcal{S}_\iota\right) d\tilde{\mathcal{H}}_\ell \, d\mathcal{S}_\iota \\ &= \mathrm{E}_{\tilde{\mathcal{H}},\mathcal{S}} \left\{ \frac{\exp\left(-\frac{\|\mathbf{y}-\tilde{\mathcal{H}}_\ell \mathcal{S}_\iota\|_{\mathrm{F}}^2}{\sigma_n^2+\sigma_{\mathbf{e}}^2\|\mathcal{S}_\iota\|_{\mathrm{F}}^2}\right)}{\pi^{N_r}\left(\sigma_n^2+\sigma_{\mathbf{e}}^2\|\mathcal{S}_\iota\|_{\mathrm{F}}^2\right)^{N_r}} \right\}, \end{aligned} \tag{5.58}$$

where $p_{\tilde{\mathcal{H}}}$ is the PDF of spatial symbols $\tilde{\mathcal{H}}$ and

$$p_{(\mathbf{Y}|\tilde{\mathcal{H}},\mathcal{S})}\left(\mathbf{y}|\tilde{\mathcal{H}}_\ell, \mathcal{S}_\iota\right) = \frac{1}{\left(\pi\left(\sigma_n^2+\sigma_{\mathbf{e}}^2\|\mathcal{S}_\iota\|_{\mathrm{F}}^2\right)\right)^{N_r}} \exp\left(\frac{-\|\mathbf{y}-\tilde{\mathcal{H}}_\ell \mathcal{S}_\iota\|_{\mathrm{F}}^2}{\sigma_n^2+\sigma_{\mathbf{e}}^2\|\mathcal{S}_\iota\|_{\mathrm{F}}^2}\right). \tag{5.59}$$

Note, by plugging (5.48) in (5.17), the received vector can be written as

$$\mathbf{y} = \mathcal{H}_\ell \mathcal{S}_\iota + \mathbf{n} = (\tilde{\mathcal{H}}_\ell + \bar{\mathbf{e}})\mathcal{S}_\iota + \mathbf{n} = \tilde{\mathcal{H}}_\ell \mathcal{S}_\iota + \bar{\mathbf{e}}\mathcal{S}_\iota + \mathbf{n}, \tag{5.60}$$

where $\bar{\mathbf{e}}$ is a subset of $\mathbf{e}$ generated depending on the used SMT, in the same way, $\mathcal{H}_\ell$ is generated. Note, both $\mathbf{n}$ and $\bar{\mathbf{e}}$ are complex Gaussian RVs with zero-mean and σ_n^2 and $\sigma_{\mathbf{e}}^2$ variances, respectively.

Plugging (5.58) in (5.57) leads to

$$\begin{aligned} H(\mathbf{Y}) &= N_r \log_2(\pi) \\ &\quad - \mathrm{E}_{\mathbf{Y}} \left\{ \log_2 \left(\mathrm{E}_{\tilde{\mathcal{H}},\mathcal{S}} \left\{ \frac{\exp\left(-\frac{\|\mathbf{y}-\tilde{\mathcal{H}}_\ell \mathcal{S}_\iota\|_{\mathrm{F}}^2}{\sigma_n^2+\sigma_{\mathbf{e}}^2\|\mathcal{S}_\iota\|_{\mathrm{F}}^2}\right)}{\pi^{N_r}\left(\sigma_n^2+\sigma_{\mathbf{e}}^2\|\mathcal{S}_\iota\|_{\mathrm{F}}^2\right)^{N_r}} \right\} \right) \right\}. \end{aligned} \tag{5.61}$$

The entropy of $\mathbf{Y}$, given the spatial and signal constellation diagrams, $\mathcal{H}$ and $\mathcal{S}$, is given by

$$\begin{aligned} H\left(\mathbf{Y}|\tilde{\mathcal{H}},\mathcal{S}\right) &= \mathop{\mathrm{E}}_{\tilde{\mathcal{H}},\mathcal{S}} \left\{ -\mathrm{E}_{\mathbf{Y}|\mathcal{H},\mathcal{S}} \left\{ \log_2 \left(p_{(\mathbf{Y}|\mathcal{H},\mathcal{S})}\left(\mathbf{y}|\tilde{\mathcal{H}}_\ell, \mathcal{S}_\iota\right) \right) \right\} \right\} \\ &= \mathrm{E}_{\mathcal{S}} \left\{ N_r \log_2 \left(\pi \left(\sigma_n^2+\sigma_{\mathbf{e}}^2\|\mathcal{S}_\iota\|_{\mathrm{F}}^2\right) \exp(1) \right) \right\}. \end{aligned} \tag{5.62}$$

where $p_{(\mathbf{Y}|\tilde{\mathcal{H}},\mathcal{S})}$ is given in (5.59).

Finally, from (5.61), (5.62), and (5.56), the mutual information of SMTs in the presence of CSE is given by

$$\begin{aligned} I\left(\tilde{\mathcal{H}},\mathcal{S};\mathbf{Y}\right) &= -\mathrm{E}_{\mathcal{S}} \left\{ N_r \log_2 \left(\left(\sigma_n^2+\sigma_{\mathbf{e}}^2\|\mathcal{S}_\iota\|_{\mathrm{F}}^2\right) \exp(1) \right) \right\} \\ &\quad - \mathrm{E}_{\mathbf{Y}} \left\{ \log_2 \left(\mathrm{E}_{\tilde{\mathcal{H}},\mathcal{S}} \left\{ \frac{\exp\left(-\frac{\|\mathbf{y}-\tilde{\mathcal{H}}_\ell \mathcal{S}_\iota\|_{\mathrm{F}}^2}{\sigma_n^2+\sigma_{\mathbf{e}}^2\|\mathcal{S}_\iota\|_{\mathrm{F}}^2}\right)}{\pi^{N_r}\left(\sigma_n^2+\sigma_{\mathbf{e}}^2\|\mathcal{S}_\iota\|_{\mathrm{F}}^2\right)^{N_r}} \right\} \right) \right\}. \end{aligned} \tag{5.63}$$

Comparing the mutual information of SMTs in the presence of CSE (5.63) to that with no CSE in (5.20) and to the mutual information of SMX in the presence of CSE in (5.55), the CSE in SMTs not only affect the SNR, where the noise increases by a power proportional to the power of the transmitted signal symbol, S_t, but it also impacts the spatial constellation symbols. As explained earlier, in the presence of CSE, the spatial constellations search space is generated from the estimated channel, $\tilde{\mathbf{H}}$, and not the perfect channel, $\mathbf{H}$. Yet, if the correct ℓth spatial symbol is chosen, there will be no error.

5.4.2 Capacity Analysis

5.4.2.1 Spatial Multiplexing MIMO

The capacity of SMX in the presence of CSE is calculated by maximizing the mutual information in (5.47),

$$
\begin{aligned}
C &= \mathrm{E}_{\tilde{\mathbf{H}}}\left\{\max_{p_{\mathbf{X}}}\left\{I(\mathbf{X};\mathbf{Y}|\tilde{\mathbf{H}})\right\}\right\} \\
&= \mathrm{E}_{\tilde{\mathbf{H}}}\left\{\max_{p_{\mathbf{X}}}\left\{H\left(\mathbf{Y}|\tilde{\mathbf{H}}\right) - H\left(\mathbf{Y}|\mathbf{X},\tilde{\mathbf{H}}\right)\right\}\right\} \\
&= \mathrm{E}_{\tilde{\mathbf{H}}}\left\{\max_{p_{\mathbf{X}}}\left\{H\left(\mathbf{Y}|\tilde{\mathbf{H}}\right)\right\} - H\left(\mathbf{Y}|\mathbf{X},\tilde{\mathbf{H}}\right)\right\},
\end{aligned}
\tag{5.64}
$$

where

(1) the maximization is done over the choices of the PDF of possible transmitted vector $\mathbf{X}$, $p_{\mathbf{X}}$,
(2) and $H(\mathbf{Y}|\mathbf{X},\tilde{\mathbf{H}})$ does not depend on $p_{\mathbf{X}}$. Therefore, the maximization is reduced to the maximization of $H(\mathbf{Y}|\tilde{\mathbf{H}})$.

From [3], and as discussed in previous sections, the distribution that maximizes the entropy is the zero mean complex Gaussian distribution. As such, $H(\mathbf{Y}|\tilde{\mathbf{H}})$ is maximized if $\mathbf{Y} \sim \mathcal{CN}(\mathbf{0}_{N_r}, \Sigma^2_{\mathbf{Y}})$. Thus

$$
p_{\mathbf{Y}|\tilde{\mathbf{H}}}\left(\mathbf{y}|\tilde{\mathbf{H}}\right) = \frac{\exp\left(-\mathbf{y}^H\left|\frac{1}{N_t}\tilde{\mathbf{H}}\tilde{\mathbf{H}}^H + \sigma^2_{\mathbf{e}}\mathbf{I}_{N_r} + \sigma^2_n\mathbf{I}_{N_r}\right|^{-1}\mathbf{y}\right)}{\pi^{N_r}\left|\frac{1}{N_t}\tilde{\mathbf{H}}\tilde{\mathbf{H}}^H + \sigma^2_{\mathbf{e}}\mathbf{I}_{N_r} + \sigma^2_n\mathbf{I}_{N_r}\right|},
\tag{5.65}
$$

where $\Sigma^2_{\mathbf{Y}} = \frac{1}{N_t}\tilde{\mathbf{H}}\tilde{\mathbf{H}}^H + \sigma^2_{\mathbf{e}}\mathbf{I}_{N_r} + \sigma^2_n\mathbf{I}_{N_r}$, and $\mathrm{E}\left\{\tilde{\mathbf{H}}\mathbf{x}_t\right\} = \frac{1}{N_t}\tilde{\mathbf{H}}\tilde{\mathbf{H}}^H$, $\mathrm{E}\{\mathbf{e}\mathbf{x}_t\} = \sigma^2_{\mathbf{e}}\mathbf{I}_{N_r}$ are assumed.

Following the same steps as in (5.7)

$$
\begin{aligned}
\max_{p_{\mathbf{X}}}\left\{H\left(\mathbf{Y}|\tilde{\mathbf{H}}\right)\right\} &= \max_{p_{\mathbf{X}}}\left\{-\mathrm{E}_{\mathbf{Y}|\tilde{\mathbf{H}}}\left\{\log_2 p_{\mathbf{Y}|\tilde{\mathbf{H}}}(\mathbf{y}|\tilde{\mathbf{H}})\right\}\right\} \\
&= \log_2\left(\pi^{N_r}\left|\frac{1}{N_t}\tilde{\mathbf{H}}\tilde{\mathbf{H}}^H + \sigma^2_{\mathbf{e}}\mathbf{I}_{N_r} + \sigma^2_n\mathbf{I}_{N_r}\right|\right) \\
&\quad + N_r\log_2(\exp(1))
\end{aligned}
$$

$$= N_r \log_2 \left(\pi \left(\sigma_{\mathbf{e}}^2 + \sigma_n^2\right) \exp(1)\right)$$
$$+ \log_2 \left| \mathbf{I}_{N_r} + \frac{1}{\sigma_{\mathbf{e}}^2 + \sigma_n^2} \frac{1}{N_t} \tilde{\mathbf{H}}\tilde{\mathbf{H}}^H \right|. \tag{5.66}$$

Plugging (5.66) and (5.54) in (5.64) gives [152]

$$C = \mathrm{E}_{\tilde{\mathbf{H}}} \left\{ N_r \log_2 \left(\pi \left(\sigma_n^2 + \sigma_{\mathbf{e}}^2\right) \exp(1)\right) + \log_2 \left| \mathbf{I}_{N_r} + \frac{1}{\sigma_{\mathbf{e}}^2 + \sigma_n^2} \frac{1}{N_t} \tilde{\mathbf{H}}\tilde{\mathbf{H}}^H \right| \right.$$
$$\left. - \mathrm{E}_{\mathbf{X}} \left\{ N_r \log_2 \left(\pi \left(\sigma_n^2 + \sigma_{\mathbf{e}}^2 \|\mathbf{x}_t\|_{\mathrm{F}}^2\right) \exp(1)\right) \right\} \right\}$$
$$= \mathrm{E}_{\mathbf{X}} \left\{ N_r \log_2 \left(\frac{\sigma_n^2 + \sigma_{\mathbf{e}}^2}{\sigma_n^2 + \sigma_{\mathbf{e}}^2 \|\mathbf{x}_t\|_{\mathrm{F}}^2} \right) \right\}$$
$$+ \mathrm{E}_{\tilde{\mathbf{H}}} \left\{ \log_2 \left| \mathbf{I}_{N_r} + \frac{1}{\sigma_{\mathbf{e}}^2 + \sigma_n^2} \frac{1}{N_t} \tilde{\mathbf{H}}\tilde{\mathbf{H}}^H \right| \right\}. \tag{5.67}$$

Comparing the capacity of SMX in the presence of CSE (5.67) to that with no CSE in (5.24), it can be seen that (5.24) can be easily deduced from (5.67) by setting $\sigma_{\mathbf{e}}^2 = 0$. Furthermore, SMX capacity in the presence of CSE depends on the power of the transmitted vector, and it is distribution where as can be seen there is an averaging over $\mathbf{X}$ in the left side of (5.67).

Finally, from (5.50), and knowing that the noise vector $\mathbf{n} \sim \mathcal{CN}(\mathbf{0}_{N_r}, \mathbf{I}_{N_r}\sigma_n^2)$, the received vector follows complex Gaussian distribution, if the summation $(\tilde{\mathbf{H}}\mathbf{x}_t + \mathbf{e}\mathbf{x}_t)$ is complex Gaussian distributed. Consequently, $\tilde{\mathbf{H}}\mathbf{x}_t$ and $\mathbf{e}\mathbf{x}_t$ each has to be complex Gaussian distributed, where the sum of two complex Gaussian distributions is complex Gaussian distributed. However, this is not possible, since $\tilde{\mathbf{H}}$ is assumed to be deterministic. Hence, the transmitted vector $\tilde{\mathbf{H}}\mathbf{x}_t$ is complex Gaussian distributed, if $\mathbf{x}_t$ follows a complex Gaussian distribution. However, $\mathbf{e}$ is a complex Gaussian distribution RV and if $\mathbf{x}_t$ follows a complex Gaussian distribution, $\mathbf{e}\mathbf{x}_t$ cannot be a complex Gaussian-distributed RV. This is because the multiplication of two complex Gaussian RVs is not a complex Gaussian RV. Thereby, the capacity for SMX in the presence of CSE cannot be achieved. Furthermore, the distribution of transmitted symbols that maximize (5.66) is needed to calculate the left-hand side of (5.67). However, and from the previous discussion, there is no distribution that maximizes (5.66). Thus, the capacity in (5.67) besides not achievable, it cannot be calculated.

Assuming that the CSE noise variance is proportional with the channel noise, $\sigma_{\mathbf{e}}^2 \propto \sigma_n^2$, [192, 215], the capacity of SMX in the presence of CSE in (5.67) can be lower bounded by [152]

$$C_{\mathrm{lower}} \geq \mathrm{E}_{\tilde{\mathbf{H}}} \left\{ \log_2 \left| \mathbf{I}_{N_r} + \frac{1}{N_t} \frac{1}{\sigma_{\mathbf{e}}^2 + \sigma_n^2} \tilde{\mathbf{H}}\tilde{\mathbf{H}}^H \right| \right\}. \tag{5.68}$$

The lower bound capacity in (5.68) can be achieved at high SNR by using maximum-likelihood (ML) receiver and complex Gaussian-distributed symbols, which are the requirements to achieve capacity neglecting CSE.

Figure 5.11 compares the lower capacity bound in (5.68) to the simulated mutual information of SMX in the presence of CSE over a MISO Rayleigh fading channels, with $\eta = 20$ bits, 1024 complex Gaussian-distributed symbols, and $N_t = 2$. It can be seen that simulated mutual information follows C_{lower} until 18 bits, where the simulated mutual information saturates at 20 bits. The saturation is because a complex Gaussian random variable has an infinite number of values, but in this example, only 1024 symbols are used. At low SNR, both $\mathbf{ex}_t$ and $\mathbf{n}$ are dominant in (5.50) as they have larger power than $\tilde{\mathbf{H}}\mathbf{x}_t$, and therefore, the received vector is not complex Gaussian distribution. Note, as discussed earlier with $\mathbf{x}_t$ and $\mathbf{e}$ being complex Gaussian distributed their multiplication is not complex Gaussian distributed. Therefore, at low SNR, C_{lower} is a lower bound. In Figure 5.11, C_{lower} is 0.16 bits lower than the simulated mutual information. As the SNR increases, both $\mathbf{ex}_t$ and $\mathbf{n}$ decrease, and the complex Gaussian-distributed transmitted vector becomes dominant. Consequently, the received vector becomes complex Gaussian dis-

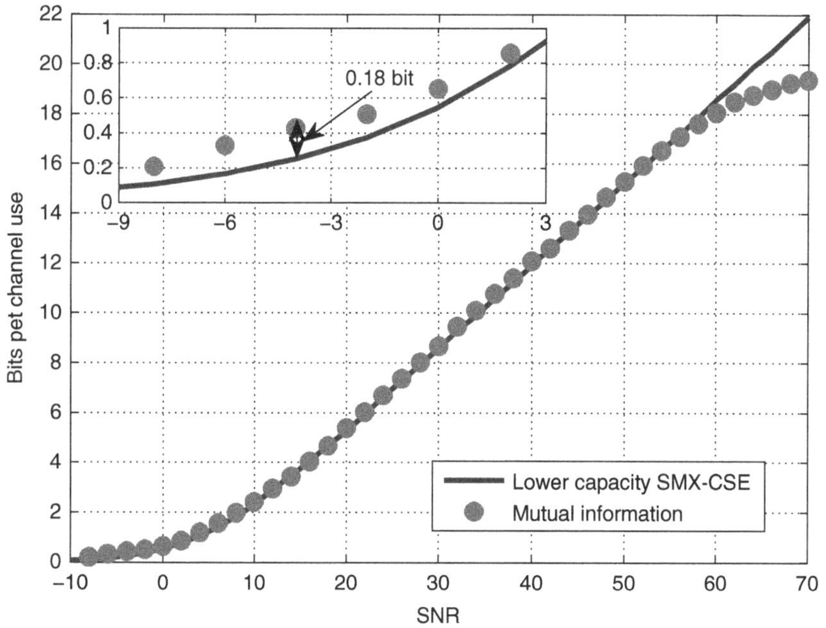

Figure 5.11 The lower capacity of SMX in the presence of CSE compared to the simulated mutual information of SMX in the presence of CSE over MISO Rayleigh fading channels, 1024 complex Gaussian distributed symbols, $\sigma_{\mathbf{e}}^2 = \sigma_n^2$ and $N_t = 2$.

tributed. In Figure 5.11 at high SNR, the simulated mutual information closely follow C_{lower}.

5.4.2.2 SMTs

The capacity is derived by maximizing the mutual information in (5.56) as

$$C = \max_{p_{\tilde{\mathcal{H}}}, p_{\mathcal{S}}} \left\{ I\left(\tilde{\mathcal{H}}, \mathcal{S}; \mathbf{Y}\right) \right\} = \max_{p_{\tilde{\mathcal{H}}}, p_{\mathcal{S}}} \{H(\mathbf{Y})\} - H\left(\mathbf{Y}|\tilde{\mathcal{H}}, \mathcal{S}\right), \tag{5.69}$$

where $H(\mathbf{Y}|\tilde{\mathcal{H}}, \mathcal{S})$ does not depend on $p_{\tilde{\mathcal{H}}}$ and $p_{\mathcal{S}}$, and therefore, the maximization was reduced to $H(\mathbf{Y})$. Moreover, as discussed in Section 5.1.2, there is no averaging over the channel as the channel is used as a way to convey information.

As discussed in Section 5.2.2.2, the entropy $H(\mathbf{Y})$ is maximized when $\mathbf{Y} \sim \mathcal{CN}(0, \sigma_{\mathbf{y}}^2)$. Hence

$$\begin{aligned} \max_{p_{\tilde{\mathcal{H}}}, p_{\mathcal{S}}} \{H(\mathbf{Y})\} &= N_r \log_2\left(\pi \sigma_{\mathbf{y}}^2 \exp(1)\right) \\ &= N_r \log_2\left(\pi\left(1 + \sigma_{\mathbf{e}}^2 + \sigma_n^2\right) \exp(1)\right) \end{aligned} \tag{5.70}$$

where $\sigma_{\mathbf{y}}^2 = 1 + \sigma_{\mathbf{e}}^2 + \sigma_n^2$, $\mathrm{E}\{\|\tilde{\mathcal{H}}_\ell \mathcal{S}_\iota\|_{\mathrm{F}}^2\} = \mathbf{I}_{N_r}$, and $\mathrm{E}\{\|\mathbf{e}\mathcal{S}_\iota\|_{\mathrm{F}}^2\} = \sigma_{\mathbf{e}}^2 \mathbf{I}_{N_r}$ is assumed.

Plugging (5.70) and (5.62) in (5.69), the capacity of SMTs in the presence of CSE is

$$\begin{aligned} C &= N_r \log_2\left(\pi\left(1 + \sigma_{\mathbf{e}}^2 + \sigma_n^2\right) \exp(1)\right) \\ &\quad - \mathrm{E}_{\mathcal{S}}\left\{N_r \log_2\left(\pi\left(\sigma_n^2 + \sigma_{\mathbf{e}}^2 \|\mathcal{S}_\iota\|_{\mathrm{F}}^2\right) \exp(1)\right)\right\} \\ &= \mathrm{E}_{\mathcal{S}}\left\{N_r \log_2\left(\frac{1 + \sigma_{\mathbf{e}}^2 + \sigma_n^2}{\sigma_n^2 + \sigma_{\mathbf{e}}^2 \|\mathcal{S}_\iota\|_{\mathrm{F}}^2}\right)\right\}. \end{aligned} \tag{5.71}$$

As in the case of perfect channel knowledge at the receiver side, the capacity of SMTs in the presence of CSE is a single theoretical equation that does not depend on the fading channel. However, in the presence of CSE, the capacity of SMT depends on the distribution of the signal constellation symbols.

As $\mathbf{n}$ is complex Gaussian distributed, $\mathbf{Y}$ is complex Gaussian distributed, and the capacity in (5.71) is achievable if

$$\mathcal{R}_{\ell,\iota} = \left(\tilde{\mathcal{H}}_\ell + \bar{\mathbf{e}}\right) \mathcal{S}_\iota \sim \mathcal{CN}\left(\mathbf{0}_{N_r}, \left(1 + \sigma_{\mathbf{e}}^2\right) \mathbf{I}_{N_r}\right). \tag{5.72}$$

This can be achieved by tailoring the signal constellation symbols for each fading channel, such that the PDF of the tailored signal constellation symbols, assuming for simplicity and without loss of generality $N_r = 1$, solves

$$\frac{1}{\pi} |\mathcal{R}_{\ell,\iota}| \exp\left(-|\mathcal{R}_{\ell,\iota}|_{\mathrm{F}}^2\right) = \int_\Omega \frac{p_{(\tilde{\mathcal{H}}_\ell + \mathbf{e})}(\Omega)}{|\Omega|} p_{\mathcal{S}_\iota}\left(\frac{\mathcal{R}_{\ell,\iota}}{\Omega}\right) d\Omega, \tag{5.73}$$

where $p_{(\tilde{\mathcal{H}}_\ell+\mathbf{e})}$ is the PDF of the summation of the two RVs $\tilde{\mathcal{H}}_\ell$ and $\mathbf{e}$,

$$\begin{aligned}
p_{(\tilde{\mathcal{H}}+\mathbf{e})}(\Omega) &= \int_{\tilde{\mathcal{H}}} p_{\tilde{\mathcal{H}}}\left(\tilde{\mathcal{H}}_\ell\right) p_{\mathbf{e}}\left(\Omega - \tilde{\mathcal{H}}_\ell\right)\, d\tilde{\mathcal{H}}_\ell \\
&= \int_{\tilde{\mathcal{H}}} p_{\tilde{\mathcal{H}}}\left(\tilde{\mathcal{H}}_\ell\right) \frac{1}{\pi\sigma_{\mathbf{e}}^2} \exp\left(\frac{\|\Omega - \tilde{\mathcal{H}}_\ell\|_{\mathrm{F}}^2}{\sigma_n^2}\right) d\tilde{\mathcal{H}}_\ell \\
&= \frac{1}{\pi\sigma_{\mathbf{e}}^2} \mathrm{E}_{\tilde{\mathcal{H}}}\left\{\exp\left(\frac{\|\Omega - \tilde{\mathcal{H}}_\ell\|_{\mathrm{F}}^2}{\sigma_n^2}\right)\right\},
\end{aligned} \tag{5.74}$$

and $p_{\mathbf{e}}$ is the PDF of the CSE noise, $\mathbf{e}$, which is assumed to be complex Gaussian distributed. Note, the PDF of the sum of two RVs is given by the convolution of the PDFs of the two RVs [214].

Substituting (5.74) in (5.73) gives

$$|\mathcal{R}_{\ell,\iota}| \exp\left(-|\mathcal{R}_{\ell,\iota}|_{\mathrm{F}}^2\right) = \int_\Omega \frac{\mathrm{E}_{\tilde{\mathcal{H}}}\left\{\exp\left(\frac{\|\Omega - \tilde{\mathcal{H}}_\ell\|_{\mathrm{F}}^2}{\sigma_n^2}\right)\right\}}{\sigma_{\mathbf{e}}^2|\Omega|} p_{S_\iota}\left(\frac{\mathcal{R}_{\ell,\iota}}{\Omega}\right) d\Omega. \tag{5.75}$$

Different to SMX, SMTs can achieve capacity in the presence of CSE as long as the used constellation diagram solves (5.75). Note, the distribution of the constellation diagram that solves (5.75) is the one to use to calculate the capacity in (5.71). In the following section, examples of how to achieve capacity for SSK and SM in the presence of CSE are given.

5.4.3 Achieving SMTs Capacity

5.4.3.1 SSK

SSK does not have signal symbols, $\mathcal{S} = \{1\}$. Therefore, to achieve the capacity, $\tilde{\mathcal{H}} \sim \mathcal{CN}(\mathbf{0}_{N_r}, \mathbf{I}_{N_r})$. Note, the sum of two independent complex normal distributed RVs is also a complex normal distributed RV [214]. Hence and in the presence of complex Gaussian distributed CSE noise, SSK can achieve the capacity only when the channel is a large-scale Rayleigh fading channel. The mutual information performance of SSK system over Rayleigh fading channel for different number of transmit antennas, $N_t = 64,\ 256$, and 1024 is depicted in Figure 5.12. The derived theoretical capacity curve in (5.71) for SMTs is depicted as well. In all results, $N_r = 2$ is assumed.

From Figure 5.12, it can be seen that the larger the number of transmit antennas, the closer the mutual information to the theoretical capacity curve. For $N_t = 1024$, the mutual information for SSK over Rayleigh fading channel and in the presence CSE is shown to follow the capacity very closely up to 5 bits. After 5 bits, the mutual information deviates because the number of transmit

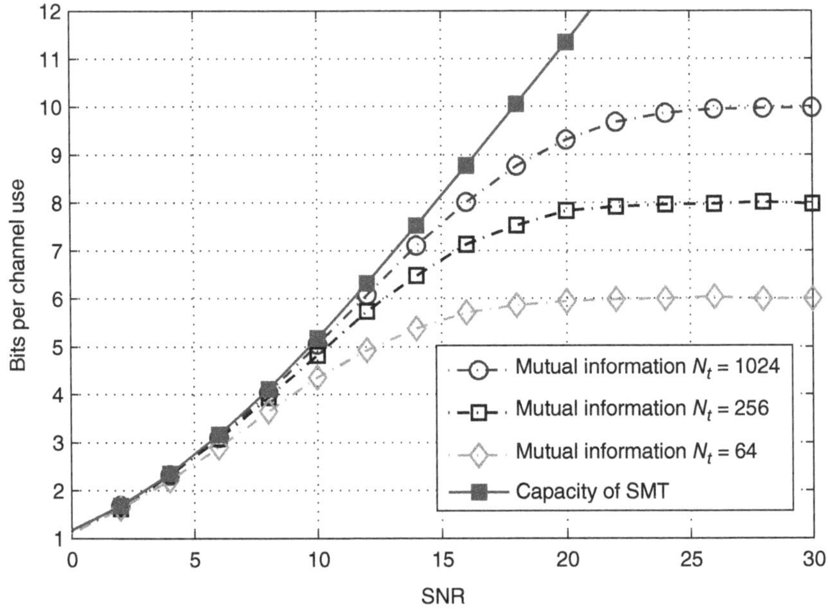

Figure 5.12 The capacity of SMTs compared to simulated mutual information of SSK over Rayleigh fading channel in the presence of CSE for $N_t = 64,\ 256$, and 1024, and $N_r = 2$.

antennas is finite, whereas a Rayleigh distributed RV continues with an infinite number of values, or in this case infinite number of transmit antennas.

5.4.3.2 SM

Following similar discussion as in the previous section and Section 5.3.2, SM would achieve the capacity if large signal and spatial constellation symbols are considered over Rayleigh fading channels while using PSK constellation diagram.

An example is shown in Figure 5.13, where the mutual information performance for SM system over Rayleigh fading channels in the presence of CSE for large-scale MIMO with $N_t = 256$, $N_r = 3$ and with 32-PSK constellation diagram and Gaussian-distributed signal constellation are depicted. The theoretical capacity in (5.71) is shown in the figure as well. The results show that CU distribution, which is obtainable through 32-PSK constellation diagram, is required to achieve the theoretical capacity limit. This again contradicts the belief that complex Gaussian-distributed symbols are needed to achieve the SMTs capacity. It is illustrated in Figure 5.13 that using Gaussian-distributed signal constellation symbols, even though not practically possible to generate, perform 0.5 bit less than that of CU distribution symbols at low SNR and 0.8 bit

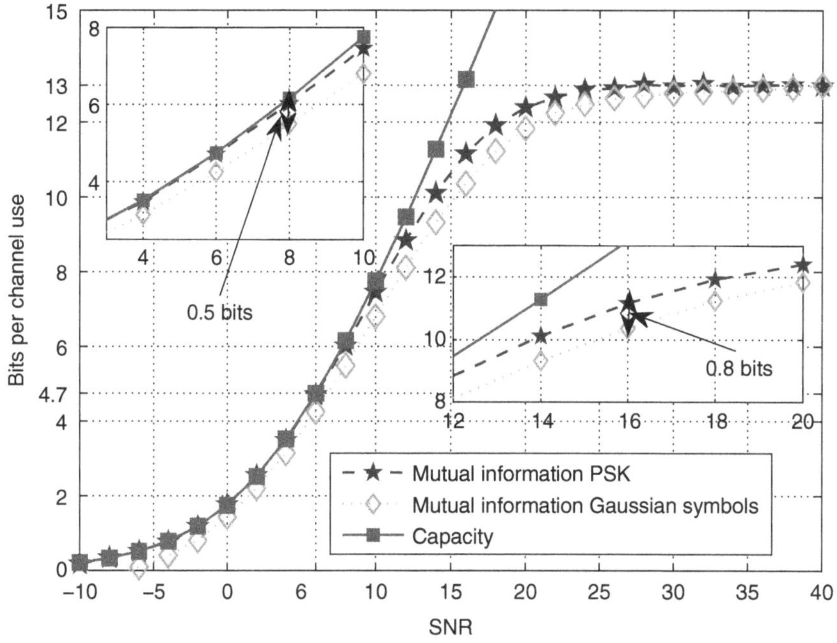

Figure 5.13 The capacity of SMT compared to the simulated mutual information of SM using PSK constellations, and Gaussian-distributed constellations, over Rayleigh fading channel in the presence of CSE, where $M = 32$, $N_t = 256$, and $N_r = 3$.

less at high SNR. These results highlight that SM systems can achieve the capacity in the presence of CSE with large number of transmit antennas and large PSK constellation diagram.

Furthermore, and from Figure 5.13, it can be seen that SM using PSK follows the theoretical capacity curve up to 4.7 bits at SNR= 6 dB. Beyond that value, it started to slowly deviate mainly due to the limiting number of spatial and signal symbols as discussed before. To achieve the capacity for the entire range of SNR, continuous distributions with infinite number of symbols are required, which is not attainable.

Finally, from Figures 5.12 and 5.13, and comparing SMT to SMX, different to SMX, SMT can achieve the capacity in the presence of CSE. Furthermore, SMT capacity in the presence of CSE can be achieved easily by

(1) increasing the number of transmit antennas, which as explained in Chapter 3 comes at a very small cost, where SMTs need only one radio frequency (RF)-chain;
(2) and in the case of SM by using CU distributed signal symbols, which are easily obtained by using PSK, as for other SMT systems by using constellation symbols with a distribution that solves (5.75).

5.5 Mutual Information Performance Comparison

A mutual information performance comparison between SM, QSM, and SMX is presented in Figures 5.14 and 5.15 for Rayleigh and Nakagami-$m = 4$ fading channels, respectively, with $\eta = 4$ and 8 and $N_t = N_r = 4$. Moreover, the theoretical capacity of SMT is depicted for reference. It can be seen that SM offers 0.7 bit performance gain for $\eta = 4$. However, for larger spectral spectral efficiency, $\eta = 8$, SMX outperforms SM by about 1.2 bits and performs 0.6 bits better than QSM. The better performance of SMX can be attributed to the use of smaller signal constellation diagram in comparison with SM and QSM, 93.75% and 75% smaller constellation size, respectively. The same can be seen for QSM and SM, where QSM offers 0.5 bit better performance than SM, as QSM modulates more information bits in the spatial domain than SM, it requires smaller signal constellation diagram.

In Figure 5.15 for Nakagami-m fading channel, different to Rayleigh fading channels, SM is shown to offer nearly the same performance as SMX for $\eta = 8$ where 0.3 bit difference is noticed at low SNR but diminishes as SNR increases.

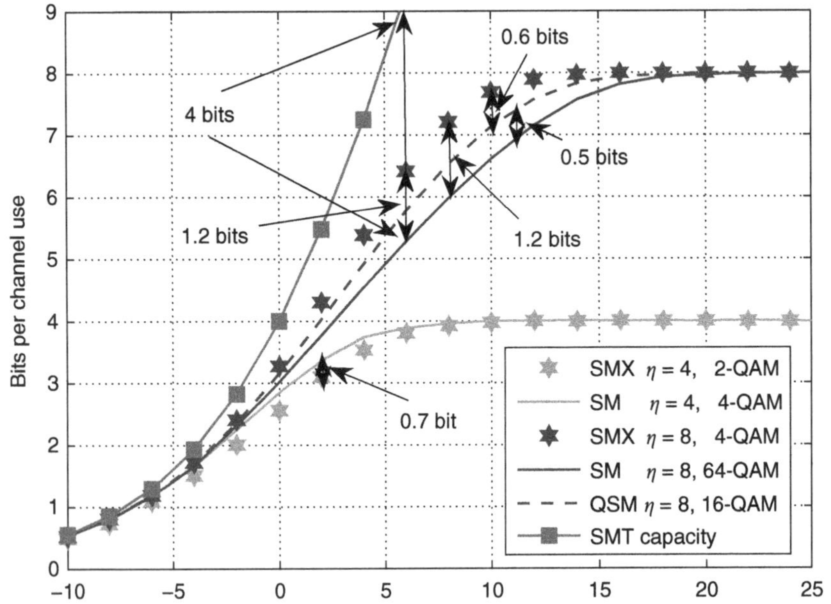

Figure 5.14 The capacity of SMT compared to the simulated mutual information of SM, QSM, and SMX over Rayleigh fading channels for different spectral efficiency, where $\eta = 4$ and 8 and $N_t = N_r = 4$.

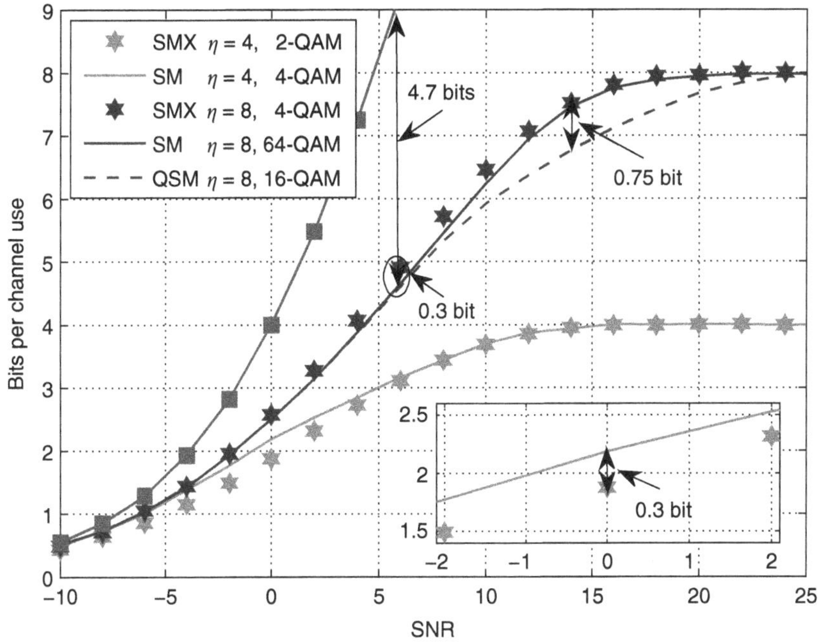

Figure 5.15 The capacity of SMT compared to the simulated mutual information of SM, QSM, and SMX over Nakagami-m = 4 fading channels for different spectral efficiencies, where $\eta = 4$ and 8 and $N_t = N_r = 4$.

Also, SM is shown to outperform QSM by 0.75 bit. Though, for Rayleigh fading channels and for $\eta = 4$ as shown in Figure 5.14, SM demonstrates better performance than SMX by about 0.3 bit.

It is also important to note that SM and QSM mutual information curves are far below the theoretical capacity limit for such systems, nearly 4 and 4.7 bits degradation can be observed for Rayleigh and Nakagami-m channels, respectively, in Figures 5.14 and 5.15. Enhancing their performance and diminishing this gap can be achieved through proper design of the signal constellation diagram for each channel statistics as discussed earlier.

The capacity of SMT is compared to the simulated mutual information of SM, QSM, and SMX in the presence of CSE with $\sigma_{\mathbf{e}}^2 = \sigma_n^2$ in Figure 5.16, where $\eta = 4$ and 8 and $N_t = N_r = 4$. Again for small spectral efficiency, SM performs 0.25 bit better performance than SMX. However, for larger spectral efficiency, SMX offers 1 bit better performance than SM and 0.5 bit better than QSM and reaches maximum mutual information 6 dB earlier than SM and QSM. Yet, in Figure 5.16, SM is performing 3.8 bits less than the maximum SMT capacity, as the used signal constellation diagram is not tailored for SM with $N_t = 4$.

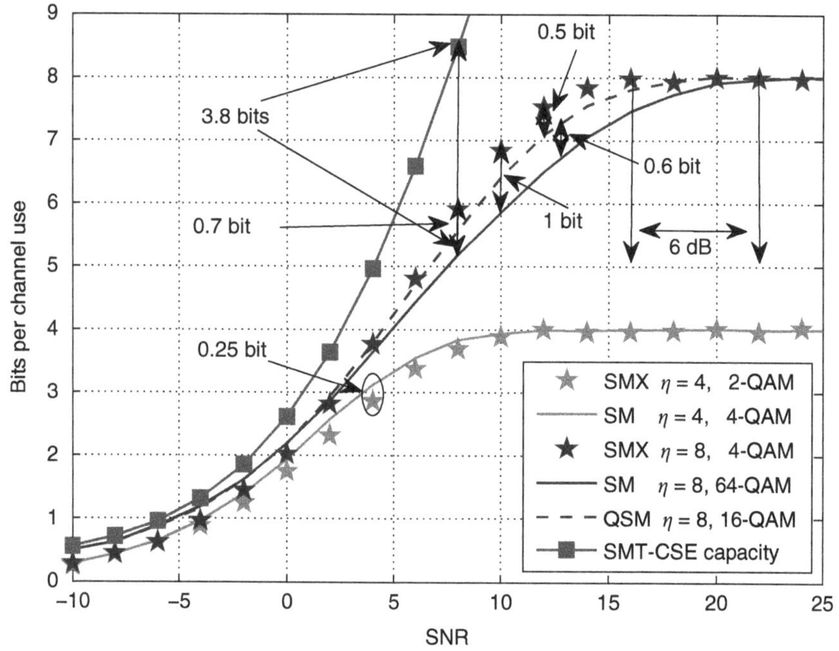

Figure 5.16 The capacity of SMT compared to the simulated mutual information of SM, QSM, and SMX over Rayleigh fading channels in the presence of CSE with $\sigma_{\mathbf{e}}^2 = \sigma_n^2$ for different spectral efficiencies, where $\eta = 4$ and 8 and $N_t = N_r = 4$.

Therefore, it is anticipated that SM and QSM would offer better performance than SMX if their signal constellations diagrams are designed for SM and QSM with $N_t = 4$ in accordance with (5.75). Finally, because QSM modulates more bits in the spatial domain than SM, in Figure 5.16, QSM offers 0.6 bit better performance than SM.

6

Cooperative SMTs

Cooperative communications create collaboration through distributed transmission/processing by allowing different nodes in a wireless network to share resources. The information for each user is sent out not only by the user but also by other collaborating users. This includes a family of configurations in which the information can be shared among transmitters and relayed to reach final destination in order to improve the systems overall capacity and coverage [209, 216–230]. As such, cooperative technologies have made their way toward wireless standards, such as IEEE 802.16 (WiMAX) [231] and long-term evolution (LTE) [232] and have been incorporated into many modern wireless applications, such as cognitive radio and secret communications.

Driven by the several advantages of space modulation techniques (SMTs) and cooperative communication technologies, cooperative SMTs have been extensively investigated in the past few years. Reported results promise significant enhancements in spectral efficiency and network coverage [98, 206, 233–244].

In this chapter, cooperative SMTs are studied and analyzed. In particular, amplify and forward (AF), decode and forward (DF), and two-way relaying (2WR) will be considered.

6.1 Amplify and Forward (AF) Relaying

In cooperative AF relaying, a source (S) and destination (D) nodes are communicating, and multiple or single AF relays participate in the communication protocol as illustrated in Figure 6.1.

At each particular time instant, η bits are to be transmitted using any of the previously studied SMTs. In the first phase, the source, which is equipped with N_t transmit antennas, applies an arbitrary SMT scheme and transmits the vector $\mathbf{x}_t$ from the available transmit antennas as discussed in previous chapters. The transmitted signal is received by both the destination and the relay node

Space Modulation Techniques, First Edition. Raed Mesleh and Abdelhamid Alhassi.

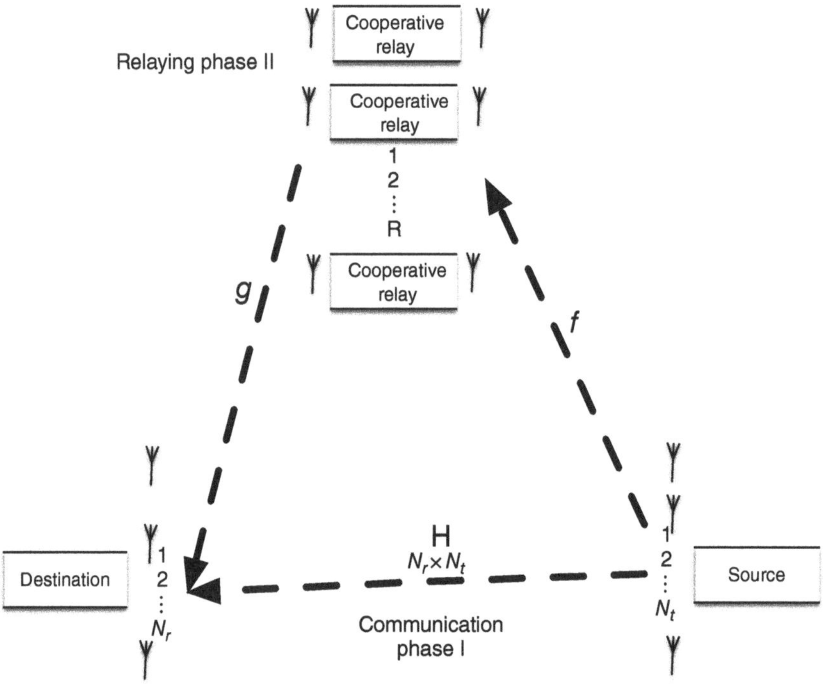

Figure 6.1 AF cooperative SMT system model. A system with N_t transmit antennas at the source, N_r receive antennas at the destination, and with R AF relays are considered.

R. Let $\mathbf{H}$ denotes the $N_r \times N_t$ multiple-input multiple-output (MIMO) channel matrix between the source and the destination. The received signal at the destination is then given by

$$\mathbf{y}_{d,s} = \mathbf{H}\mathbf{x}_t + \mathbf{n}_{d,s} = \mathcal{H}_{\ell_t}\mathcal{S}_{\iota_t} + \mathbf{n}_{d,s}, \tag{6.1}$$

where depending on the considered SMT, $\mathcal{H}_\ell \in \mathcal{H}$ is the ℓth spatial symbol chosen from the spatial constellation diagram $\mathcal{H}$, where $\mathcal{H}$ is generated form $\mathbf{H}$ as discussed in Chapter 3. Moreover, $\mathcal{S}_\iota \in \mathcal{S}$ is the ιth signal constellation symbol chosen from the signal constellation diagram $\mathcal{S}$. Finally, $\mathbf{n}_{d,s}$ is an N_r-length additive white Gaussian noise (AWGN) vector with zero mean and variance σ_n^2. Note, for simplicity, $\mathrm{E}\{\|\mathbf{H}\mathbf{x}_t\|_\mathrm{F}^2\} = N_r$, and therefore, the signal-to-noise-ratio (SNR) is $\mathrm{SNR} = 1/\sigma_n^2$.

In the depicted scenario in Figure 6.1, R single-antenna AF relays are assumed. Hence, the signal received at the kth relay in the first time slot is given by

$$y_{k,s} = \mathbf{f}_k\mathbf{x}_t + n_{k,s} = \mathcal{F}_{\ell,k}\mathcal{S}_\iota + n_{k,s}, \tag{6.2}$$

where $\mathbf{f}_k$ is the N_t–dimensional multiple-input single-output (MISO) channel vector between the source and the kth relay, which has similar characteristics as $\mathbf{H}$, and $\mathcal{F}_{\ell,k} \in \mathcal{F}_k$ denotes the ℓth spatial symbol chosen from the spatial constellation diagram $\mathcal{F}_k$, where $\mathcal{F}_k$ is generated, depending on the used SMT, from $\mathbf{f}_k$. Finally, $n_{k,s}$ is an AWGN with zero mean and variance σ_n^2 seen at the input of the kth relay.

In conventional AF relaying, all the relays participate in the second phase by retransmitting the source signal to the destination in a predetermined orthogonal time slots. Therefore, $R + 1$ time slots are needed for each symbol transmission. The relayed signal is an amplified version of the received signal at the relay node. As such, the amplification process is performed in the analog domain without further processing. Hence, the received signal at the destination can be written as

$$\mathbf{y}_{d,k} = \underbrace{A\mathbf{g}_k\mathcal{F}_{k,\ell}\mathcal{S}_\iota}_{\text{Signal part}} + \underbrace{A\mathbf{g}_k n_{k,s} + \mathbf{n}_{d,k}}_{\text{Noise part}} = A\mathbf{g}_k\mathcal{F}_{k,\ell}\mathcal{S}_\iota + \ddot{\mathbf{n}}_k, \tag{6.3}$$

where $\mathbf{g}_k$ denotes an N_r-dimensional single–input multiple–output (SIMO) channel vector between the kth relay and the destination, A is the amplification factor, and $\ddot{\mathbf{n}}_k$ is a colored Gaussian noise vector, $\ddot{\mathbf{n}}_k \sim \mathcal{CN}(\mathbf{0}_{N_r}, \sigma_n^2\boldsymbol{\Upsilon}_k)$, where $\boldsymbol{\Upsilon}_k = (A^2\mathbf{g}_k\mathbf{g}_k^H + \mathbf{I}_{N_r})$, and $\mathbf{0}_N$ and $\mathbf{I}_N$ are an N-length all zeros vector and N dimensional identity square matrix, respectively.

It is assumed that the receiver has full channel state information (CSI). Hence, the optimum maximum–likelihood (ML) detector assuming perfect time synchronization is given by

$$[\hat{\ell}, \hat{\iota}] = \arg\min_{\ell,\iota}\left(\|\mathbf{y}_{d,s} - \mathcal{H}_\ell\mathcal{S}_\iota\|_\mathrm{F}^2 + \sum_{k=1}^{R}\left\|\boldsymbol{\Upsilon}_k^{-\frac{1}{2}}(\mathbf{y}_{d,k} - A\mathbf{g}_k\mathcal{F}_{k,\ell}\mathcal{S}_\iota)\right\|_\mathrm{F}^2\right), \tag{6.4}$$

where the search is over $\ell \in \{1, \ldots, 2^{\eta_\mathcal{H}}\}$ and $\iota \in \{1, \ldots, 2^{\eta_\mathcal{S}}\}$, with $\eta_\mathcal{H}$ and $\eta_\mathcal{S}$ denoting the number of bits modulated in the spatial and signal domains, respectively.

6.1.1 Average Error Probability Analysis

The ML receiver in (6.4) can be rewritten as

$$\hat{\mathbf{x}} = \arg\min_{\mathbf{x}\in\mathbf{X}}\left(\|\mathbf{y}_{d,s} - \mathbf{H}\mathbf{x}\|_\mathrm{F}^2 + \sum_{k=1}^{R}\left\|\boldsymbol{\Upsilon}_k^{-\frac{1}{2}}(\mathbf{y}_{d,k} - A\mathbf{g}_k\mathbf{f}_k\mathbf{x})\right\|_\mathrm{F}^2\right), \tag{6.5}$$

where $\mathbf{X}$ denoting a space containing all possible transmitted vectors.

The pairwise error probability (PEP) of an AF cooperative system is given by

$$\begin{aligned}
\Pr_{\text{error}} &= \Pr\left(\mathbf{x}_t \rightarrow \mathbf{x}|\mathbf{H}, \mathbf{g}_1, \mathbf{f}_1, \ldots, \mathbf{g}_R, \mathbf{f}_R\right)\\
&= \Pr\left(\left(\|\mathbf{y}_{d,s} - \mathbf{H}\mathbf{x}_t\|_F^2 + \sum_{k=1}^{R}\left\|\boldsymbol{\Upsilon}_k^{-\frac{1}{2}}(\mathbf{y}_{d,k} - A\mathbf{g}_k\mathbf{f}_k\mathbf{x}_t)\right\|_F^2\right)\right.\\
&\qquad \left.> \left(\|\mathbf{y}_{d,s} - \mathbf{H}\mathbf{x}\|_F^2 + \sum_{k=1}^{R}\left\|\boldsymbol{\Upsilon}_k^{-\frac{1}{2}}(\mathbf{y}_{d,k} - A\mathbf{g}_k\mathbf{f}_k\mathbf{x})\right\|_F^2\right)\right)\\
&= \Pr\left(\left(\|\mathbf{n}_{d,s}\|_F^2 + \sum_{k=1}^{R}\left\|\boldsymbol{\Upsilon}_k^{-\frac{1}{2}}\ddot{\mathbf{n}}_k\right\|_F^2\right)\right.\\
&\qquad \left.> \left(\|\mathbf{H}(\mathbf{x}_t - \mathbf{x}) + \mathbf{n}_{d,s}\|_F^2 + \sum_{k=1}^{R}\left\|\boldsymbol{\Upsilon}_k^{-\frac{1}{2}}(A\mathbf{g}_k\mathbf{f}_k(\mathbf{x}_t - \mathbf{x}) + \ddot{\mathbf{n}}_k)\right\|_F^2\right)\right)\\
&= \Pr\left(\left(\|\mathbf{n}_{d,s}\|_F^2 + \sum_{k=1}^{R}\left\|\boldsymbol{\Upsilon}_k^{-\frac{1}{2}}\ddot{\mathbf{n}}_k\right\|_F^2\right)\right.\\
&\qquad \left.> \left(\|\mathbf{H}\boldsymbol{\Psi} + \mathbf{n}_{d,s}\|_F^2 + \sum_{k=1}^{R}\left\|\boldsymbol{\Upsilon}_k^{-\frac{1}{2}}(\mathbf{F}_k\boldsymbol{\Psi} + \ddot{\mathbf{n}}_k)\right\|_F^2\right)\right),
\end{aligned} \tag{6.6}$$

where $\boldsymbol{\Psi} = (\mathbf{x}_t - \mathbf{x})$, and $\mathbf{F}_k = A\mathbf{g}_k\mathbf{f}_k$.

Now,

$$\begin{aligned}
\left\|\boldsymbol{\Upsilon}_k^{-\frac{1}{2}}(\mathbf{F}_k\boldsymbol{\Psi} + \ddot{\mathbf{n}}_k)\right\|_F^2 &= \left\|\boldsymbol{\Upsilon}_k^{-\frac{1}{2}}\mathbf{F}_k\boldsymbol{\Psi}\right\|_F^2 - 2\,\mathrm{Re}\left\{\left(\boldsymbol{\Upsilon}_k^{-\frac{1}{2}}\mathbf{F}_k\boldsymbol{\Psi}\right)^H\boldsymbol{\Upsilon}_k^{-\frac{1}{2}}\ddot{\mathbf{n}}_k\right\}\\
&\quad + \left\|\boldsymbol{\Upsilon}_k^{-\frac{1}{2}}\ddot{\mathbf{n}}_k\right\|_F^2\\
&= \left\|\boldsymbol{\Upsilon}_k^{-\frac{1}{2}}\mathbf{F}_k\boldsymbol{\Psi}\right\|_F^2 - 2\,\mathrm{Re}\left\{\left(\boldsymbol{\Upsilon}_k^{-\frac{1}{2}}\mathbf{F}_k\boldsymbol{\Psi}\right)^H\bar{\bar{\mathbf{n}}}_k\right\}\\
&\quad + \|\bar{\bar{\mathbf{n}}}_k\|_F^2,
\end{aligned} \tag{6.7}$$

where $\bar{\bar{\mathbf{n}}}_k = \boldsymbol{\Upsilon}_k^{-\frac{1}{2}}\ddot{\mathbf{n}}_k$ is a white Gaussian noise with zero mean and covariance given by

$$\begin{aligned}
\mathrm{E}\{\bar{\bar{\mathbf{n}}}_k\} &= \mathrm{E}\left\{\boldsymbol{\Upsilon}_k^{-\frac{1}{2}}\ddot{\mathbf{n}}_k\left(\boldsymbol{\Upsilon}_k^{-\frac{1}{2}}\ddot{\mathbf{n}}_k\right)^H\right\} = \mathrm{E}\left\{\boldsymbol{\Upsilon}_k^{-\frac{1}{2}}\ddot{\mathbf{n}}_k\ddot{\mathbf{n}}_k^H\left(\boldsymbol{\Upsilon}_k^{-\frac{1}{2}}\right)^H\right\}\\
&= \boldsymbol{\Upsilon}_k^{-\frac{1}{2}}\mathrm{E}\{\ddot{\mathbf{n}}_k\ddot{\mathbf{n}}_k^H\}\left(\boldsymbol{\Upsilon}_k^{-\frac{1}{2}}\right)^H = \boldsymbol{\Upsilon}_k^{-\frac{1}{2}}(\sigma_n^2\boldsymbol{\Upsilon}_k)\left(\boldsymbol{\Upsilon}_k^{-\frac{1}{2}}\right)^H = \sigma_n^2\mathbf{I}_{N_r}.
\end{aligned} \tag{6.8}$$

Plugging (6.7) in (6.6) and following the same steps as in (4.46),

$$\begin{aligned}
\Pr_{\text{error}} &= \Pr\left(2\text{Re}\left\{\mathbf{n}_{d,s}^{H}\mathbf{H}\boldsymbol{\Psi} + \sum_{k=1}^{R} \overline{\overline{\mathbf{n}}}_{k}^{H}\boldsymbol{\Upsilon}_{k}^{-\frac{1}{2}}\mathbf{F}_{k}\boldsymbol{\Psi}\right\}\right. \\
&\quad > \left.\left(\|\mathbf{H}\boldsymbol{\Psi}\|_{\text{F}}^{2} + \sum_{k=1}^{R}\left\|\boldsymbol{\Upsilon}_{k}^{-\frac{1}{2}}\mathbf{F}_{k}\boldsymbol{\Psi}\right\|_{\text{F}}^{2}\right)\right) \\
&= Q\left(\sqrt{\frac{\|\mathbf{H}\boldsymbol{\Psi}\|_{\text{F}}^{2} + \sum_{k=1}^{R}\left\|\boldsymbol{\Upsilon}_{k}^{-\frac{1}{2}}\mathbf{F}_{k}\boldsymbol{\Psi}\right\|_{\text{F}}^{2}}{2\sigma_{\mathbf{n}}^{2}}}\right) \\
&= Q\left(\sqrt{\gamma_{d,s} + \sum_{k=1}^{R}\gamma_{d,k}}\right) \\
&= \frac{1}{\pi}\int_{0}^{\pi/2}\exp\left(-\frac{\gamma_{d,s} + \sum_{k=1}^{R}\gamma_{d,k}}{2\sin^{2}\theta}\right)d\theta \\
&= \frac{1}{\pi}\int_{0}^{\pi/2}\exp\left(-\frac{\gamma_{d,s}}{2\sin^{2}\theta}\right)\prod_{k=1}^{R}\exp\left(-\frac{\gamma_{d,k}}{2\sin^{2}\theta}\right)d\theta,
\end{aligned} \tag{6.9}$$

where

$$\gamma_{d,s} = \frac{\|\mathbf{H}\boldsymbol{\Psi}\|_{\text{F}}^{2}}{2\sigma_{n}^{2}}, \tag{6.10}$$

$$\gamma_{d,k} = \frac{\left\|\boldsymbol{\Upsilon}_{k}^{-\frac{1}{2}}\mathbf{F}_{k}\boldsymbol{\Psi}\right\|_{\text{F}}^{2}}{2\sigma_{n}^{2}}. \tag{6.11}$$

The average PEP is then computed by taking the expectation of (6.9),

$$\text{E}\left\{\Pr_{\text{error}}\right\} = \frac{1}{\pi}\int_{0}^{\frac{\pi}{2}}\mathcal{M}_{\gamma_{d,s}}\left(-\frac{1}{2\sin^{2}\theta}\right)\prod_{k=1}^{R}\mathcal{M}_{\gamma_{d,k}}\left(-\frac{1}{2\sin^{2}\theta}\right) \tag{6.12}$$

$$\leq \frac{1}{2}\mathcal{M}_{\gamma_{d,s}}\left(-\frac{1}{2}\right)\prod_{k=1}^{R}\mathcal{M}_{\gamma_{d,k}}\left(-\frac{1}{2}\right), \tag{6.13}$$

where $\mathcal{M}_{\gamma_{d,s}}(s)$ and $\mathcal{M}_{\gamma_{d,k}}(s)$ are the moment-generation functions (MGFs) of $\gamma_{d,s}$ and $\gamma_{d,k}$, respectively. Note, the upper bound in (6.13) is obtained by using $\theta = \pi/2$ in the integral in (6.12) [244].

Finally, the average bit error ratio (ABER) of AF cooperative system is

$$\text{ABER} \leqslant \frac{1}{2^\eta} \sum_{\mathbf{x}_t \in \mathbb{X}} \sum_{\mathbf{x} \in \mathbb{X}} \frac{e_{\mathbf{x}}^{\mathbf{x}_t}}{\eta} \frac{1}{2} \mathcal{M}_{\gamma_{d,s}}\left(-\frac{1}{2}\right) \prod_{k=1}^{R} \mathcal{M}_{\gamma_{d,k}}\left(-\frac{1}{2}\right), \tag{6.14}$$

where $e_{\mathbf{x}}^{\mathbf{x}_t}$ is the number of bits in error associated with the corresponding PEP event.

For $N_r = 1$, then,

$$\begin{aligned}
\gamma_{d,k} &= \frac{\left\|\mathbf{\Upsilon}_k^{-\frac{1}{2}} \mathbf{F}_k \mathbf{\Psi}\right\|_{\mathrm{F}}^2}{2\sigma_n^2} = \frac{\left\|\mathbf{\Upsilon}_k^{-\frac{1}{2}} A g_k \mathbf{f}_k \mathbf{\Psi}\right\|_{\mathrm{F}}^2}{2\sigma_n^2} \\
&= \frac{\left\|(1 + A^2 |g_k|^2)^{-\frac{1}{2}} A g_k \mathbf{f}_k \mathbf{\Psi}\right\|_{\mathrm{F}}^2}{2\sigma_n^2} \\
&= \frac{A^2 |g_k|^2}{1 + A^2 |g_k|^2} \frac{\|\mathbf{f}_k \mathbf{\Psi}\|_{\mathrm{F}}^2}{2\sigma_n^2} = A_k^2 |g_k|^2 \frac{\|\mathbf{f}_k \mathbf{\Psi}\|_{\mathrm{F}}^2}{2\sigma_n^2} = A_k^2 \varphi_k \gamma_k,
\end{aligned} \tag{6.15}$$

where $\varphi_k = |g_k|^2$, $\gamma_k = \|\mathbf{f}_k \mathbf{\Psi}\|_{\mathrm{F}}^2/(2\sigma_n^2)$, and $A_k = A/\sqrt{1 + A^2 |g_k|^2}$.

Assuming Rayleigh fading channels, the random variables (RVs) φ_k and γ_k are exponential RVs with means $\bar{\varphi}_k$ and $\bar{\gamma}_R$, respectively, with $\bar{\varphi}_k = 1$ and

$$\bar{\gamma}_R = \frac{\|\mathbf{\Psi}\|_{\mathrm{F}}^2}{2\sigma_n^2}. \tag{6.16}$$

The cumulative distribution function (CDF) of the RV $\gamma_{d,k}$, which is the result of the multiplication of the RVs A_k, φ_k and γ_k, is given by [123, 245, 246],

$$F_{\gamma_{d,k}}(\gamma) = 1 - 2\sqrt{\frac{\gamma}{\bar{\gamma}_R}} \exp\left(-\frac{\gamma}{\bar{\gamma}_R}\right) K_1\left(2\sqrt{\frac{\gamma}{\bar{\gamma}_R}}\right), \tag{6.17}$$

and the probability distribution function (PDF) is given by [246]

$$p_{\gamma_{d,k}}(\gamma) = \frac{2}{\bar{\gamma}_R} \exp\left(-\frac{\gamma}{\bar{\gamma}_R}\right) \left[\sqrt{\frac{\gamma}{\bar{\gamma}_R}} K_1\left(2\sqrt{\frac{\gamma}{\bar{\gamma}_R}}\right) + K_0\left(2\sqrt{\frac{\gamma}{\bar{\gamma}_R}}\right)\right], \tag{6.18}$$

where $K_a(\cdot)$ is the ath-order modified Bessel function of the second kind.

Finally, the MGF of $\gamma_{d,k}$ is given by [246]

$$\mathcal{M}_{\gamma_{d,k}}(s) = \frac{1}{1 + s\bar{\gamma}_R} + \frac{\bar{\gamma}_R}{(1 + s\bar{\gamma}_R)^2} s \exp\left(\frac{1}{1 + s\bar{\gamma}_R}\right) E_1\left(\frac{1}{1 + s\bar{\gamma}_R}\right), \tag{6.19}$$

where $E_1(\cdot)$ is the exponential integral function.

The MGF of the exponential RV $\gamma_{d,s}$ is given by [123],

$$\mathcal{M}_{\gamma_{d,s}}(s) = \frac{1}{1 + s\bar{\gamma}_R}. \tag{6.20}$$

The ABER of AF SMTs over MISO Rayleigh fading channels can be calculated by substituting (6.19) and (6.20) in (6.14).

6.1.1.1 Asymptotic Analysis

At asymptotically high SNR values, and using Taylor series, the PDF of $\gamma_{d,k}$ is simplified as

$$p_{\gamma_{d,k}}(\gamma) \approx \gamma \left(\frac{1}{\bar{\gamma}_R} + \frac{1}{\bar{\gamma}_R} \left[\psi(1) - \log \left(\frac{1}{\bar{\gamma}_R} \right) \right] \right), \tag{6.21}$$

where $\psi(\cdot)$ is the digamma function with $\psi(1) = -0.57721$.

Thus, the average PEP can be computed as [247]

$$\mathrm{E}\left\{ \Pr_{\text{error}} \right\} \leq \frac{1}{\bar{\gamma}_R} \frac{2^R \Gamma(R+0.5)}{\sqrt{\pi}\Gamma(R+1)} \left(\frac{1}{\bar{\gamma}_R} + \frac{1}{\bar{\gamma}_R} \left[\psi(1) - \log \left(\frac{1}{\bar{\gamma}_R} \right) \right] \right)^R. \tag{6.22}$$

The diversity gain of R is clearly seen in (6.22). Finally, the asymptotic ABER of AF SMTs over MISO Rayleigh fading channels can be computed by plugging (6.22) in (6.14).

6.1.1.2 Numerical Results

In the first results shown in Figure 6.2, the performance of AF space shift keying (SSK) system is evaluated through Monte-Carlo simulations and analytical formulas for $N_t = 2$, $N_r = 1$, and variable number of AF relays from $R = 1, 3$, and 5. Results reveal that increasing the number of relays significantly enhances the

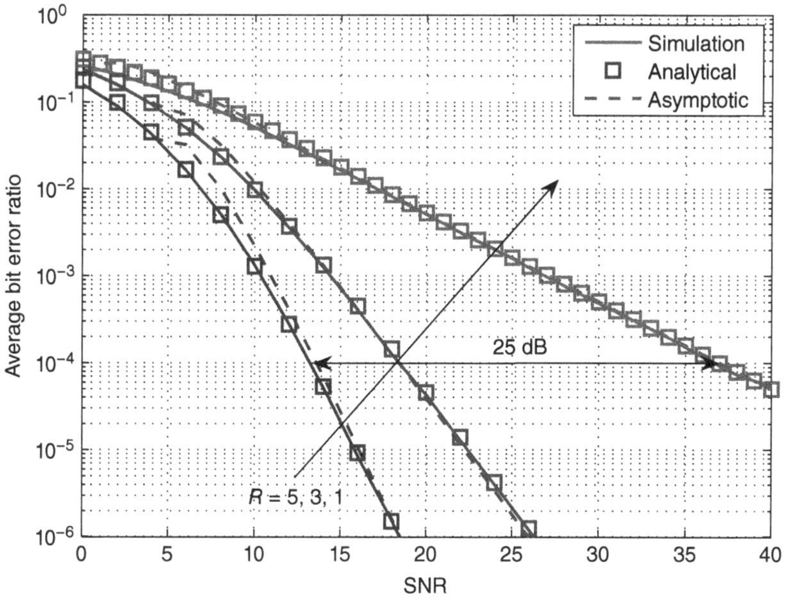

Figure 6.2 Simulation, analytical, and asymptotic results for AF SSK system with $N_t = 2$, $N_r = 1$, and variable $R = 1, 3,$ and 5.

ABER performance. Also, analytical and asymptotic curves are shown to closely match Monte Carlo simulation results for a wide range of SNR values and for the different number of relays. An increase in diversity with the increase of R is also clear from the figure. Such increase in diversity gain is shown to provide about 25 dB gain in SNR at an ABER of 10^{-4}. It should be noted, though, that the spectral efficiency is not identical for the depicted curves even though they all transmit the same number of data bits. This is because the number of needed orthogonal time slots for the AF scheme is $R + 1$, which increases with increasing R. Therefore, an AF scheme with $R = 5$ requires six time slots to convey source data to destination, whereas a system with $R = 1$ requires only two time slots.

Results for AF quadrature spatial modulation (QSM) system with $N_t = 2$, $N_r = 1$, and $R = 3 \rightarrow 5$ while considering 4-quadrature amplitude modulation (QAM) are shown in Figure 6.3. Again, higher R value results in better error performance, and analytical and simulation results are shown to match closely for a wide range of SNR values. Also and as discussed for the previous results in Figure 6.2, increasing the value of R enhances the performance and degrades the spectral efficiency as $R + 1$ time slots are needed to convey the source information bits. A performance comparison between spatial modulation (SM) and

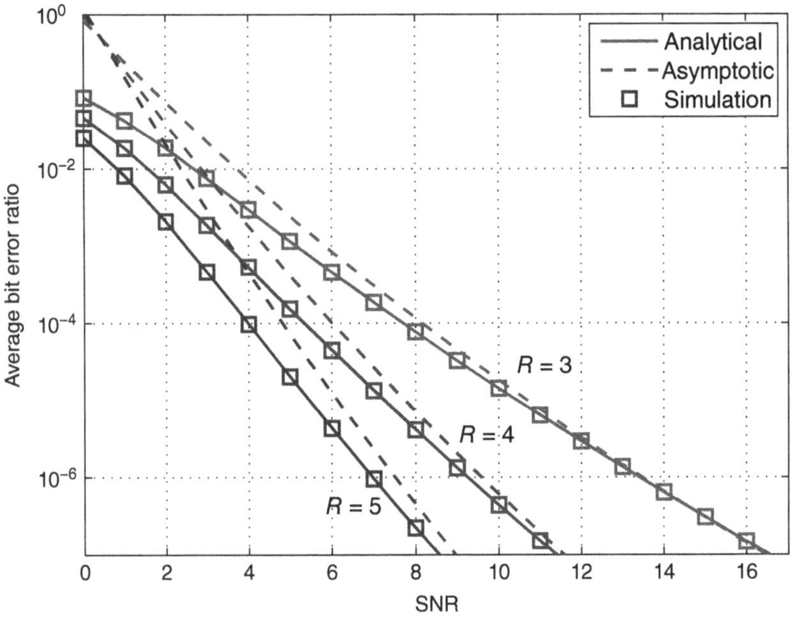

Figure 6.3 Analytical, simulation, and asymptotic results for an AF QSM system with $N_t = 2$, $N_r = 1$, and 4-QAM modulation while varying $R = 2 \rightarrow 5$.

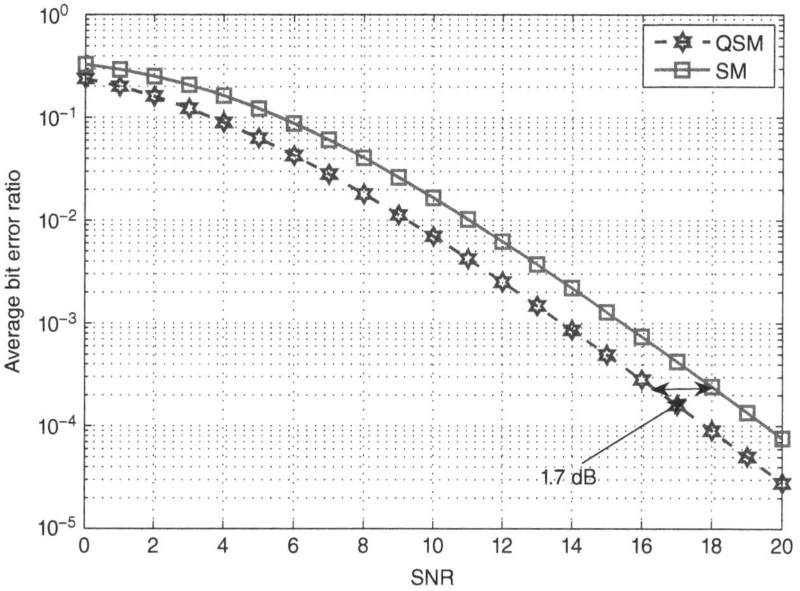

Figure 6.4 AF cooperative QSM and SM performance comparison with $\eta = 4$ bits and with 4-QAM modulation assuming $N_t = 2$ for QSM and $N_t = 8$ for SM and $N_r = 1$ and $R = 1$.

QSM with single AF relay and for a spectral efficiency of $\eta = 4$ bits is depicted in Figure 6.4 while assuming 4-QAM modulation. QSM system is implemented with $N_t = 2$ while SM considers $N_t = 8$ to achieve the target spectral efficiency. Results show that QSM outperforms SM performance by about 1.7 dB. Similar comparison between SM and QSM but for a spectral efficiency of $\eta = 6$ bits is shown in Figure 6.5. The target spectral efficiency is achieved by considering 4-QAM for both schemes and assuming $N_t = 4$ for QSM and $N_t = 16$ for SM systems. A single AF cooperative relay is also considered. Again, QSM demonstrates better performance and a gain of about 1.7 dB can be clearly seen from the figure.

6.1.2 Opportunistic AF Relaying

Previous conventional AF relaying scheme requires $R + 1$ time slots to convey the source message to the destination. To enhance the spectral efficiency, opportunistic relaying can be considered, where only the best relay participates in the relaying process. This relay is chosen by selecting the indirect link from the source to the kth relay and then to the destination, S–R_k–D, that gives the minimum instantaneous error probability.

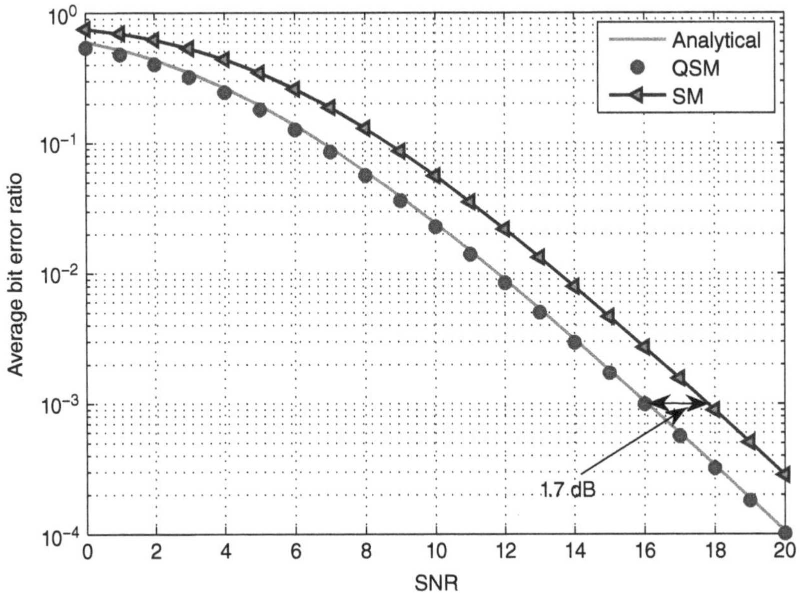

Figure 6.5 AF cooperative QSM and SM performance comparison with $\eta = 6$ bits and with 4-QAM modulation assuming $N_t = 4$ for QSM and $N_t = 16$ for SM and $N_r = 1$.

Following similar steps as in Section 6.1.1, the PEP for the kth relay link can be written as

$$\Pr_{\text{error}} = Q(\sqrt{\gamma_{d,k}}) = Q\left(\sqrt{A_k^2|g_k|^2\frac{\|\mathbf{f}_k\boldsymbol{\Psi}\|_{\mathrm{F}}^2}{2\sigma_n^2}}\right). \tag{6.23}$$

The relay that minimizes (6.23) is selected to participate in the retransmission process. Hence, the chosen relay is formulated as

$$\begin{aligned} k_{\text{Sel}} &= \min_k\left(Q\left(\sqrt{A_k^2|g_k|^2\frac{\|\mathbf{f}_k\boldsymbol{\Psi}\|_{\mathrm{F}}^2}{2\sigma_n^2}}\right)\right) \\ &= \max_k\left(A_k^2|g_k|^2\frac{\|\mathbf{f}_k\boldsymbol{\Psi}\|_{\mathrm{F}}^2}{2\sigma_n^2}\right). \end{aligned} \tag{6.24}$$

Please note that since only a single relay participates in the retransmission process, only two time slots are needed regardless of the number of relays in the network, and proper communication between the relays is assumed.

The ML receiver using opportunistic relaying is then given by

$$[\hat{\ell},\hat{\imath}] = \arg\min_{\ell,\imath}\left(\|\mathbf{y}_{s,d}-\mathcal{H}_\ell \mathcal{S}_\imath\|_F^2+\left\|\mathbf{\Upsilon}_{k_{\text{Sel}}}^{-\frac{1}{2}}\left(\mathbf{y}_{k_{\text{Sel}},d}-A\mathbf{g}_{k_{\text{Sel}}}\mathcal{F}_{k_{\text{Sel}},\ell}\mathcal{S}_\imath\right)\right\|_F^2\right). \tag{6.25}$$

6.1.2.1 Average Error Probability Analysis

In opportunistic relaying, only one relay is retransmitting. Thus and following similar steps as in Section 6.1.1, the average PEP using opportunistic relaying can be written as

$$\mathrm{E}\left\{\Pr_{\text{error}}\right\} = \frac{1}{2}\mathcal{M}_{\gamma_{d,s}}\left(-\frac{1}{2}\right)\mathcal{M}_{\gamma_{d,k_{\text{Sel}}}}\left(-\frac{1}{2}\right), \tag{6.26}$$

where

$$\boldsymbol{\gamma}_{d,k_{\text{Sel}}} = A_{k_{\text{Sel}}}^2 |g_{k_{\text{Sel}}}|^2 \frac{\|\mathbf{f}_{k_{\text{Sel}}}\mathbf{\Psi}\|_F^2}{2\sigma_n^2}. \tag{6.27}$$

The CDF of $\boldsymbol{\gamma}_{d,k_{\text{Sel}}}$ can be written as [248]

$$\begin{aligned} F_{\boldsymbol{\gamma}_{d,k_{\text{Sel}}}}(\gamma) &= \Pr(\boldsymbol{\gamma}_{d,k_{\text{Sel}}} \leq \gamma) = \Pr(\max_k \boldsymbol{\gamma}_{d,k} \leq \gamma) \\ &= \prod_{k=1}^{R} \Pr(\boldsymbol{\gamma}_{d,k} \leq \gamma) = \prod_{k=1}^{R} F_{\gamma_{d,k}} \\ &= \prod_{k=1}^{R}\left(1-2\sqrt{\frac{\gamma}{\bar{\gamma}_R}}\exp\left(-\frac{\gamma}{\bar{\gamma}_R}\right)K_1\left(2\sqrt{\frac{\gamma}{\bar{\gamma}_R}}\right)\right), \end{aligned} \tag{6.28}$$

where $F_{\gamma_{d,k}}$ is given in (6.17).

Furthermore, the PDF is given by

$$p_{\gamma_{d,k_{\text{Sel}}}}(\gamma) = \prod_{k=1}^{R}\frac{2}{\bar{\gamma}_R}\exp\left(-\frac{\gamma}{\bar{\gamma}_R}\right)\left[\sqrt{\frac{\gamma}{\bar{\gamma}_R}}K_1\left(2\sqrt{\frac{\gamma}{\bar{\gamma}_R}}\right)+K_0\left(2\sqrt{\frac{\gamma}{\bar{\gamma}_R}}\right)\right]. \tag{6.29}$$

Thus, the MGF of $\gamma_{d,k}$ is given by

$$\underset{\gamma_{d,k_{\text{Sel}}}}{M}(s) = \prod_{k=1}^{R}\left(\frac{1}{1+s\bar{\gamma}_R}+\frac{\bar{\gamma}_R}{(1+s\bar{\gamma}_R)^2}s\exp\left(\frac{1}{1+s\bar{\gamma}_R}\right)E_1\left(\frac{1}{1+s\bar{\gamma}_R}\right)\right). \tag{6.30}$$

Plugging (6.30) in (6.26) results in the same average PEP as derived for conventional AF SMTs in (6.13). Hence, opportunistic AF relaying offers the same ABER performance as conventional AF relaying, but with an enhanced spectral efficiency.

6.1.2.2 Asymptotic Analysis

At high SNR values, and using Taylor series, the PDF of $\gamma_{d,k_{\text{Sel}}}$ in (6.29) can simplified to

$$p_{\gamma_{d,k_{\text{Sel}}}}(\gamma) = \frac{\gamma^{R-1}}{\Gamma(R)}\left(\frac{1}{\bar{\gamma}_R} + \frac{1}{\bar{\gamma}_R}\left[\psi(1) - \log\left(\frac{1}{\bar{\gamma}_R}\right)\right]\right)^R. \tag{6.31}$$

Hence, and using (6.26), the average PEP can be computed as

$$\mathrm{E}\left\{\Pr_{\text{error}}\right\} \leq \frac{1}{2}\frac{1}{\bar{\gamma}_R}\frac{2^{R-1}\Gamma(R+0.5)}{\sqrt{\pi}\Gamma(R+1)}\left(\frac{1}{\bar{\gamma}_R} + \frac{1}{\bar{\gamma}_R}\left[\psi(1) - \log\left(\frac{1}{\bar{\gamma}_R}\right)\right]\right)^R. \tag{6.32}$$

As in conventional AF relays, a diversity gain of R is clearly seen in (6.32) for opportunistic relaying.

6.2 Decode and Forward (DF) Relaying

In DF relaying, all existing relays process the received signal and decode the transmitted information. It is generally assumed that an error detection mechanism is available and the relay can tell if the decoded bits are correct or not. If the relay decodes the source signal correctly, it participates in the second phase by forwarding the message to the destination. However, if an error is detected in the retrieved data at the relay, it remains silent at this particular time instant. The relays that will participate in the second phase are grouped in a set C and allocated orthogonal slots. Therefore, the spectral efficiency will decay by a factor of $R_C + 1$, and required synchronization and signaling between relays is needed, with R_C denoting the number of relays participating in the relaying phase. In practical DF systems, as in IEEE 802.16j standard [231] and other literature [248, 249], error detection techniques [250] are used and the relay participates in the cooperative phase if it detects the whole packet correctly. However, the simplified assumption made here facilitates the derived analysis of the error probability and commonly assumed in the literature (see [235, 237, 239, 248, 249, 251–253]). Besides the conducted analysis provides a benchmark for all practical systems.

6.2.1 Multiple single-antenna DF relays

The source is equipped with multiple antennas and applies a specific SMT. However, in the first scenario considered here, single-antenna DF relays are considered as shown in Figure 6.1. Similar to previous discussion for AF relaying, η bits are to be transmitted by the source at each particular time instant. The signal received at the destination node through the direct link can be written as in (6.1). Also, the signal received at the kth relay in the first time slot is given by (6.2).

The relays apply the specific SMT-ML decoder to decode the received source signal and retrieve the transmitted bits. If the retrieved bits are correct, the relay participates in the retransmission process. However, the relay is equipped with single antenna. Hence, the participating relay will forward the following message to the destination, $\mathbf{f}_k\mathbf{x}_t$. As such, the received signal at the destination from kth cooperative relay, $k \in \{C\}$, is given by

$$\mathbf{y}_{d,k} = \mathbf{g}_k\mathbf{f}_k\mathbf{x}_t + \mathbf{n}_{d,k}. \tag{6.33}$$

At the destination, the ML optimum detector is considered to jointly decode the received signals from the source and the cooperating relays:

$$\hat{x} = \arg\min_{x\in\mathbf{X}} \left(\|\mathbf{y}_{d,s} - \mathbf{H}\mathbf{x}\|_F^2 + \sum_{k=1}^{R_C} \|\mathbf{y}_{d,k} - \mathbf{g}_k\mathbf{f}_k\mathbf{x}\|_F^2 \right). \tag{6.34}$$

6.2.2 Single DF Relay with Multiple Antennas

The previous DF system considers multiple DF relays each equipped with single transmit and receive antennas. Alternatively, DF relays with multiple transmit antennas can be considered as illustrated in Figure 6.6. In such case, the relay will apply the same SMT considered at the source and transmits identical data. A DF relay with N_t antennas exists and participates in the retransmission phase if it detects the source signal correctly. The source and the destination nodes are the same as discussed before. The received signal at the destination is the same as given in (6.1). The signal received at the relay is given by

$$\mathbf{y}_{r,s} = \mathbf{F}\mathbf{x}_t + \mathbf{n}_{r,s}, \tag{6.35}$$

where $\mathbf{F}$ is the $N_t \times N_t$ square channel matrix between relay and the source, and $\mathbf{n}_{r,s}$ is an N_t-length AWGN vector with zero-mean and σ_n^2 variance.

The relay decodes the signal using the ML decoder as

$$\hat{x} = \arg\min_{\mathbf{x}\in\mathbf{X}} \|\mathbf{y}_{r,s} - \mathbf{F}\mathbf{x}\|_F^2. \tag{6.36}$$

If the signal is decoded correctly, the relay retransmits the decoded symbol vector $\mathbf{x}_t$ using the same SMT used at the transmitter. Therefore, the received signal at the destination in the cooperative phase is given by

$$\mathbf{y}_{d,r} = \mathbf{G}\mathbf{x}_t + \mathbf{n}_{d,r}, \tag{6.37}$$

where $\mathbf{G}$ is an $N_r \times N_t$ fading channel matrix between the relay and the destination, and $\mathbf{n}_{d,r}$ is an N_r-length AWGN vector with zero-mean and σ_n^2 variance.

The destination node combines the received signals from the direct link and the cooperative link to detect the source signal as

$$\hat{x} = \arg\min_{\mathbf{x}\in\mathbf{X}} \left(\|\mathbf{y}_{d,s} - \mathbf{H}\mathbf{x}\|_F^2 + \|\mathbf{y}_{d,r} - \mathbf{G}\mathbf{x}\|_F^2 \right). \tag{6.38}$$

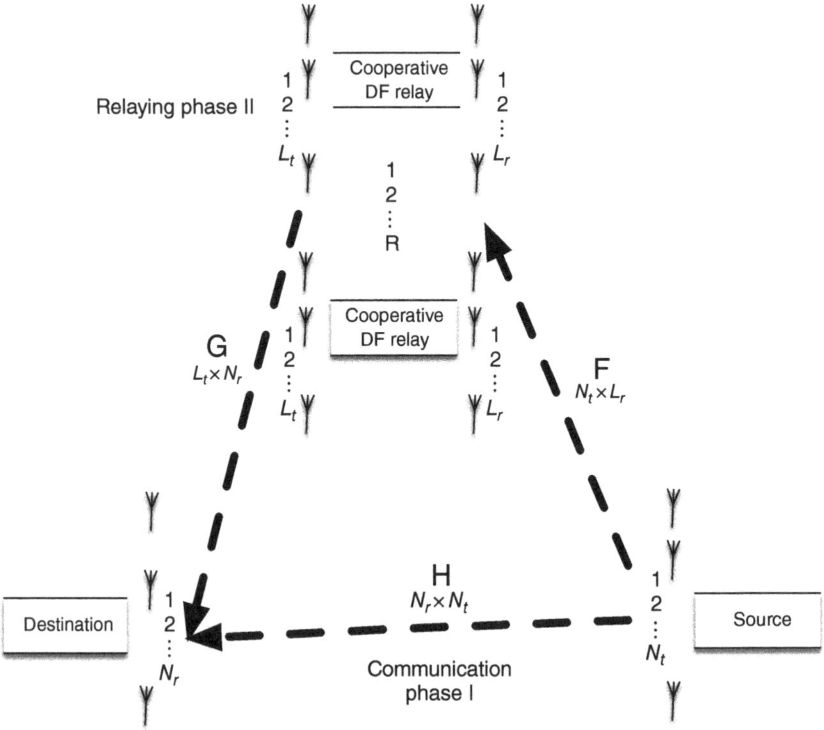

Figure 6.6 DF cooperative SMTs system model. A system with N_t transmit antennas at the source, N_r receive antennas at the destination, and with L_t transmit and L_r receive antennas at the DF relays are considered.

6.2.3 Average Error Potability Analysis

6.2.3.1 Multiple Single-Antenna DF Relays

In multiple single-antenna DF relaying, the transmitted message is received via a direct link and through all relays that detected the transmitted SMT signal correctly. The average PEP that the kth relay detects the SMT signal incorrectly, and thus being off, is given by

$$P_{\text{off}} = \text{E}\left\{\Pr\left(\mathbf{x}_t \to \mathbf{x}|\mathbf{f}_k\right)\right\} = \text{E}\left\{\Pr\left(\|\mathbf{y}_{k,s} - \mathbf{f}_k\mathbf{x}_t\|_\text{F}^2 > \|\mathbf{y}_{k,s} - \mathbf{f}_k\mathbf{x}\|_\text{F}^2\right)\right\}$$
$$= \text{E}\left\{Q\left(\sqrt{\frac{\|\mathbf{f}_k\mathbf{\Psi}\|_\text{F}^2}{2\sigma_n^2}}\right)\right\} = \text{E}\left\{Q\left(\sqrt{\gamma_k}\right)\right\}. \tag{6.39}$$

As defined earlier in Section 6.1.1, $\gamma_k = \|\mathbf{f}_k\boldsymbol{\Psi}\|_{\mathrm{F}}^2/2\sigma_n^2$ is an exponential RV with a mean $\bar{\gamma}_R$ given in (6.16). Hence, and from (4.6),

$$P_{\text{off}} = \frac{1}{2}\left(1 - \sqrt{\frac{\bar{\gamma}_R/2}{1+\bar{\gamma}_R/2}}\right). \tag{6.40}$$

In the considered multiple single-antenna DF relays, different scenarios can be defined. In the first scenario, none of the available relays decoded the source message correctly. Therefore, all relays will be off and will not participate in the relaying phase. As such, the destination has to rely on the direct link signal to decode the message. In the second scenario, part of the relays decoded the signal correctly and are grouped in the set C. Finally, all relays can decode the signal correctly and $R_C = R$. Let the PEP given that the rth scenario occurs be $\Pr_{e|\text{Scn}_r}$. Then, the average $\Pr_{\text{error}}$ for all scenarios is given by

$$\Pr_{\text{error}} = \mathrm{E}\left\{\Pr_{e|\text{Scn}_k}\right\} = \sum_{r=0}^{R}\binom{R}{r}P_{\text{off}}^{R-r}(1-P_{\text{off}})^r \Pr_{e|\text{Scn}_r}. \tag{6.41}$$

In the first scenario, all relays are off and the receiver decodes only the direct link signal. Hence, the PEP of this scenario is

$$\begin{aligned}\Pr_{e|\text{Scn}_0} &= \Pr(\mathbf{x}_t \to \mathbf{x}|\mathbf{H})\\ &= \Pr(\|\mathbf{y}_{s,d} - \mathbf{H}\mathbf{x}_t\|_{\mathrm{F}}^2 > \|\mathbf{y}_{s,d} - \mathbf{H}\mathbf{x}\|_{\mathrm{F}}^2)\\ &= Q(\sqrt{\gamma_{d,s}}),\end{aligned} \tag{6.42}$$

where $\gamma_{d,s}$ is given in (6.10).

In the other two scenarios, $0 < r \leq R$ relays detected the signal correctly and retransmitted it to the destination. Hence, the receiver will be receiving the signal from the direct link and from r relaying links. Thus, and following the same steps as in Section 6.1.1, the PEP for the rth scenario is given by

$$\begin{aligned}\Pr_{e|\text{Scn}_r} &= \Pr\left(\mathbf{x}_t \to \mathbf{x}|\mathbf{H}, \mathbf{g}_1, \mathbf{f}_1, \ldots, \mathbf{g}_r, \mathbf{f}_r\right)\\ &= \Pr\left(\left(\|\mathbf{y}_{d,s} - \mathbf{H}\mathbf{x}_t\|_{\mathrm{F}}^2 + \sum_{k=1}^{r}\|\mathbf{y}_{d,k} - \mathbf{g}_k\mathbf{f}_k\mathbf{x}_t\|_{\mathrm{F}}^2\right)\right.\\ &\quad \left.> \left(\|\mathbf{y}_{d,s} - \mathbf{H}\mathbf{x}\|_{\mathrm{F}}^2 + \sum_{k=1}^{r}\|\mathbf{y}_{d,k} - \mathbf{g}_k\mathbf{f}_k\mathbf{x}\|_{\mathrm{F}}^2\right)\right)\\ &= Q\left(\sqrt{\gamma_{d,s} + \sum_{k=1}^{r}\gamma_{d,k}^{\text{DF}_R}}\right),\end{aligned} \tag{6.43}$$

where

$$\gamma_{d,k}^{\mathrm{DF}_R} = \frac{\|\mathbf{g}_k\mathbf{f}_k\boldsymbol{\Psi}\|_\mathrm{F}^2}{2\sigma_n^2} = \|\mathbf{g}_\mathrm{k}\|_\mathrm{F}^2\frac{\|\mathbf{f}_k\boldsymbol{\Psi}\|_\mathrm{F}^2}{2\sigma_n^2} = \varphi_k\gamma_k, \tag{6.44}$$

where φ_k and γ_k are, as defined earlier in Section 6.1.1, exponential RVs with means $\bar{\varphi}_k = 1$ and $\bar{\gamma}_R$ given in (6.16).

Plugging (6.42) and (6.43) in (6.41) gives

$$\Pr_{\text{error}} = P_{\text{off}}^R Q\left(\sqrt{\gamma_{d,s}}\right) + \sum_{r=1}^{R}\binom{R}{r}P_{\text{off}}^{R-r}(1-P_{\text{off}})^r Q\left(\sqrt{\gamma_{d,s} + \sum_{k}^{r}\gamma_{d,k}^{\mathrm{DF}_R}}\right). \tag{6.45}$$

Taking the expectation of the PEP in (6.45), using (4.6) and following the same steps as in Section 6.1.1,

$$\begin{aligned}
\mathrm{E}\left\{\Pr_{\text{error}}\right\} &= P_{\text{off}}^R \mathrm{E}\left\{Q\left(\sqrt{\gamma_{d,s}}\right)\right\} \\
&\quad + \sum_{r=1}^{R}\binom{R}{r}P_{\text{off}}^{R-r}(1-P_{\text{off}})^r \mathrm{E}\left\{Q\left(\sqrt{\gamma_{d,s} + \sum_{k=1}^{r}\gamma_{d,k}^{\mathrm{DF}_R}}\right)\right\} \\
&\leq \frac{1}{2}P_{\text{off}}^R\left(1-\sqrt{\frac{\bar{\gamma}_R/2}{1+\bar{\gamma}_R/2}}\right) \\
&\quad + \frac{1}{2}\sum_{r=1}^{R}\binom{R}{r}P_{\text{off}}^{R-r}(1-P_{\text{off}})^r \mathcal{M}_{\gamma_{d,s}}\left(-\frac{1}{2}\right)\prod_{k=1}^{r}\mathcal{M}_{\gamma_{d,k}^{\mathrm{DF}_R}}\left(-\frac{1}{2}\right).
\end{aligned} \tag{6.46}$$

Note, $\mathcal{M}_{\gamma_{d,s}}$ is given in (6.20).

From [198], the PDF of $\gamma_{d,k}^{\mathrm{DF}_R}$, which is the result of the multiplication of two exponential RVs φ_k and γ_k, is given by

$$p_{\gamma_{d,k}^{\mathrm{DF}_R}}(\gamma) = \frac{2}{\bar{\gamma}_R}K_0\left(2\sqrt{\frac{\gamma}{\bar{\gamma}_R}}\right). \tag{6.47}$$

The MGF of $\gamma_{d,k}^{\mathrm{DF}_R}$ is then given by

$$\mathcal{M}_{\gamma_{d,k}^{\mathrm{DF}_R}}(s) = \frac{1}{\bar{\gamma}_R s}\exp\left(-\frac{1}{\bar{\gamma}_R s}\right)\Gamma\left(0, \frac{1}{\bar{\gamma}_R s}\right), \tag{6.48}$$

where $\Gamma(0, \cdot)$ is the incomplete Gamma function.

6.2.3.2 Single DF Relay with Multiple-Antennas

From (6.46), the average PEP for a single DF relay is given by

$$\begin{aligned} \mathrm{E}\left\{\Pr_{\text{error}}\right\} &= P_{\text{off}}\mathrm{E}\left\{Q\left(\sqrt{\gamma_{d,s}}\right)\right\} + (1-P_{\text{off}})\,\mathrm{E}\left\{Q\left(\sqrt{\gamma_{d,s}+\gamma_{d,r}^{\text{DF}}}\right)\right\} \\ &\leq \frac{1}{2}P_{\text{off}}\left(1-\sqrt{\frac{\bar{\gamma}_R/2}{1+\bar{\gamma}_R/2}}\right) \\ &\quad + \frac{1}{2}(1-P_{\text{off}})\,\mathcal{M}_{\gamma_{d,s}}\left(-\frac{1}{2}\right)\mathcal{M}_{\gamma_{d,r}^{\text{DF}}}\left(-\frac{1}{2}\right), \end{aligned} \tag{6.49}$$

where $\gamma_{d,r}^{\text{DF}} = \|\mathbf{g}_k\boldsymbol{\Psi}\|_{\text{F}}^2/(2\sigma_n^2)$ is an exponential RV with a mean $\bar{\gamma}_R$. Hence, the MGF of $\gamma_{d,r}^{\text{DF}}$ is equal to the MGF of $\gamma_{d,s}$ given in (6.20). Therefore, the average PEP in (6.49) can be rewritten as

$$\mathrm{E}\left\{\Pr_{\text{error}}\right\} \leq \frac{1}{2}P_{\text{off}}\left(1-\sqrt{\frac{\bar{\gamma}_R/2}{1+\bar{\gamma}_R/2}}\right) + \frac{1}{2}(1-P_{\text{off}})\left(\frac{1}{1-\bar{\gamma}_R/2}\right)^2. \tag{6.50}$$

From (6.39), and considering that the relay has N_t receive antennas, the PEP of the relay detecting the transmitted SMT signal incorrectly and being off is

$$P_{\text{off}} = \mathrm{E}\left\{\sqrt{\gamma_r}\right\}. \tag{6.51}$$

Different to single-antenna relay, γ_r in (6.51) is a Chi-squared RV and from [123, 191], the average PEP in (6.51) is given by

$$P_{\text{off}} = \alpha_a^{N_t}\sum_{i=0}^{N_t-1}\binom{N_t-1+i}{i}[1-\alpha_a]^i, \tag{6.52}$$

where $\alpha_a = \frac{1}{2}\left(1-\sqrt{\frac{\bar{\gamma}_R/2}{1+\bar{\gamma}_R/2}}\right)$.

6.2.3.3 Numerical Results

Simulation results for SM system with DF relays are illustrated in Figure 6.7, where $N_t = 2, N_r = 1$, and $R = 1 \rightarrow 4$ while considering binary phase shift keying (BPSK) modulation. Similar behavior as noted for AF relaying can be seen here as well. Increasing the number of relays significantly enhances the error performance due to the increase of diversity gain. Increasing R from one to four enhances the SNR performance by about 12 dB at an ABER of 10^{-4}. Such gain is achieved while degrading the spectral efficiency, as noted for AF system, since a maximum of $R + 1$ time slots will be needed to convey the source information bits through R DF relays. It is shown in the figure, as well, that analytical and simulation results closely match for a wide range of system parameters.

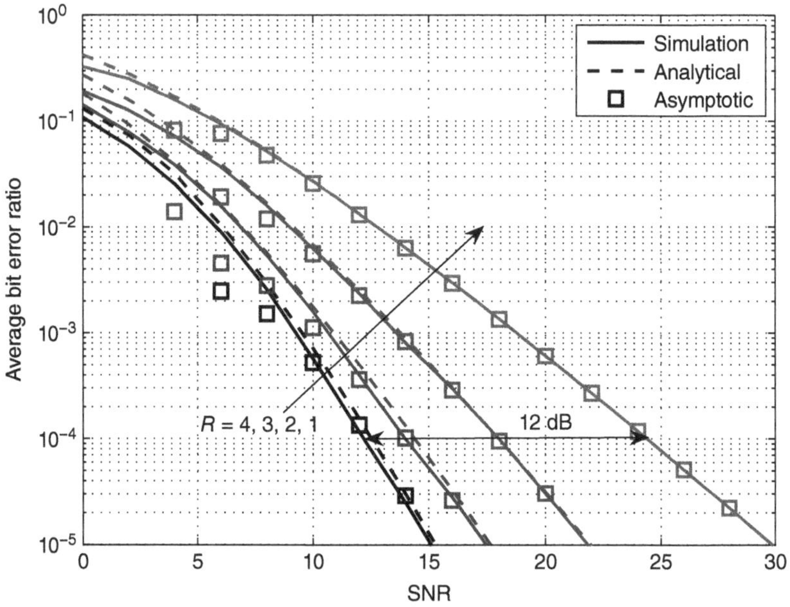

Figure 6.7 Simulation, analytical and asymptotic results for cooperative DF SM system with $N_t = 2$, $N_r = 1$, BPSK modulation and variable $R = 1 \rightarrow 4$.

Increasing the modulation order to quadrature phase shift keying (QPSK) is shown to degrade the DF SM performance by about $1 \rightarrow 2$ dB as shown in Figure 6.8. Furthermore, and as in previous results, Figure 6.8 shows that increasing the number of relays enhances the performance. From the figure, compared to no relays, using five relays offers a 20 dB gain in the SNR.

6.3 Two-Way Relaying (2WR) SMTs

In all previously discussed cooperative networks, multiple orthogonal time slots are needed to broadcast information from a source node to a destination node. However, an enhanced spectral-efficiency relaying algorithm that attracted significant interest in literature is 2WR [236, 242, 254–256]. In 2WR scheme, two source nodes are allowed to simultaneously transmit their data blocks toward a relay node. The relay node retrieves data bits from both nodes and applies network coding principle on the decoded messages. The new generated coded data block is then forwarded to both nodes. To receive the data from the other node, each node reverses the coding operation applied at the relay node.

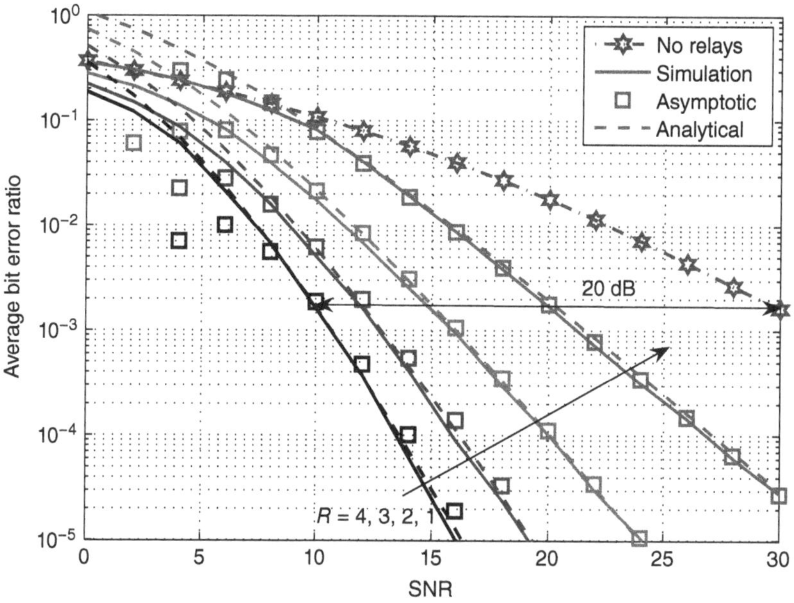

Figure 6.8 Simulation, analytical, and asymptotic results for cooperative DF SM system with $N_t = 2$, $N_r = 1$, and QPSK modulation and variable $R = 1 \rightarrow 4$.

A system model for 2WR protocol is illustrated in Figure 6.9, which consists of two source nodes, A and B, that exchange information with the aid of a relay node R. The number of transmit antennas is, respectively, given by N_A, N_B, and N_R for A, B and R nodes. Similarly, the number of receive antennas is denoted by L_A, L_B, and L_R.

Data transmission is performed in two consecutive phases, namely, transmission phase and relaying phase. In the transmission phase, both A and B nodes concurrently transmit η bits toward the relay node using an SMT. A DF relay decodes the received data from both nodes and obtains an estimate for the 2η bits. In the relaying phase, the relay precodes the estimated bits to generate a new message with η bits that will be forwarded to both nodes using any of the discussed SMTs.

In what follows, the transmission and reception protocols in both phases are described in detail.

6.3.1 The Transmission Phase

In this phase, both nodes A and B concurrently transmit η bits toward the relay node, R, using a specific SMT. In order to keep the same block length (i.e., η bits)

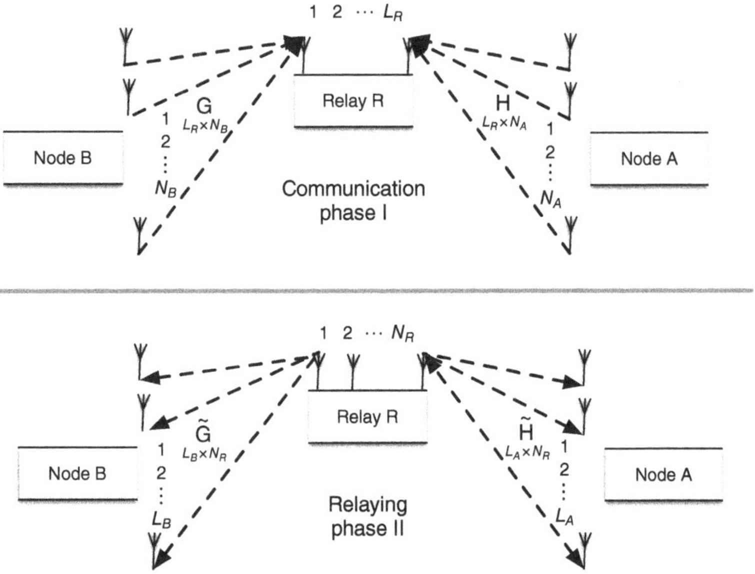

Figure 6.9 A two-way relaying system model applying SMTs at any transmitting node. It is assumed that the sources A and B are equipped, respectively, with N_A and N_B transmit antennas and L_A and L_B receive antennas, and the relay has N_R transmit and L_R receive antennas.

from both nodes, different nodes with unequal number of transmit antennas should use different modulation orders.

The transmitted block from A is denoted by $\mathbf{a} = [a_1, a_2, \ldots, a_\eta]$, while the transmitted block from B is denoted by $\mathbf{b} = [b_1, b_2, \ldots, b_\eta]$. Also, the transmitted vectors from nodes A and B are denoted by $\mathbf{x}_t^A$ and $\mathbf{x}_t^B$, respectively, that are generated from their corresponding data blocks (i.e., $\mathbf{a}$ and $\mathbf{b}$).

The received signal vector at the relay R is denoted by $\mathbf{y}_R$ and is given as follows:

$$\mathbf{y}_R = \mathbf{H}\mathbf{x}_t^A + \mathbf{G}\mathbf{x}_t^B + \mathbf{n}_R, \tag{6.53}$$

where $\mathbf{H}$ and $\mathbf{G}$ are, respectively, the $N_A \times L_R$ and $N_B \times L_R$ MIMO channel matrices between node A and the relay and node B and the relay. Moreover, the vector $\mathbf{n}_R$ is an L_R-length AWGN vector with zero mean and σ_n^2 variance.

The ML detector is considered at the relay node to retrieve the transmitted vectors $\hat{\mathbf{x}}^A$ and $\hat{\mathbf{x}}^B$ as

$$[\hat{\mathbf{x}}^A, \hat{\mathbf{x}}^B] = \underset{\substack{\mathbf{x}^A \in \mathbf{X}^A \\ \mathbf{x}^B \in \mathbf{X}^B}}{\arg\min} \|\mathbf{y}_R - \mathbf{H}\mathbf{x}^A - \mathbf{G}\mathbf{x}^B\|_F^2, \tag{6.54}$$

where $\mathbf{X}^A$ and $\mathbf{X}^B$ represent the sets of all possible transmission vectors from nodes A and B, respectively. The detected vectors $\hat{\mathbf{x}}^A$ and $\hat{\mathbf{x}}^B$ are then, respectively, mapped back to their corresponding data blocks $\hat{\mathbf{a}}$ and $\hat{\mathbf{b}}$.

6.3.2 The Relaying Phase

Upon obtaining the blocks $\hat{\mathbf{a}}$ and $\hat{\mathbf{b}}$, the relay node processes them to obtain a third data block of size η bits, denoted by $\mathbf{r}$. This is usually performed through a simple XOR operation as

$$\mathbf{r} = \hat{\mathbf{a}} \oplus \hat{\mathbf{b}} = [r_1, r_2, \ldots, r_\eta]. \tag{6.55}$$

The $\mathbf{r}$ bits block is then used to obtain the transmission vector $\mathbf{x}_t^R$, which is transmitted from the relay node in this phase.

The received vectors at nodes A and B are

$$\mathbf{y}_A = \widetilde{\mathbf{H}}\mathbf{x}_t^R + \mathbf{n}_A, \tag{6.56}$$

$$\mathbf{y}_B = \widetilde{\mathbf{G}}\mathbf{x}_t^R + \mathbf{n}_B, \tag{6.57}$$

where $\widetilde{\mathbf{H}}$ and $\widetilde{\mathbf{G}}$ are, respectively, the $N_R \times L_A$ and $N_R \times L_B$ MIMO channel matrices between the relay and the nodes A and B, and $\mathbf{n}_A$ and $\mathbf{n}_B$ are the L_A and L_B length AWGN vectors at A and B nodes, respectively.

At nodes A and B, the ML detection is applied on the received signals in order to estimate the detected vectors $\hat{\mathbf{x}}^R$ as follows:

$$\hat{\mathbf{x}}^R = \arg\min_{\mathbf{x}^R \in \mathbf{X}^R} \left\| \mathbf{y}_A - \widetilde{\mathbf{H}}\mathbf{x}^R \right\|_F^2, \tag{6.58}$$

and

$$\hat{\mathbf{x}}^R = \arg\min_{\mathbf{x}^R \in \mathbf{X}^R} \left\| \mathbf{y}_B - \widetilde{\mathbf{G}}\mathbf{x}^R \right\|_F^2, \tag{6.59}$$

where $\mathbf{X}^R$ is the set of all possible transmitted vectors from the relay node R.

At the end of this phase, each node maps its detected vector into its corresponding bit block. Let $\hat{\mathbf{r}}_A$ denotes the obtained block at node A, and $\hat{\mathbf{r}}_B$ is the obtained bits block at node B. Consequently, each node can extract a corrupted version of the transmitted bits from the other node by performing an XOR operation with its own transmitted bits as

$$\mathbf{d}_A = \hat{\mathbf{r}}_A \oplus \mathbf{a} \tag{6.60}$$

and

$$\mathbf{d}_B = \hat{\mathbf{r}}_B \oplus \mathbf{b}, \tag{6.61}$$

where $\mathbf{d}_A$ being the corrupted version of $\mathbf{b}$ obtained at node A, and $\mathbf{d}_B$ is the corrupted version of $\mathbf{a}$ obtained at node B.

At the end of both phases, each node has received η bits from the other node. Compared to the conventional one-way relaying systems, as discussed previously, 2WR can double the spectral efficiency. However, the cost of the improvement in the spectral efficiency is paid in the overall error performance at both nodes, which will be analyzed and discussed hereinafter.

6.3.3 Average Error Probability Analysis

The error performance at either nodes, A or B, is identical. Therefore, only the error performance at node A is considered. Starting from (6.60), the received block $\mathbf{d}_A$ can be expressed in terms of $\mathbf{b}$ as follows:

$$\begin{aligned}\mathbf{d}_A &= \hat{\mathbf{r}}_A \oplus \mathbf{a}\\ &= \mathbf{r} \oplus \mathbf{e}_R \oplus \mathbf{a}\\ &= \hat{\mathbf{a}} \oplus \hat{\mathbf{b}} \oplus \mathbf{e}_R \oplus \mathbf{a}\\ &= \mathbf{a} \oplus \mathbf{e}_A \oplus \mathbf{b} \oplus \mathbf{e}_B \oplus \mathbf{e}_R \oplus \mathbf{a}\\ &= \mathbf{0} \oplus \mathbf{e}_A \oplus \mathbf{b} \oplus \mathbf{e}_B \oplus \mathbf{e}_R\\ &= \mathbf{b} \oplus \mathbf{e}_A \oplus \mathbf{e}_B \oplus \mathbf{e}_R,\end{aligned} \tag{6.62}$$

where $\mathbf{e}_A$, $\mathbf{e}_B$, and $\mathbf{e}_R$ are the error vectors in the transmitted block from A, B, and R nodes, respectively, and all are of η length. A specific bit in an error vector is 1 if the corresponding bit has been received in error. Otherwise, it is 0. For example, a specific bit in $\mathbf{e}_A$ is 1 if the corresponding bits in $\mathbf{a}$ and $\hat{\mathbf{a}}$ are not equal. The second line in (6.62) is obtained by substituting $\hat{\mathbf{r}}_A = \mathbf{r} \oplus \mathbf{e}_R$, the third line is obtained by substituting $\mathbf{r} = \hat{\mathbf{a}} \oplus \hat{\mathbf{b}}$, the fourth line is obtained by substituting $\hat{\mathbf{a}} = \mathbf{a} \oplus \mathbf{e}_A$ and $\hat{\mathbf{b}} = \mathbf{b} \oplus \mathbf{e}_B$, the fifth line is obtained by substituting $\mathbf{a} \oplus \mathbf{a} = \mathbf{0}$ where $\mathbf{0}$ is an all zeros vector, and the last line is obtained using $\mathbf{0} \oplus \mathbf{b} = \mathbf{b}$.

Now, the last line in (6.62) can be rewritten on a bit base as

$$d_A = b \oplus e_A \oplus e_B \oplus e_R, \tag{6.63}$$

where d_A is an arbitrary bit in $\mathbf{d}_A$, and b, e_A, e_B, and e_R represent the corresponding bits in $\mathbf{b}$, $\mathbf{e}_A$, $\mathbf{e}_B$, and $\mathbf{e}_R$, respectively. The bit d_A is correct if it is equal to b. Hence, the bit error rate at A, denoted by ζ, can be expressed as follows:

$$\begin{aligned}\zeta &= \Pr[d_A \neq b]\\ &= \Pr[e_A \oplus e_B \oplus e_R = 1].\end{aligned} \tag{6.64}$$

Notice that $e_A \oplus e_B \oplus e_R = 1$ occurs if only one of them is 1 or all of them are 1's. Thus, (6.64) can be expanded to

$$\begin{aligned}\zeta = &\Pr[e_A = 1 \cap e_B = 0 \cap e_R = 0]\\ &+ \Pr[e_A = 0 \cap e_B = 1 \cap e_R = 0]\\ &+ \Pr[e_A = 1 \cap e_B = 1 \cap e_R = 1]\\ &+ \Pr[e_A = 0 \cap e_B = 0 \cap e_R = 1],\end{aligned} \tag{6.65}$$

which can be further simplified to

$$\begin{aligned}\zeta &= \Pr[e_A = 1 \cap e_B = 0]\Pr[e_R = 0] \\ &\quad + \Pr[e_A = 0 \cap e_B = 1]\Pr[e_R = 0] \\ &\quad + \Pr[e_A = 1 \cap e_B = 1]\Pr[e_R = 1] \\ &\quad + \Pr[e_A = 0 \cap e_B = 0]\Pr[e_R = 1] \\ &= (\alpha_{10} + \alpha_{01})(1-\beta) + (\alpha_{00} + \alpha_{11})\beta, \end{aligned} \tag{6.66}$$

where

$$\alpha_{10} = \Pr[e_A = 1 \cap e_B = 0] \tag{6.67}$$

$$\alpha_{01} = \Pr[e_A = 0 \cap e_B = 1] \tag{6.68}$$

$$\alpha_{11} = \Pr[e_A = 1 \cap e_B = 1] \tag{6.69}$$

$$\alpha_{00} = \Pr[e_A = 0 \cap e_B = 0] \tag{6.70}$$

and $\beta = \Pr[e_R = 1]$.

Also, it can be easily verified that $\alpha_{00} = 1 - \alpha_{11} - \alpha_{10} - \alpha_{01}$. Therefore, (6.66) can be further simplified to

$$\zeta = (\alpha_{10} + \alpha_{01})(1-\beta) + (1 - \alpha_{10} - \alpha_{10})\beta. \tag{6.71}$$

The parameter α_{10} represents the ABER in the received block at the relay for the source A given that the relay received the block from B correctly. Similarly, α_{01} is the average ABER in the received block from B at the relay given that the relay received the A block correctly, and β represents the average ABER in the received block from the relay at node A.

The computation of the average ABER can be obtained through the union bound technique. As such, α_{10} is formulated as

$$\alpha_{10} = \frac{1}{2^\eta}\sum_{i=1}^{2^\eta}\sum_{j=1}^{2^\eta}\frac{\delta_{ij}}{\eta}\Pr\left[\mathbf{x}^{A_i} \neq \mathbf{x}^{A_j} \cap \mathbf{x}^{B_i} = \mathbf{x}^{B_j}\right], \tag{6.72}$$

where δ_{ij} is the hamming distance between the two bit blocks corresponding to $\mathbf{x}^{A_i}$ and $\mathbf{x}^{A_j}$. The probability $\Pr\left[\mathbf{x}^{A_i} \neq \mathbf{x}^{A_j} \cap \mathbf{x}^{B_i} = \mathbf{x}^{B_j}\right]$ can be expressed using (6.58) as

$$\Pr\left[\mathbf{x}^{A_i} \neq \mathbf{x}^{A_j} \cap \mathbf{x}^{B_i} = \mathbf{x}^{B_j}\right] = \Pr\left[\|\mathbf{n}_A\|_F^2 > \left\|\mathbf{H}\left(\mathbf{x}^{A_i} - \mathbf{x}^{A_j}\right) + \mathbf{n}_A\right\|_F^2\right]. \tag{6.73}$$

For a given $\mathbf{H}$, and as shown previously and in Chapter 4, the above probability can be expressed by means of Q-function as

$$\Pr[\mathbf{x}^{A_i} \neq \mathbf{x}^{A_j} \cap \mathbf{x}^{B_i} = \mathbf{x}^{B_j}|\mathbf{H}] = Q\left(\sqrt{\frac{\|\mathrm{H}\,\Delta_A\|_F^2}{2\sigma_n^2}}\right), \tag{6.74}$$

where $\Delta_A = \mathbf{x}^{A_i} - \mathbf{x}^{A_j}$.

Assuming Rayleigh fading channels, the probability in (6.74) can be averaged and upper bounded as

$$\Pr\left[\mathbf{x}^{A_i} \neq \mathbf{x}^{A_j} \cap \mathbf{x}^{B_i} = \mathbf{x}^{B_j}\right] \leq \frac{1}{2}\frac{1}{\left|\mathbf{I}_{L_R N_A} + \frac{1}{4\sigma_n^2}\mathbf{\Psi}_A\right|}, \tag{6.75}$$

where $\mathbf{\Psi}_A = \mathbf{I}_{L_R} \bigotimes \mathbf{\Delta}_A \mathbf{\Delta}_A^H$ and $\bigotimes$ denotes the Kronecker product. A proof of (6.75) can be easily obtained considering the ABER derivations in Chapter 4.

Substituting (6.75) in (6.72), the probability α_{10} can be upper bounded as

$$\alpha_{10} \leqslant \frac{1}{2^\eta}\sum_{i=1}^{2^\eta}\sum_{j=1}^{2^\eta}\frac{\delta_{ij}}{\eta}\frac{1}{2}\frac{1}{\left|\mathbf{I}_{L_R N_A} + \frac{1}{4\sigma_n^2}\mathbf{\Psi}_A\right|}. \tag{6.76}$$

Following similar steps as used to obtain (6.78), α_{01} can be derived and is given by

$$\alpha_{01} = \frac{1}{2^\eta}\sum_{i=1}^{2^\eta}\sum_{j=1}^{2^\eta}\frac{\varsigma_{ij}}{\eta}\Pr\left[\mathbf{x}^{A_i} = \mathbf{x}^{A_j} \cap \mathbf{x}^{B_i} \neq \mathbf{x}^{B_j}\right], \tag{6.77}$$

where ς_{ij} is the hamming distance between the bit blocks corresponding to $\mathbf{x}^{B_i}$ and $\mathbf{x}^{B_j}$. Following the same procedure in (6.72)–(6.76), the probability α_{01} can be upper bounded by

$$\alpha_{01} = \frac{1}{2^\eta}\sum_{i=1}^{2^\eta}\sum_{j=1}^{2^\eta}\frac{\varsigma_{ij}}{\eta}\frac{1}{2}\frac{1}{\left|\mathbf{I}_{L_R N_B} + \frac{1}{4\sigma_n^2}\mathbf{\Psi}_B\right|}, \tag{6.78}$$

where $\mathbf{\Psi}_B = \mathbf{I}_{L_R} \bigotimes \Delta_B \Delta_B^H$ and $\Delta_B = \mathbf{x}^{B_i} - \mathbf{x}^{B_j}$. Please note that both α_{10} and α_{01} are identical (but not equal) due to the assumption that both **H** and **G** have same statistical fading model.

The last probability to be estimated in (6.71) is β, which is given by

$$\beta = \frac{1}{2^\eta}\sum_{i=1}^{2^\eta}\sum_{j=1}^{2^\eta}\frac{\nu_{ij}}{\eta}\Pr\left[\mathbf{x}^{R_i} \neq \mathbf{x}^{R_j}\right], \tag{6.79}$$

where ν_{ij} represents the hamming distance between the bit blocks corresponding to $\mathbf{x}^{R_i}$ and $\mathbf{x}^{R_j}$.

Now using (6.58), and similar to (6.74), the probability $\Pr[\mathbf{x}^{R_i} \neq \mathbf{x}^{R_j}]$ for a given $\widetilde{\mathbf{H}}$ can be expressed as

$$\Pr\left[\mathbf{x}^{R_i} \neq \mathbf{x}^{R_j}|\widetilde{\mathbf{H}}\right] = Q\left(\sqrt{\frac{\|\widetilde{\mathbf{H}}\Delta_R\|_{\mathrm{F}}^2}{2\sigma_n^2}}\right), \tag{6.80}$$

where $\Delta_R = \mathbf{x}^{R_i} - \mathbf{x}^{R_j}$ and (6.80) can be rewritten as

$$\Pr\left[\mathbf{x}^{R_i} \neq \mathbf{x}^{R_j}\right] \leqslant \frac{1}{2}\frac{1}{\left|\mathbf{I}_{L_A N_R} + \frac{1}{4\sigma_n^2}\mathbf{\Psi}_R\right|}, \tag{6.81}$$

where $\mathbf{\Psi}_R = \mathbf{I}_{L_A} \bigotimes \Delta_R \Delta_R^H$.

Therefore, β can be computed by substituting (6.81) into (6.79) as follows:

$$\beta \leq \frac{1}{2^\eta}\sum_{i=1}^{2^\eta}\sum_{j=1}^{2^\eta}\frac{\nu_{ij}}{\eta}\frac{1}{2}\frac{1}{\left|\mathbf{I}_{L_A N_R} + \frac{1}{4\sigma_n^2}\mathbf{\Psi}_R\right|}. \tag{6.82}$$

Finally, the average ABER at node A (or node B) can be expressed by substituting (6.76), (6.78), and (6.82) into (6.71).

6.3.3.1 Numerical Results

The ABER for 2WR QSM system is depicted in Figure 6.10 for $N_A = N_B = N_R = 2$, $M_A = M_B = M_R = 4$-QAM, and $L_A = L_B = L_R$ is varied from $1 \rightarrow 3$. Considering this setup, each node will deliver $\eta = 4$ bits by the end of the two phases. As illustrated in the figure, increasing the number of receive antennas

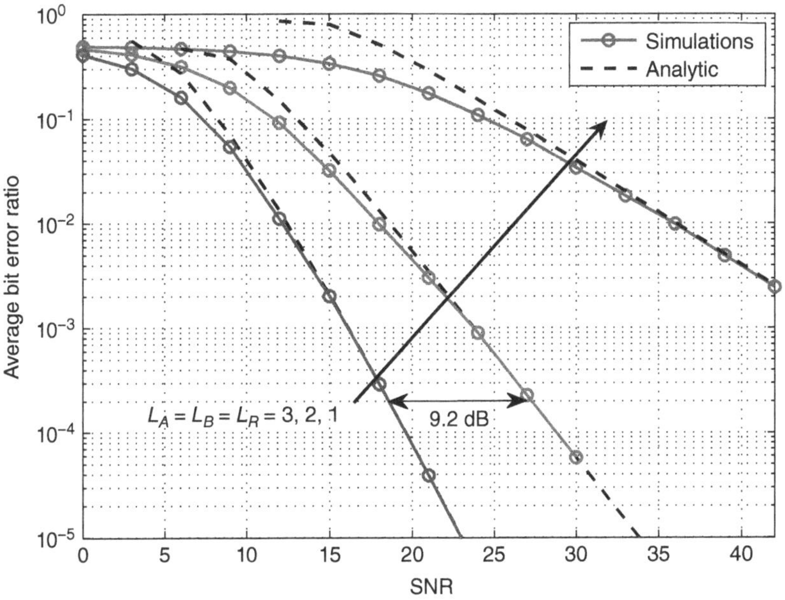

Figure 6.10 Simulation and analytical results for the ABER versus the average SNR for 2WR QSM MIMO system. The block length is $\eta = 4$ bits per channel use for each transmitting node. The nodes are assumed to have two transmit antennas and using 4-QAM modulation order. The number of received antennas is varied from $1 \rightarrow 3$.

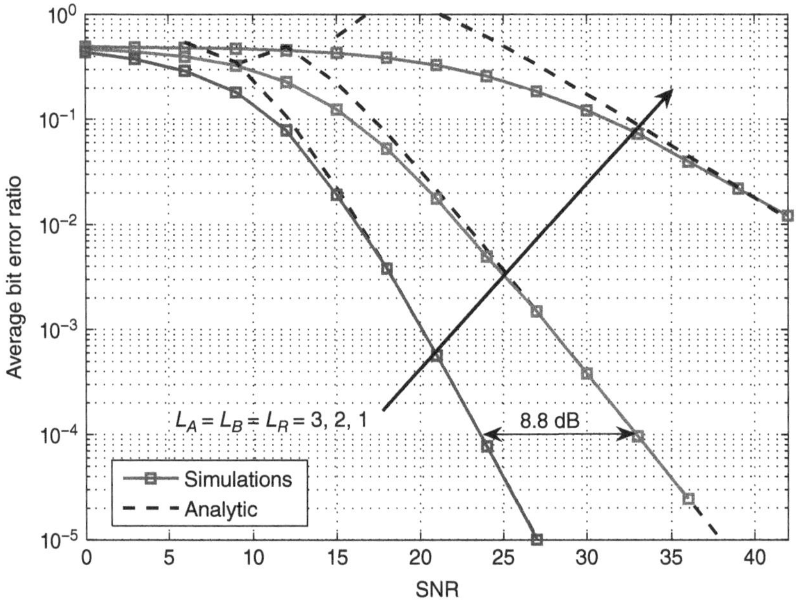

Figure 6.11 Simulation and analytical results for the ABER versus the average SNR for 2WR QSM MIMO system. The block length is $\eta = 5$ bits per channel use from each transmitting node. Each node is assumed to be equipped with two transmit antennas and transmits an 4-QAM symbol. The number of received antennas is varied from 1 → 3.

at all nodes significantly improves the overall performance and a gain of about 9.2 dB can be noticed at an ABER of 10^{-4} when having three receive antennas instead of two. In addition, depicted analytical and simulation results are shown to match closely and for the different depicted curves with variant number of receive antennas.

Another setup considering $N_A = N_B = N_R = 2$, $M_A = M_B = M_R = 8$-QAM and the receive antennas are varied from $1 \rightarrow 3$ is evaluated, and the results are shown in Figure 6.11. With these configurations, the spectral efficiency increases to $\eta = 5$ bits for each transmitting node. Similar conclusions as drawn in the previous figure can be concluded here as well.

7

SMTs for Millimeter-Wave Communications

Millimeter-wave (mmWave) communications offer a plenty of frequency spectrum, ranging from 30 to 300 GHz, that can be exploited to achieve multi-gigabits per second (Gb/s) data rates [257–259]. The E-band at 70/80 GHz offers 1 Gb s^{-1} up to 10 Gb s^{-1} for typical distances of 3 km with an available worldwide low-cost license [260]. Also, it has been demonstrated in [260–263] that wave propagation at E-band has negligible atmospheric attenuation (less than 0.5 dB km^{-1}) and is unaffected by dust, snow, and any other channel deterioration. Although heavy rain is shown to significantly impact the performance of E-band mmWave systems. However, heavy rain usually occurs in limited part of the world [264], and the radio link can be designed to overcome the attenuation resulted from heavy rain [260]. As such, mmWave technology is one of the promising techniques for 5G and beyond wireless standards, and it has been adopted in several recent standards such as mmWave WPAN (IEEE 802.15.3c-2009) [265], WiGig (IEEE 802.11ad) [266], and WirelessHD [267].

The use of mmWave for multiple–input multiple–output (MIMO) systems requires careful design of propagation characteristics of radio signals. The signals at mmWave frequencies propagate in line-of-sight (LOS) environment and do not penetrate solid materials very well. Several experimental studies on mmWave outdoor channel modeling proved that the channel can be safely modeled as LOS or near LOS links [268–273]. The performance of several space modulation technique (SMT) systems over LOS mmWave channel has been studied in [51, 56, 210, 274–277]. Another experimental work conducted by New York University (NYU) Wireless Lab developed a three-dimensional (3D) modeling of the mmWave channel [258, 259, 278–280]. The 3D channel models are comprehensive and fit the conducted measurements. The performance of quadrature spatial modulation (QSM) over the 3D mmWave channel model was studied in [55], and a performance comparison between spatial modulation (SM) and spatial multiplexing (SMX) systems over the 3D mmWave model is presented in [53, 54].

Space Modulation Techniques, First Edition. Raed Mesleh and Abdelhamid Alhassi.

In this chapter, we will present both mmWave channel models, LOS and 3D, and discuss the performance of SMTs over such channel models.

7.1 Line of Sight mmWave Channel Model

In an LOS channel model, it is generally assumed that the antenna arrays at both the transmitter and the receiver are uniformly spaced and aligned to the broadside of each other. As such, the ℓth column of the $N_r \times N_t$ LOS MIMO channel matrix, $\mathbf{H}$, can be modeled as [270, 272]

$$\mathbf{h}_\ell = \frac{\lambda}{4\pi R} \exp\left(-j\frac{2\pi}{\lambda}\mathbf{d}_\ell\right), \tag{7.1}$$

where R is the distance between the transmitter antenna array and the receiver antenna array, $\mathbf{h}_\ell$ denotes the ℓth transmit antenna vector containing all channel paths between the ℓ antenna and all receive antennas, $\mathbf{d}_\ell$ is the distance vector between ℓ transmit antenna and all receive antennas, and $\lambda = c/f$ is the carrier wavelength with f being the carrier frequency and c denoting the speed of light. Furthermore, the signal-to-noise-ratio (SNR) is defined as $\text{SNR} = E_s/N_o = (\lambda/(4\pi R\sigma_n))^2$, where $E_s = \text{E}\{||\mathbf{H}\mathbf{x}_t||_\text{F}^2\} = N_r(\lambda/(4\pi R))^2$.

7.1.1 Capacity Analysis

It is shown in Chapter 5 that the capacity of SMT systems is given by

$$C = N_r \log_2(1 + \text{SNR}), \tag{7.2}$$

where N_r is the number of receiver antennas. It is also demonstrated that achieving the capacity depends on the channel, where the distribution of the multiplication of the spatial, $\mathcal{H}_\ell$, and signal, $\mathcal{S}_\iota$, symbols has to follow a complex Gaussian distribution $\mathcal{R}_{\ell,\iota} = \mathcal{H}_\ell \mathcal{S}_\iota \sim \mathcal{CN}(\mathbf{0}_{N_r}, \sigma_r^2 \mathbf{I}_{N_r})$, where $\mathbf{0}_N$ is an N-length of all zeros vector, and $\mathbf{I}_N$ is an $N \times N$ identity matrix.

In what follows and for illustration purposes, the conditions under which the capacity of SM and QSM can be achieved for LOS mmWave channel are derived and discussed.

7.1.1.1 SM

In SM, the spatial and signal symbols are $\mathcal{H}_\ell = \mathbf{h}_\ell$ and $\mathcal{S}_\iota = s_\iota$, where s_ι is a symbol drawn from a complex constellation diagram such as quadrature amplitude modulation (QAM) or phase shift keying (PSK).

The spatial symbol generated from channel coefficients in (7.1) can be rewritten as

$$\mathbf{h}_\ell = \frac{\lambda}{4\pi R} \exp(j\mathbf{\Theta}_\ell), \tag{7.3}$$

where $\boldsymbol{\Theta}_\ell = -(2\pi/\lambda)\mathbf{d}_\ell$. Hence, the transmitted SM symbol can be written as

$$\mathcal{R}_{\ell,\iota} = \mathbf{h}_\ell s_\iota = \frac{\lambda}{4\pi R} s_\iota \exp\left(j\boldsymbol{\Theta}_\ell\right). \tag{7.4}$$

To achieve the capacity, $\mathcal{R}_{\ell,\iota}$ has to follow a complex Gaussian distribution. Knowing that the complex Gaussian distribution has Rayleigh distribution amplitude and uniform distribution phase, the capacity can be achieved if the phase of the spatial symbols is $\boldsymbol{\Theta}_\ell \sim \mathcal{U}[-\pi, \pi]$, and the signal symbols are either Rayleigh distributed or complex Gaussian distributed. Note, the sum of two random phases with $\mathcal{U}[-\pi, \pi]$ distribution is also a $\mathcal{U}[-\pi, \pi]$.

7.1.1.2 QSM

For QSM, the spatial and signal symbols are $\mathcal{H}_\ell = [\mathbf{h}_{\ell\Re}, \mathbf{h}_{\ell\Im}]$ and $\mathcal{S}_\iota = [\mathrm{Re}\{s_\iota\}, \mathrm{Im}\{s_\iota\}]^T$, respectively. Hence, and from (7.3), the transmitted QSM symbols are

$$\mathcal{R}_{\ell,\iota} = \frac{\lambda}{4\pi R}\mathrm{Re}\{s_\iota\} \exp\left(j\boldsymbol{\Theta}_{\ell\Re}\right) + \frac{\lambda}{4\pi R}\mathrm{Im}\{s_\iota\} \exp\left(j\boldsymbol{\Theta}_{\ell\Im}\right). \tag{7.5}$$

From (7.5) and from the discussion in the previous section for SM, the capacity can be achieved if the phase of the spatial symbol is uniformly distributed in the range $[-\pi, \pi]$, and the signal symbols are Rayleigh distributed.

7.1.1.3 Randomly Spaced Antennas

It is shown in previous sections that the capacity can be achieved if the different N_t channel vectors follow a circular uniform distribution. [1] In other words, to achieve the capacity, the different N_t channel vectors have to be i) not equal; ii) chosen equally likely; iii) and have an absolute value equal to the channel gain. The last two conditions are already fulfilled, because 1) the different N_t transmit antennas are chosen equally likely since no precoding is assumed and 2) the considered LOS mmWave channel is a phase shift with a constant amplitude as given in (7.1).

The first condition aims at decorrelating the MIMO channel matrix and is of significant research interest. Therefore, it has been intensively studied in literature, and an algorithm called optimally spaced antennas (OSA) is considered in [270, 272, 274, and references therein]. In OSA, the distance between the neighbor transmit antennas, d_t, and the neighbor receive antennas, d_r, is chosen in a way that achieves an orthogonal channel as

$$d_t d_r = \frac{R\lambda}{N_r}. \tag{7.6}$$

However, it has been shown in [282] that the assumption of orthogonal channel paths in (7.6) is only valid for a number of receive antennas that is larger or equal

1 Note, if u is uniformly distributed over the range $[-\pi, \pi]$, then $r = \exp(u)$ follows a circular uniform distribution [281].

to the number of transmit antennas, $N_r \geq N_t$. But, for $N_r < N_t$, the channel will not be orthogonal, and the first condition will not be fulfilled. For example, for $R = 5$ m, $f = 60$ GHz, $N_r = 2 < N_t = 4$, and $d_t = d_r = 11.18$ cm the channel matrix $\mathbf{H}$ is

$$\mathbf{H} = \begin{bmatrix} \overbrace{\begin{matrix} 0.796 \\ -j0.796 \end{matrix}}^{\mathbf{h}_1} & \overbrace{\begin{matrix} -j0.796 \\ 0.796 \end{matrix}}^{\mathbf{h}_2} & \overbrace{\begin{matrix} 0.796 \\ -j0.796 \end{matrix}}^{\mathbf{h}_3} & \overbrace{\begin{matrix} -j0.796 \\ 0.796 \end{matrix}}^{\mathbf{h}_4} \end{bmatrix} \cdot 10^{-4}. \tag{7.7}$$

It can be clearly seen from (7.7) that the channel paths to the first receive antenna ($\mathbf{h}_1$) and to the third receive antenna ($\mathbf{h}_3$) are equal. Similarly, the channel paths to the second and fourth receive antennas, $\mathbf{h}_2$ and $\mathbf{h}_4$, are also equal. Thus, capacity cannot be achieved as the N_t channel paths are highly correlated.

To overcome the limitations of OSA in an unbalanced MIMO configuration, a method called randomly spaced antennas (RSA) was proposed in [51]. In RSA, the transmit and receive antennas are randomly distributed along the broadside of each other according to uniform distributions in the ranges of $[0, D_{\text{Tx}}]$ and $[0, D_{\text{Rx}}]$, respectively, where D_{Tx} and D_{Rx} are the maximum transmit and receive array lengths. This guarantees that the N_t channel paths are not equal. D_{Tx} and D_{Rx} are chosen to be equal to the maximum transmit and receive array lengths if OSA was used,

$$D_{\text{Tx}} = (N_t - 1)\sqrt{R\lambda/N_r}, \tag{7.8}$$

$$D_{\text{Rx}} = (N_r - 1)\sqrt{R\lambda/N_r}. \tag{7.9}$$

Using the same system setup for the example in (7.7), the maximum lengths of the transmit and receive antenna arrays are $D_{\text{Tx}} = 33.5$ cm and $D_{\text{Rx}} = 11.2$ cm. Based on the maximum lengths, the distances of the transmit and receive antennas to the start of the antenna array are chosen randomly according to uniform distribution in the ranges of [0, 33.5 cm] and [0, 11.2 cm], respectively. The resultant distances for the four transmit antennas are 2.7, 17.9, 26, and 30.4 cm from the start of the transmit array, and the two receive antennas are 2.6 and 8.3 cm from the start of the receive array. The channel matrix is then obtained as,

$$\mathbf{H} = \underbrace{\begin{bmatrix} \overbrace{\begin{matrix} 0.80 + j0.00 \\ 0.74 - j0.30 \end{matrix}}^{\mathbf{h}_1} & \overbrace{\begin{matrix} -0.78 - j0.17 \\ 0.32 - j0.73 \end{matrix}}^{\mathbf{h}_2} & \overbrace{\begin{matrix} 0.67 + j0.45 \\ -5.40 + j0.59 \end{matrix}}^{\mathbf{h}_3} & \overbrace{\begin{matrix} -0.78 + j0.17 \\ 0.78 + j0.14 \end{matrix}}^{\mathbf{h}_4} \end{bmatrix}}_{\times 10^{-4}}. \tag{7.10}$$

It is evident from (7.10) that unlike (7.7), the four channel paths are not equal and the first condition is fulfilled even for $N_r \leq N_t$.

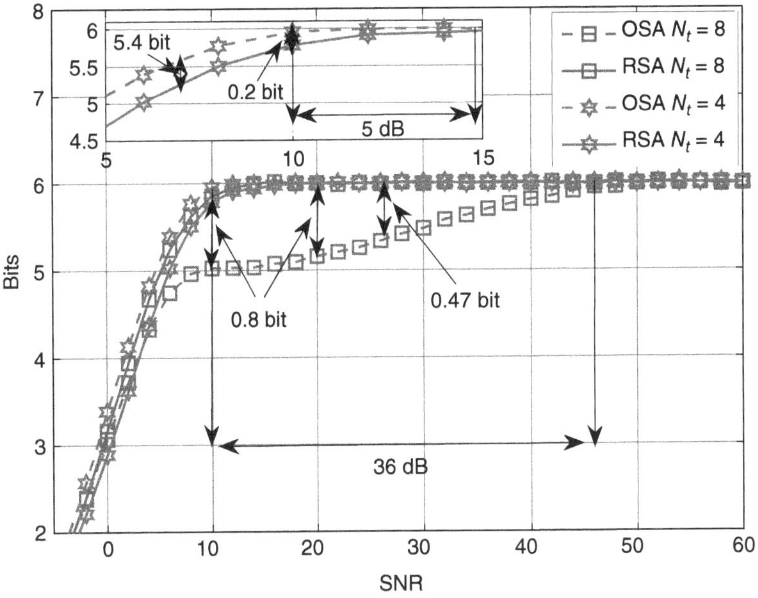

Figure 7.1 Simulated mutual information comparison for mmWave-SM over OSA and RSA channels, with f = 60 GHz, R = 5 m, m = 6, N_t = 4 and 8, and N_r = 4.

Figure 7.1 compares the mutual information of mmWave-SM for $\eta = 6$ bits, $N_t = 4$, and 8, and $N_r = 4$, while considering RSA and OSA channel design algorithms. It is seen in the figure that for unbalanced MIMO systems, where $N_t > N_r$, the performance of RSA algorithm is much better than the OSA method. Mutual information gains of (0.8 and 0.47) bits at SNR= (10 and 26) dB, respectively, can be clearly noticed in the figure. Besides it is demonstrated in Figure 7.1 that SM with RSA reaches the maximum mutual information of $\eta = 6$ bits, 36 dB earlier than OSA. These enhancements are because RSA algorithm insures that the channel follows a circular uniform distribution for unbalanced MIMO systems. However, in Figure 7.1, OSA is shown to have a better performance as compared to RSA for $N_r \geq N_t$, where OSA offers 5.4, and 0.2 bits at SNR= (7, and 10) dB, respectively. Moreover, SM with OSA reaches the maximum mutual information 5 dB earlier than RSA. That is because circular uniform distribution is a continuous distribution, and RSA for small number of antennas offers a discrete circular uniform distribution. Hence, for RSA to perform better even for $N_r \geq N_t$, a very large number of transmit antennas should be used, i.e. large-scale-MIMO. Fortunately, this is not a problem with SMT as the computational complexity does not increase much with the number of transmit antennas. In addition, using mmWave allows more transmit antennas to be deployed in a small space.

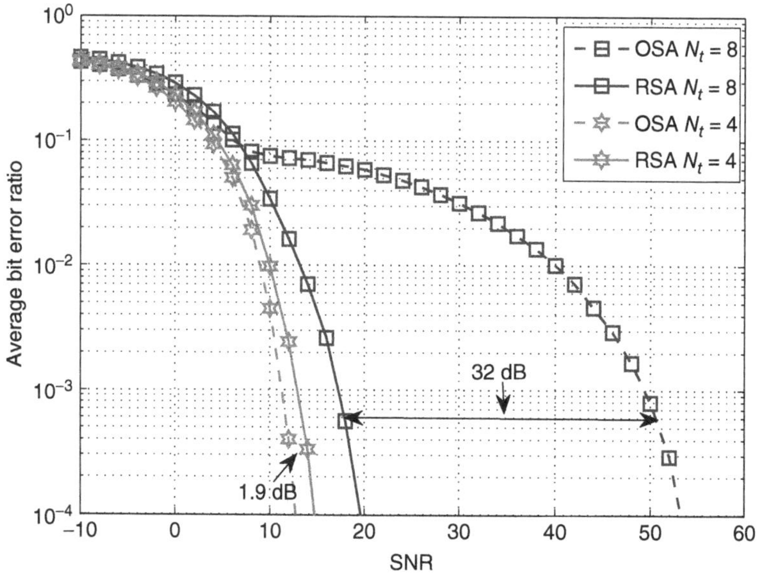

Figure 7.2 ABER performance comparison for mmWave-QSM over OSA and RSA channels, with $f = 60$ GHz, $R = 5$ m, $\eta = 6$ bits, $N_t = 4$, and 8, and $N_r = 4$.

Yet, RSA by itself would not achieve the capacity as the signal symbol has to be especially shaped in accordance with the used SMT system to achieve the capacity. In Figure 7.1, uniformly distributed QAM symbols were used.

The performance enhancement offered by RSA for $N_r < N_t$ can also be seen in the average bit error ratio (ABER) curves depicted in Figure 7.2. It can be seen that RSA offers 32 dB gain in the SNR for $N_t = 8 > N_r = 4$. However, for $N_r \geq N_t$, OSA offers better ABER performance than RSA, 3.5 and 2 dB can be noticed for $N_t = 2$ and $N_t = 4$, respectively.

7.1.1.4 Capacity Performance Comparison

Monte Carlo simulation results for the mutual information of SM, QSM, and SMX over LOS mmWave channel for $\eta = 4$ and 8 and $N_t = N_r = 4$ are depicted in Figure 7.3. It is shown that for $\eta = 4$, SM offers 0.4 bit higher mutual information than SMX. However, for higher spectral efficiency, $\eta = 8$, SMX outperforms QSM and SM, where it reaches the maximum mutual information 4 and 8 dB earlier.

Figure 7.4 depicts the simulated mutual information for SM with $N_t = 8$ and SMX with $N_t = 2$, where $\eta = 6$ bits and $N_r = 8$. It can be seen that using a large number of transmit antennas for SM enhances the performance, where a gain of about 0.8 bit and 2 dB better than SMX is reported. Thus, large-scale SMTs promises significant gains for LOS mmWave communication systems.

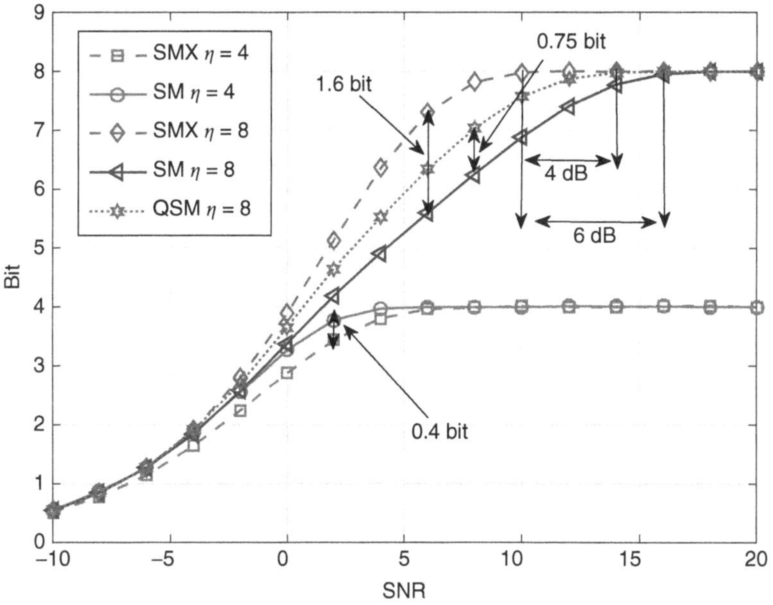

Figure 7.3 Simulated mutual information comparison between SM, QSM, and SMX over LOS mmWave channel using OSA, with $f = 60$ GHz, $R = 5$ m, $\eta = 4$ and 8 bits, and $N_t = N_r = 4$.

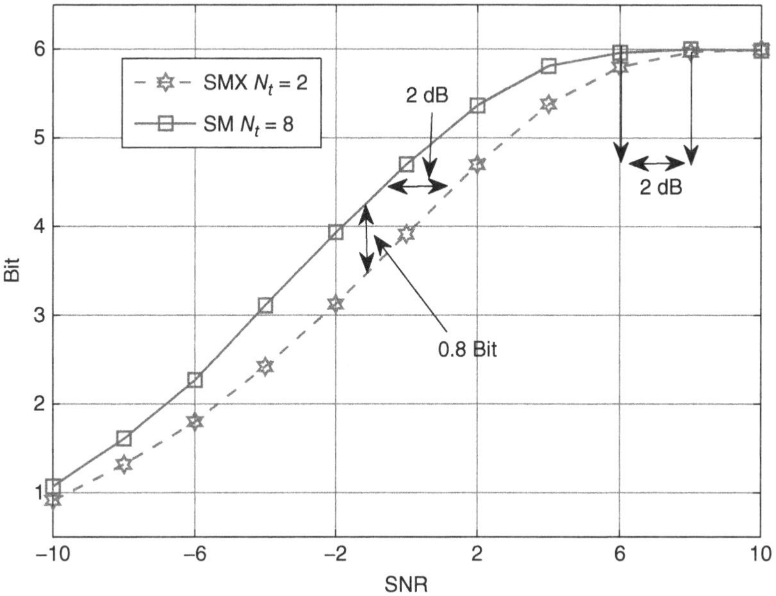

Figure 7.4 Simulated mutual information comparison between SM with $N_t = 8$ and SMX with $N_t = 2$ over LOS mmWave channel, where $f = 60$ GHz, $R = 5$ m, $\eta = 6$ bits, and $N_r = 8$.

7.1.2 Average Bit Error Rate Results

The ABER performance comparison for different spectral efficiencies and MIMO setups between SM, QSM, and SMX is depicted in Figures 7.5 and 7.6. As discussed in previous section, for $N_r < N_t$, RSA is the best to use, and for $N_r \geq N_t$, OSA is a better choice. Therefore, in Figure 7.5 where $N_r < N_t$, RSA is used, and in Figure 7.6, where $N_r = N_t$ OSA is considered.

In Figure 7.5 where $N_r = 2 < N_t = 4$, it can be seen that for $\eta = 4$ bits, SM outperforms SMX by about 2.8 dB. However, for larger spectral efficiency, $\eta = 8$, SMX performs 1.5 dB better than SM. The same can be seen in Figure 7.6 for $N_r = N_t = 4$, where for $\eta = 4$, SM performs 1.9 dB better than SMX. But, for larger spectral efficiency, $\eta = 8$, SMX performs 7.2 dB better than SM. That is because SMX uses much smaller constellation size as compared to SM at high spectral efficiencies. Furthermore and from both Figures 7.5 and 7.6, it can be seen that QSM offers 1.2 dB better performance than SMX for $N_r < N_t$. Yet, for $N_r = N_t$, QSM performs 4.2 dB worse than SMX. That is because for $N_r < N_t$, RSA is used, which is designed so that QSM would achieve near capacity.

One of the advantages offered by mmWave communications is that a large number of antennas can be installed in small spaces. Furthermore, in SMTs and

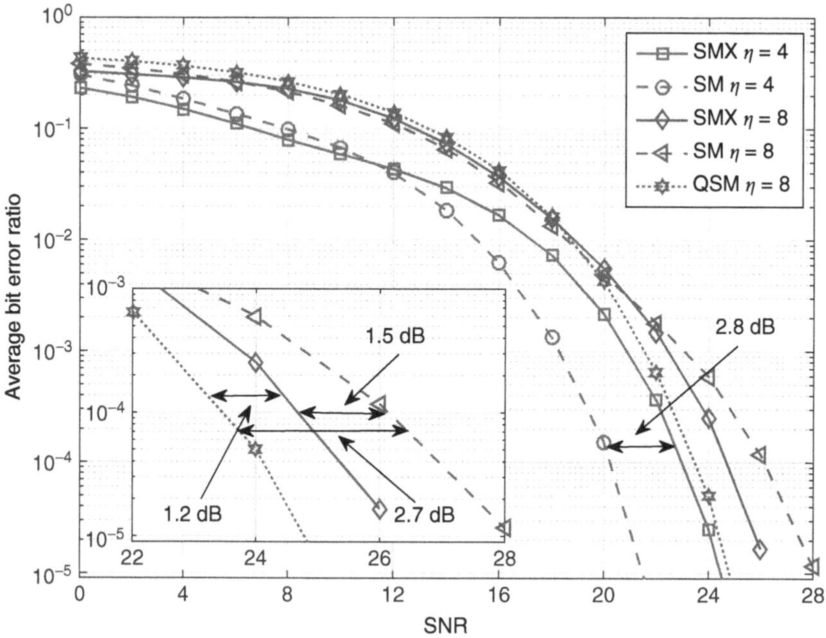

Figure 7.5 ABER performance comparison between SM, QSM and SMX over LOS mmWave channel using RSA, with $f = 60$ GHz, $R = 5$ m, $\eta = 4$ and 8 bits, $N_t = 4$, and $N_r = 2$.

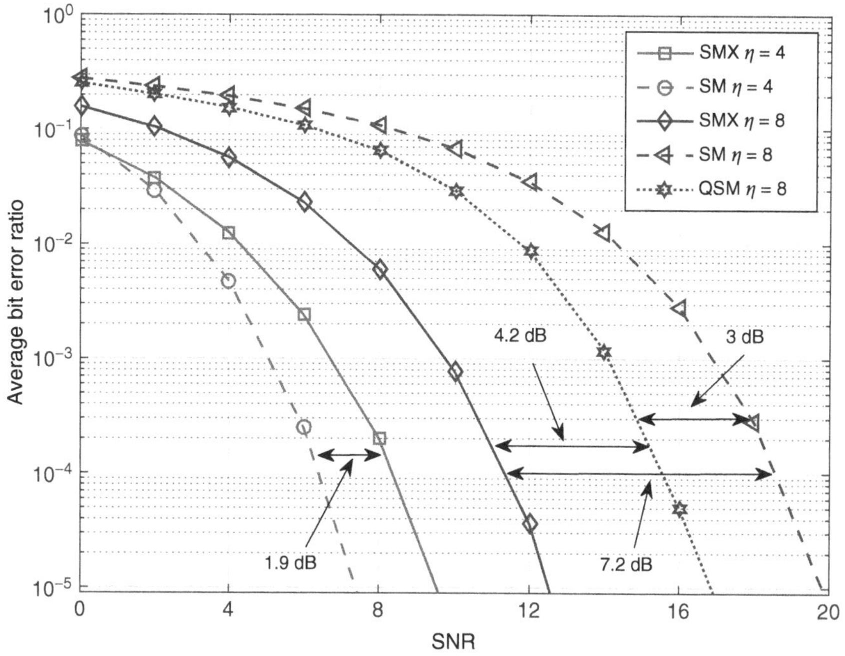

Figure 7.6 ABER performance comparison between SM, QSM and SMX over LOS mmWave channel using OSA, with $f = 60$ GHz, $R = 5$ m, $\eta = 4$ and 8 bits, $N_t = 4$, and $N_r = 4$.

unlike SMX, increasing the number of transmit antennas comes with nearly no cost, as only a maximum of single radio frequency (RF) chain is needed as illustrated earlier. Hence, large-scale SMTs are very good candidates for mmWave communications. Figure 7.7 demonstrates a comparison between SMX with $N_t = 2$ and SM with a larger number of transmit antennas $N_t = 8$, where OSA is used and $N_r = 8$. It can be seen that SM offers 2 dB better performance than SMX, where increasing the number of transmit antennas increased the size of the spatial constellation diagram and decreased the size of the needed signal constellation diagram. In conclusion, even though SMX offers better performance than SM for equal number of transmit antennas, SM would outperform SMX when using larger number of transmit antennas.

7.2 Outdoor Millimeter-Wave Communications 3D Channel Model

In [258, 259, 278–280], NYU Wireless Lab proposed a 3D model for outdoor (mmWave) channels. The proposed model is comprehensive, and therefore, it

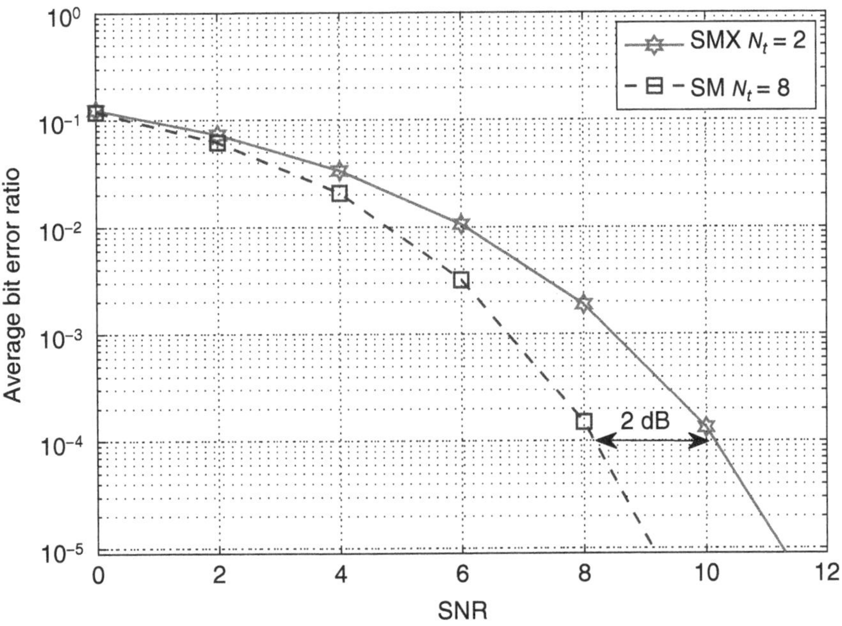

Figure 7.7 ABER performance comparison between SM with $N_t = 8$ and SMX with $N_t = 2$ over LOS mmWave channel, where $f = 60$ GHz, $R = 5$ m, $\eta = 6$ bits, and $N_r = 8$.

is adopted in this section to study the ABER and capacity performance of the SMTs over outdoor mmWave channels.

The 3D mmWave channel model and measurements in [259] consider omni directional antennas operating at mmWave frequencies. The channel impulse response $h_{n_t,n_r}(t)$ for the n_tth and n_rth transmit and receive antennas can be calculated using the double-directional channel model proposed in [283, 284] and given by

$$h(t)_{n_t,n_r} = \sum_{l=1}^{L} h^l_{n_t,n_r} a_l e^{j\varphi_l} \delta(t - \tau_l) \delta\left(\Theta - \Theta_{n_t,l}\right) \delta\left(\Phi - \Phi_{n_r,l}\right), \tag{7.11}$$

where $h^l{}_{n_t,n_r}$ is the lth subpath complex channel attenuation between the n_tth and n_rth transmit and receive antennas, a_l, φ_l, and τ_l are the amplitude, phase, and absolute propagation delay of the lth subpath, $\Theta_{n_t,l}$, and $\Phi_{n_r,l}$ are the vectors of azimuth/elevation angle of departure (AOD) and angle of arrival (AoA) for the n_tth and n_rth transmit and receive antennas, respectively, and L is the total number of multipath components.

Assuming that the antenna arrays at both the transmitter and the receiver is uniformly spaced with distance d and aligned along the z-dimension, the impulse response in (7.11) can be reduced to

$$h_{n_t,n_r}(t) = \sum_{l=1}^{L} h^l_{n_t,n_r} a_l e^{j\varphi_l} \delta(t-\tau_l) \delta\left(\theta^z - \theta^z_{n_t,l}\right) \delta\left(\phi^z - \phi^z_{n_t,l}\right), \tag{7.12}$$

where $\theta^z_{n_t,l}$ and $\phi^z_{n_r,l}$ denoting the elevation AOD and AOA for the n_tth and n_rth transmit and receive antennas, respectively.

From [285], the transfer function of the impulse response in (7.12) is given by

$$h_{n_t,n_r}(f) = \sum_{l=1}^{L} h^l_{n_t,n_r} a_k e^{j\varphi_l} e^{-j\frac{2\pi}{\lambda}d\left(n_t \sin\left(\theta^z_{n_t,l}\right)+n_r \sin\left(\phi^z_{n_r,l}\right)\right)} e^{-j2\pi f \tau_l}, \tag{7.13}$$

where λ is the carrier wavelength.

The values of a_l, φ_l, $\theta^z_{n_t,l}$, $\phi^z_{n_r,l}$, and τ_l in this chapter are generated using the 3D statistical channel model for outdoor mmWave communications derived in [259], where the frequency is 73 GHz, antenna gains are 24.5 dBi, and the distance at each particular time instance is varied equally likely in the range of [60 m – 200 m] [259].

Furthermore, let $\mathbf{H}^l$ be an $N_r \times N_t$ matrix containing all $h^l_{n_t,n_r}$ complex MIMO channel attenuations, then, from [284],

$$\mathbf{H}^l = \mathbf{R}_{\text{Rx}}^{1/2} \mathbf{H}_{\text{Rician}} \mathbf{R}_{\text{Tx}}^{1/2}, \tag{7.14}$$

where R_{Tx} and R_{Rx} are the transmitter and receiver correlation matrices, respectively, and $\mathbf{H}_{\text{Rician}}$ is a matrix whose elements obey the small-scale Rician distribution with $K = 10$ dB [71]. From [286], the correlation matrices can be calculated by

$$R_{u,v} = e^{-j\Theta}\left(0.9e^{-|u-v|d} + 0.1\right), \tag{7.15}$$

where Θ follows a uniform distribution in the range $[-\pi, \pi]$.

The histogram of the amplitude (r_h) and the phase (ϑ_h) of the 3D mmWave channel model are plotted in Figures 7.8 and 7.9. From Figure 7.8, it can be seen that the amplitude of the 3D mmWave channel model can be fitted to a log-normal with parameters in the range of $\mu = [-1.365, -1.30413]$ and $\sigma = [1.19843, 1.27692]$ given by

$$p_{r_h}(r_h) = \frac{1}{r_h \sigma_{r_h}\sqrt{2\pi}} e^{-\frac{\left(\ln r_h - \mu_{r_h}\right)^2}{2\sigma^2_{r_h}}} \coprod_{\mathbb{R}^+}(r_h), \tag{7.16}$$

where $\coprod_B(b) = 1$ if $b \in B$ and zero otherwise, and $\mathbb{R}^+$ denotes the set of all positive real numbers. Furthermore, it can be seen from Figure 7.9 that the phase of the 3D mmWave channel model can be fitted to a continues uniform distribution in the range of $[-\pi, \pi]$ as

$$p_{\vartheta_h}(\vartheta_h) = \frac{1}{2\pi} \coprod_{[-\pi,\pi]}(\vartheta_h). \tag{7.17}$$

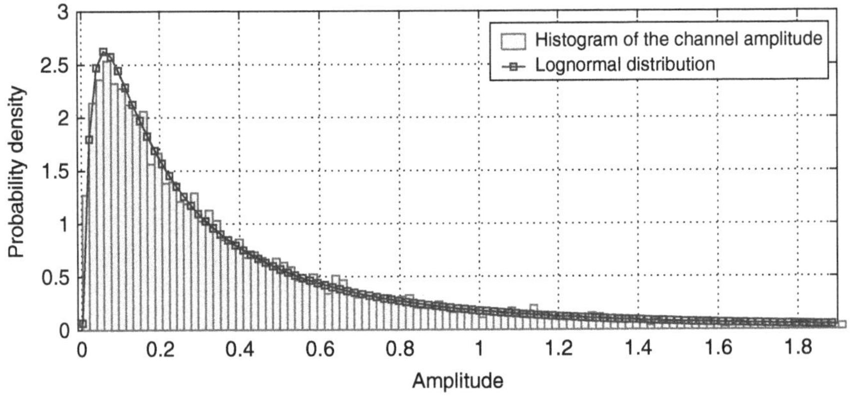

Figure 7.8 Histogram of the amplitude of the 3D mmWave Channel model fitted to a log-normal distribution.

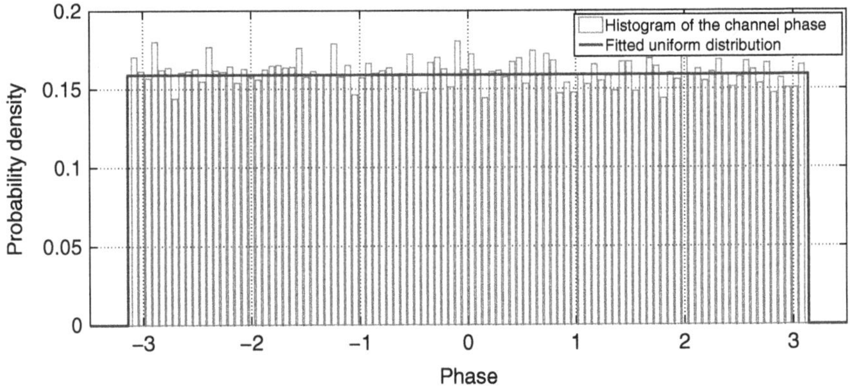

Figure 7.9 Histogram of the phase of the 3D mmWave Channel model fitted to a uniform distribution.

As the phase and amplitude of the 3D mmWave channel model are independent, the joint amplitude and phase probability distribution function (PDF) of the 3D mmWave channel model are,

$$p_h(h) = \frac{1}{2\pi} \frac{e^{-\frac{\left(\ln r_h - \mu_{r_h}\right)^2}{2\sigma_{r_h}^2}}}{r_h \sigma_{r_h} \sqrt{2\pi}} \coprod_{\mathbb{R}^+}(r_h) \coprod_{[-\pi,\pi]}(\vartheta_h). \tag{7.18}$$

Figure 7.10 illustrates the simulated mutual information of a QSM system over the 3D mmWave channel model and the lognormal channel model in (7.18) for different spectral inefficiencies $\eta = 6, 8, 12,$ and 14 bits and different MIMO setups $N_t = N_r = 2$, $N_t = 4$ and $N_r = 2$, $N_t = 2$ and $N_r = 4$,

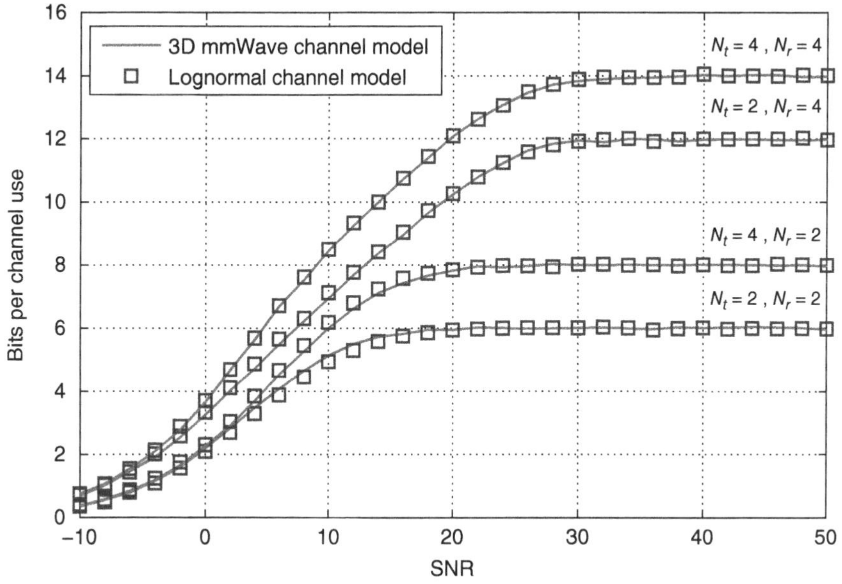

Figure 7.10 The simulated mutual information of QSM over the 3D mmWave channel model and the lognormal channel model for different spectral inefficiencies, and different MIMO setups.

and $N_t = N_r = 4$. The mutual information results for the 3D mmWave channel and for the log-normal fading channel demonstrate close match for wide and pragmatic range of SNR values and for different number of transmit and receive antennas.

7.2.1 Capacity Analysis

The capacity in (7.2) is only achievable if each element of the transmitted SMT symbol $\mathcal{R}_{\ell,\iota} = \mathcal{H}_\ell S_\iota$ follows a complex Gaussian distribution. Hence, (i) substituting (7.18) in (5.45) and (ii) without loss of generality, consider SM system with $N_r = 1$ as an example; the capacity is achieved if the signal symbols are shaped such that its PDF solves

$$p_S = \arg_{p_S} \left\{ |\mathcal{R}_{\ell,\iota}| \exp\left(-\frac{|\mathcal{R}_{\ell,\iota}|^2}{2}\right) = \int_0^\infty \frac{p_s\left(\frac{|\mathcal{R}_{\ell,\iota}|}{h_\ell}\right)}{\sigma_{r_{h_\ell}}\sqrt{2\pi}|h_\ell|^2} \exp\left(-\frac{(\ln|h_\ell| - \mu_{r_{h_\ell}})^2}{2\sigma^2_{r_{h_\ell}}}\right) dh_\ell \right\}. \tag{7.19}$$

From (7.19), it can be seen that for SM to achieve the capacity over the 3D-mmWave channel, the constellation symbols have to be shaped, such that the PDF of the constellation symbols, p_S, solves (7.19). Hence, it is a two-stage process, where in the first stage, p_S has to be found for the given mmWave channel statistics by solving (7.19). The second stage is to shape the constellation symbols to achieve p_S. Signal shaping or other methods need to be considered to shape the constellation symbols such that their distribution follows the obtained p_S. This is an open design problem that is yet unsolved for SMTs as it is mathematically involved and requires further investigations and studies.

In almost all previous studies dealing with SM capacity over mmWave channel, as in [210, and references therein], it is concluded that signal constellation symbols must be Gaussian to achieve the theoretical capacity. However, as discussed in Chapter 5, and shown in Section 7.1.1 and (7.19), complex Gaussian distribution is not always the needed distribution to achieve the theoretical capacity. For the 3D mmWave channel model and from (7.19), it is clear that complex Gaussian distribution is not the required distribution. Figure 7.11 shows the mutual information for SM system over the 3D mmWave

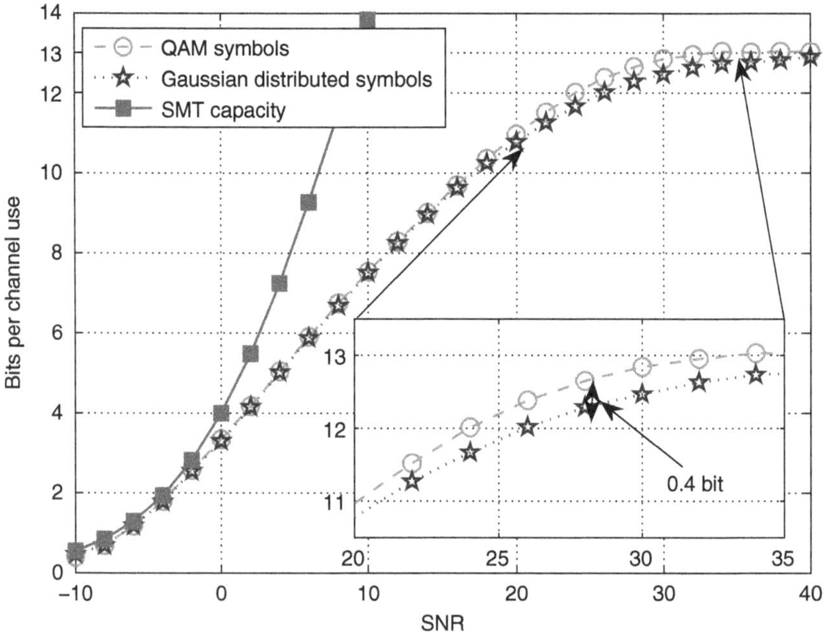

Figure 7.11 Mutual information for a SM system over 3D mmWave channel with $N_t = 8$, $N_r = 2$ and $M = 1024$ constellation diagram assuming Gaussian distributed symbols and QAM symbols.

channel model with $N_t = 8$, $N_r = 2$, and $M = 1024$ constellation size. The mutual information is computed for Gaussian distributed symbols and QAM symbols. It is clear from Figure 7.11 that Gaussian distribution does not achieve the capacity. In fact, the achieved mutual information using Gaussian distributed symbols is 0.4 bit less than the achieved mutual information using ordinary QAM symbols, which are uniformly distributed.

A comparison between simulated mutual information of SM, QSM, and SMX for different spectral efficiencies ($\eta = 4$, 8, and 12) is depicted in Figure 7.12, with $N_t = N_r = 4$. Note, the minimum number of bits QSM can send for $N_t = 4$ is 5 bits. The theoretical channel capacity for SMT and SMX is also shown. It can be seen that for low spectral efficiency, $\eta = 4$, SM and SMX have almost the same performance. Moreover, for $\eta = 4$ and also $\eta = 8$, QSM and SMX have almost the same performance. Yet, for $\eta = 12$, SMX offers higher mutual information than SM and QSM by about 1.63 bits and 1 bit higher, respectively. This enhancement can be attributed to the need of smaller constellation diagram of SMX as compared to SM and QSM systems. Please note that for $\eta = 12$, SMX constellation size is about 99.6% and 97% smaller than the SM and QSM constellation sizes, respectively. The decrease

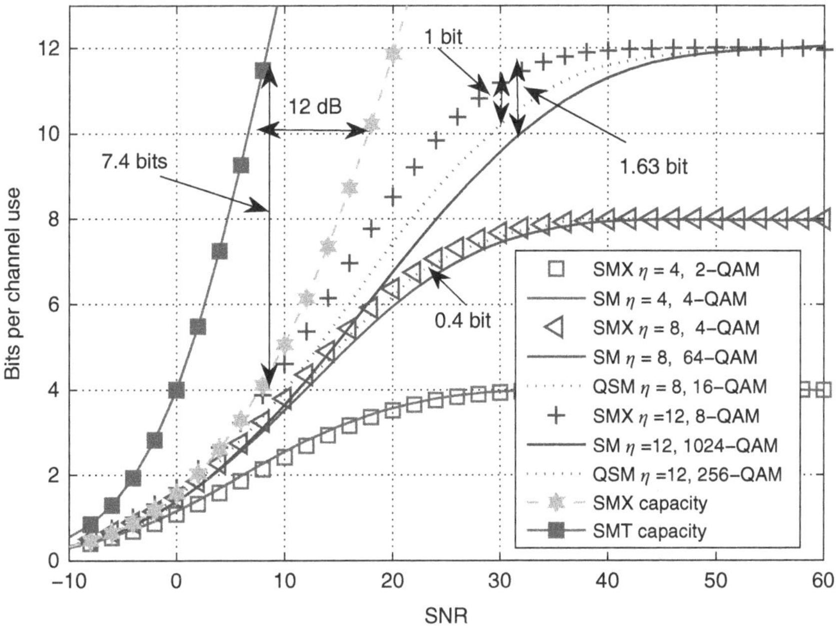

Figure 7.12 The capacity of SMT and SMX compared to the simulated mutual information of SM, QSM, and SMX over the 3D mmWave channel for different spectral efficiencies, where $\eta = 4, 8$, and 12 and $N_t = N_r = 4$.

in performance due to increasing the constellation size can also be seen when comparing QSM to SM for $\eta = 8$. QSM is shown to perform 0.4 bit better than SM, and nearly the same as SMX, where SMX constellation size is only 50% smaller than QSM, compared to 87.5% for SM.

It is also shown that the SMT capacity is 12 dB higher than the SMX capacity. Moreover, at SNR= 8 dB, the SMT capacity is 7.4 bits higher than the capacity of SMX. Thus and even though SMX outperforms SM and QSM, both SMT systems can achieve higher mutual information with proper design of the constellation symbols. Hence, solving (7.19) is of a great interest as it promises a great extension to the existing MIMO capacity.

7.2.2 Average Bit Error Rate Results

The ABER results for both SM and SMX with $N_r = N_t = 4$ are demonstrated in Figure 7.13. The results are compared for spectral efficiencies of $\eta = 4$, 8, 12, and 16. It can be seen that the performance of SM and SMX degrades with increasing the spectral efficiency. Also, SMX is shown to outperform SM performance by about 1.2, 4, 8, and 14 dB for $\eta = 4$, 8, 12, and 16, respectively. Another important observation from the figure is the significant degradation of SM performance with higher spectral efficiencies.

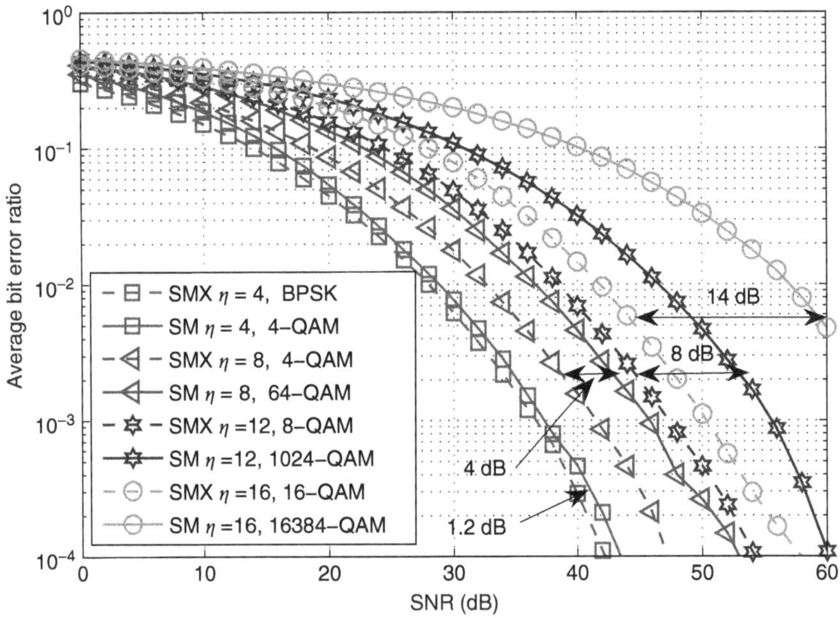

Figure 7.13 ABER performance comparison between SMX and SM for different η, where $\eta = 4, 8, 12$, and 16, $N_t = 4$, and $N_r = 4$.

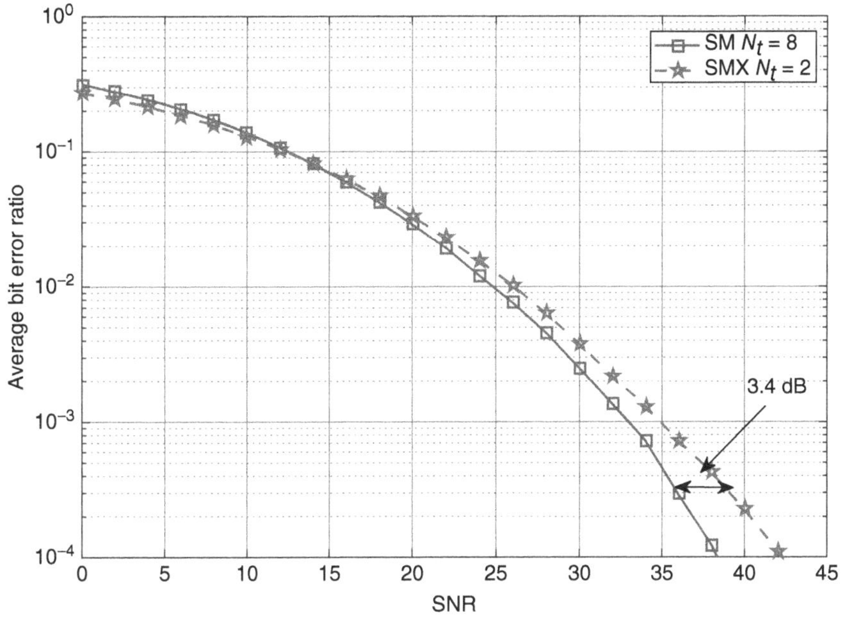

Figure 7.14 ABER performance comparison between SMX and SM for different $\eta = 6$, $N_r = 16$, $N_t = 2$ for SMX and $N_t = 16$ for SM.

For the same MIMO setup, SMX uses smaller constellation diagram than SM, and it offers better ABER performance. However, as discussed in the previous chapters, in SMT, increasing the number of transmit antennas comes with nearly no cost. Hence, large-scale SMT over mmWave communications is a cheep solution that can offer a better ABER performance than SMX. It is illustrated in Figure 7.14 for $\eta = 6$ and $N_r = 16$, SM with $N_t = 16$ offers about 3.4 dB better performance than SMX with $N_t = 2$. Note, SM with $N_t = 16$ needs only 4 signal constellation symbols, i.e. 50% less than SMX with $N_t = 2$, which needs 8 signal constellation symbols.

8

Summary and Future Directions

8.1 Summary

Space modulation techniques (SMTs) were presented, analyzed, and thoroughly discussed in this book. The book presented a comprehensive overview of different SMTs and analyzes their performance. SMTs are new and unique multiple-input multiple-output (MIMO) wireless communication systems. Most of the existing analysis in literature cannot be directly applied to SMTs. It is shown in this book that SMTs need different system designs for both transmitter and receiver. Also, the performance analysis for these systems requires different studies, which go beyond existing theory. In SMTs, new constellation diagram is defined and used to convey information. In previous theory of MIMO systems, multipath propagation channel is used to increase the channel capacity of wireless systems. Uncorrelated channel paths from different transmit antennas are utilized to transmit independent and cochannel data streams to increase the data rate. Such schemes generally achieve spatial multiplexing gain and they are called spatial multiplexing (SMX) MIMO systems. SMX systems suffer from inter-channel interference (ICI), which degrades their error performance and complicates the receiver design. Also, each transmit antenna must be driven by a complete radio frequency (RF) chain including IQ modulators, filters, power amplifier (PA), and others. These components are generally very expensive and produce good amount of heat, which require proper cooling. Also, transmit antennas must be synchronized to start the transmission simultaneously.

In SMTs, the MIMO channel is utilized in a different manner, which relaxes all previous SMX limitations. SMTs consider the multipath propagation channel as a spatial constellation diagram and use the uncorrelated channel paths from each transmit antenna to all receive antennas as constellation points in the spatial constellation diagram. As such, SMX gain is achieved, which depends on the number of available symbols in the spatial domain.

Thus, in SMTs, data bits can be transmitted through two domains, spatial and signal domains. Some SMTs as space shift keying (SSK) and generalized space shift keying (GSSK) use only the spatial domain. Other SMTs as spatial modulation (SM) and quadrature spatial modulation (QSM) use both signal and constellation domains. However, in all SMTs, single-RF carrier signal is transmitted, and the transmitted carrier is either modulated or unmodulated. Hence, a maximum of one RF-chain is needed to implement the transmitter of any SMT. It was demonstrated in this book that SMTs transmitting unmodulated carrier signal, using only the spatial domain, require no RF-chain and can be implemented through RF switches and some other basic components. Thereby, a simple and low-cost implementation of SMTs is very feasible through simple RF components such as RF switches, RF couplers, RF splitters, and RF combiners. The book evaluated the performance of different SMTs in terms of energy efficiency, implementation cost, receiver computational complexity, and average bit error ratio (ABER) and compared it to SMX. It was reported that SMTs can be designed to achieve much better performance in all these metrics as compared to SMX.

Considering the working mechanism of SMTs, analyzing their performance in different environments and over variant channel conditions is not trivial and witnessed tremendous studies in literature. The book presented a detailed performance analyses for SMTs over Rayleigh fading channel with perfect and imperfect channel knowledge at the receiver. Also, a general analytical framework for the ABER performance of SMTs was developed and shown to be accurate. Noncoherent SMTs ABER performance analyses were also discussed and developed.

In addition, the book presented the capacity analysis of SMTs and discussed the conditions under which capacity can be achieved. Again and similar to bit error analysis, SMTs are unique wireless communication systems and their capacity cannot be derived by applying existing theory. It was demonstrated in this book that the derivation of SMTs capacity should consider both spatial and signal constellation diagrams. It was revealed that the capacity can be achieved if the multiplication of the signal symbol by the spatial symbol follows a complex Gaussian distribution. This means that for a given channel distribution, the signal symbols should be shaped such that their multiplication with the spatial symbols is complex Gaussian distributed. These novel results were validated through different analysis, and it is a major contribution of this book.

The adoption of SMTs in two emerging technologies, millimeter-wave (mmWave) and cooperative communications, was presented as well in the book. mmWave is an auspicious technology for future wireless systems. Applying SMTs in mmWave systems promises significant enhancements as reported in the book. Detailed performance analysis and results were presented and discussed. Also, cooperative communications were a key-enabling technology for fourth-generation (4G) wireless standard, and it will play a significant role in fifth-generation (5G) and beyond systems. SMTs promise significant enhancements when considered in cooperative communication. Detailed system design and performance analysis for different cooperative topologies were presented and discoursed.

8.2 Future Directions

SMTs attracted significant research interests in different areas. SMTs were developed almost 10 years ago, and many emerging technologies are considering the adoption of SMTs at the moment. Hereinafter, we will summarize some possible future directions and interesting applications of SMTs.

8.2.1 SMTs with Reconfigurable Antennas (RAs)

An antenna is reconfigurable if it is possible to change its frequency, polarization or radiation characteristics. This is attained through variant techniques, which redistribute the antenna current to modify the electromagnetic fields of its effective aperture. Reconfigurable antennas (RAs) witnessed significant research interest very recently to accommodate sophisticated system requirements by modifying their geometry and electrical behavior. Such modifications are generally performed to adapt to anticipated changes in environmental conditions or system requirements [287]. RA promises several advantages including reduced number of hardware components, complexity, and cost.

The use of RA to implement SMTs was first proposed in [62] with the aim to enhance the spectral efficiency and to reduce the implementation complexity. As discussed in this book, the implementation of SMTs can be easily facilitated through RF switches, which can be easily deployed using RA. In [62], RA are considered to provide a low complexity implementation of SM system. This is achieved by modifying the radiation patterns of RA based on the sequence of incoming data bits. Each incoming data bits sequence modulates a certain polarization parameters that produces unique radiation pattern. The transmitted radiation pattern is detected at the receiver side and used to estimate the transmitted data bits. The spectral efficiency of such scheme depends on the number of possible radiation patterns, and the probability of error is determined by the uniqueness of such patterns. It is important to note that different radiation patterns will be received as a unique signal due to interaction with the available scatterers in the propagation medium [61]. Therefore, RA can be used to create a new constellation dimension and can be combined with other signal and spatial dimensions to enhance the overall spectral efficiency.

Recently, an implementation of a communication system utilizing RA for SM is proposed in [288]. The RA are based on a meander line radiating element surrounded by two L-shaped wire resonators connected to a metallic ground plane with two PIN diodes. The design system allows for a generation of four radiation patterns through switching the PIN diodes. It is reported that the cross correlation among different patterns is in the range of 11–80%. Other antenna geometries are proposed in order to increase the number of radiation patterns that can be generated while still preserving good correlation properties for practical communication applications.

The world first testbed that uses RA to realize SM is proposed in [289]. Novel RA designs are considered to encode the information bit stream and their different *energy patterns* are visualized with the aid of an innovative radio wave display that is capable of measuring the received power. The display clearly shows that distinct received energy patterns are obtained when different radiation patterns of RA are activated at the transmitter. Finally, in [290], RA are used to enhance the performance of SSK system over Rician fading channels.

8.2.2 Practical Implementation of SMTs

So far, intensive theoretical analysis of SMTs have been conducted in literature. However, a major asset for SMTs is the promise to simplify hardware designs, which promise significant enhancements in energy consumption, computational power, and hardware cost. Yet, hardware designs and practical implementation of SMTs are not yet addressed in literature. Some attempts are conducted utilizing existing hardware components, which are not tailored to the specific working mechanism of SMTs. In [39], a hardware testbed for a 4×2 SM MIMO system is developed using National Instruments (NI) PXI MIMO platform. Recently, Mesleh et al. [40] proposed transmitter designs for different SMTs with minimum number of RF chains. It is, also, reported in this book that SMTs based on this optimum design can achieve significant enhancements as compared to SMX MIMO systems. However, the impact of several hardware blocks on the performance of SMTs are yet to be studied. Some recent attempts in this direction were reported where the impact of IQ imbalance on the performance of QSM system is analyzed in [77]. Furthermore, the impact of antenna switching time on the performance of SM system is studied in [156, 291].

Therefore, the investigation of the potential of SMTs via practical implementation testbeds and under real-time channel conditions is very much needed to demonstrate their true potential. This is a significant research direction for future studies.

8.2.3 Index Modulation and SMTs

Index modulation (IM) techniques are developed based on the concept of SMTs. They have witnessed significant research interest in literature. However, their full potential is not yet explored. Noncoherent techniques for IM are to be developed. Also, IM techniques can be deployed with single transmit antennas by modulating the orthogonal frequency division multiplexing (OFDM) subcarriers. Hence, combining them with different SMTs might lead to significant enhancements in terms of spectral and energy efficiencies. The impact of different imperfections such as frequency offset, peak to average power ratio (PARP), frequency, and timing errors are to be investigated and analyzed.

8.2.4 SMTs for Optical Wireless Communications

Optical wireless communications (OWC) is another promising technology for future wireless systems [107–110]. It utilizes a huge portion of an unlicensed spectrum that promises an immense increase in the data rate. Indoor systems are generally called *visible light communications* whereas outdoor applications are named *free space optics*. Transmitting wireless signals over optical carriers is not trivial, and specific modulation techniques can only be considered. In optical systems, the intensity of the light is modulated, which must be real and unipolar and quadrature modulation is not possible. For instant, SSK system can be directly applied to OWC. Other SMTs can be applied in conjunction with multicarrier communication such as OFDM [99–104]. However, the major challenge in applying SMTs in OWC is the high spatial correlation among different channel paths from different transmit light sources to multiple receivers. Uncorrelated channel paths can be created by properly spacing transmitters and receivers, such that the spacing is larger than the fading correlation length. In particular, for OWC, the total length of the transmitters must not exceed the capture zone of the receiver. However, designing such an OWC setup may not be feasible in practical systems, since the available spacing among different transmitters may not be adequate for these requirements. A solution that attracted significant interest in literature to decorrelate the optical wireless (OW) MIMO channel paths is the imaging receiver. Imaging receivers are shown to be efficient in eliminating the effects of ambient light noise, cochannel interference, and multipath distortion. Such advantages are attained through allowing signals arriving from certain angles to be passed and all other signals are discarded.

Designing an OW system utilizing all advantages of SMTs and achieving good performance is still an open research question that needs to be studied and is a very interesting topic for future research.

A

Matlab Codes

A.1 Generating the Constellation Diagrams

A.1.1 SSK

```
function SSK_Cons_Diagram = SSK_Cons_Diagram_Gen(Nt)
% ==================================================================
% Book title: Space Modulation Techniques
% Authors:    Raed Mesleh and Abdelhamid Alhassi
% Publisher:  John Wiley & Sons, Ltd
% Date:       2017
% ==================================================================
%
% Description
%  This script generates the space shift keying (SSK) constellations
%  diagram containing all possible SSK modulated transmit vectors.
%
% Input
%  Nt                  Number of transmit antennas.
%
% Output
%  SSK_Cons_Diagram    SSK constellations diagram.
%
% Usage
%  SSK_Cons_Diagram = SSK_Cons_Diagram_Gen(Nt)

%% Generating SSK constellation matrix

SSK_Cons_Diagram =  eye(Nt) ;

end
```

Space Modulation Techniques, First Edition. Raed Mesleh and Abdelhamid Alhassi.

A.1.2 GSSK

```
function GSSK_Cons_Diagram = GSSK_Cons_Diagram_Gen(Nt,nu)
% ======================================================================
% Book title: Space Modulation Techniques
% Authors:    Raed Mesleh and Abdelhamid Alhassi
% Publisher:  John Wiley & Sons, Ltd
% Date:       2017
% ======================================================================
%
% Description
%  This script generates the generalized space shift keying (GSSK)
%  constellations diagram containing all possible GSSK modulated
%  transmit vectors.
%
% Inputs
%  Nt                  Number of transmit antennas.
%  nu                  Number of active transmit antennas.
%
% Output
%  GSSK_Cons_Diagram   GSSK constellations diagram
%
% Usage
%  GSSK_Cons_Diagram = GSSK_Cons_Diagram_Gen(Nt,nu)

%% Generating the antenna combinations

% Calculating the spectral efficiency
eta          = floor(log2(nchoosek(Nt,nu))) ;

% Generating all possible nu active antenna combinations
All_Ant_Comb = flipud(nchoosek(1:Nt,nu)) ;

% Taking a multiple of two number of the possible antenna combinations
Poss_Ant     = All_Ant_Comb(1:2^eta ,:) ;

%% Generating the GSSK constellation diagram

% Creating a matrix containing all possible spatial indexes
PossAnt_Ind = zeros(nu,2^eta) ;
for i = 1 : nu

    PossAnt_Ind(i ,:) = sub2ind([Nt 2^eta], Poss_Ant(:,i).', 1:2^eta) ;

end

% The GSSK Constellation Diagram
GSSK_Cons_Diagram                   = zeros(Nt,2^eta) ;
GSSK_Cons_Diagram(PossAnt_Ind) = 1 / sqrt(nu) ;

end
```

A.1.3 SM

```
function SM_Cons_Diagram = SM_Cons_Diagram_Gen(Nt,M)
% ==========================================================
% Book title: Space Modulation Techniques
% Authors:    Raed Mesleh and Abdelhamid Alhassi
% Publisher:  John Wiley & Sons, Ltd
% Date:       2017
% ==========================================================
%
% Description
%  This script generates the spatial modulation (SM) constellations
%  diagram containing all possible SM-modulated transmit vectors.
%
% Inputs
%  Nt              Number of transmit antennas.
%  M               Size of the signal constellation diagram.
%
% Output
%  SM_Cons_Diagram   SM constellations Diagram.
%
% Usage
%  SM_Cons_Diagram = SM_Cons_Diagram_Gen(Nt,M)

%% Generating the M-QAM signal constellation symbols

% Constructing an M-QAM modulator
alphabit        = modem.qammod('M', M) ;
% Generating signal constellation diagram
SigConsDiag     = modulate( alphabit , (0:M-1)).';
% Calculating average power of the signal constellation diagram
SigConsDiag_avg = SigConsDiag'*SigConsDiag / M ;
% Normalizing the power of the signal constellation diagram
SigConsDiag     = SigConsDiag / sqrt(SigConsDiag_avg);

%% SM modulation

% Creating a matrix containing the indexes of all possible signal
% symbols for the different spatial symbols
PossibleSignalSymbols = repmat((1:M).',Nt,1);

% Calculating the spectral efficiency
eta   = log2(Nt) + log2(M) ;

% Creating a matrix containing all possible spatial indexes
PossAnt     = reshape(repmat(1:Nt,M,1),[2^eta 1]);
PossAnt_Ind = sub2ind([Nt 2^eta], PossAnt.', 1:2^eta) ;

% The SM constellation diagram
SM_Cons_Diagram                = zeros( Nt , Nt*M ) ;
SM_Cons_Diagram(PossAnt_Ind) = SigConsDiag(PossibleSignalSymbols) ;

end
```

A.1.4 GSM

```
function GSM_Cons_Diagram = GSM_Cons_Diagram_Gen(Nt,nu,M)
% ==================================================================
% Book title: Space Modulation Techniques
% Authors:    Raed Mesleh and Abdelhamid Alhassi
% Publisher:  John Wiley & Sons, Ltd
% Date:       2017
% ==================================================================
%
% Description
%  This script generates the generalized spatial modulation (GSM)
%  constellations diagram containing all possible GSM modulated
%  transmit vectors.
%
% Inputs
%  Nt                 Number of transmit antennas.
%  nu                 Number of active transmit antennas.
%  M                  Size of the signal constellation diagram.
%
% Outputs
%  GSM_Cons_Diagram   GSM constellations Diagram
%
% Usage
%  GSM_Cons_Diagram = GSM_Cons_Diagram_Gen(Nt,nu,m)

%% Generating the M-QAM signal constellation symbols

% Constructing an M-QAM modulator
alphabit        = modem.qammod('M', M) ;
% Generating signal constellation diagram
SigConsDiag     = modulate( alphabit , (0:M-1)).';
% Calculating average power of the signal constellation diagram
SigConsDiag_avg = SigConsDiag'*SigConsDiag / M ;
% Normalizing the power of the signal constellation diagram
SigConsDiag     = SigConsDiag / sqrt(SigConsDiag_avg);

%% Generating the antenna combinations

% Number of bits transmitted in the spatial domain
SpatialBits         = floor(log2(nchoosek(Nt,nu))) ;

% Calculating the spectral efficiency
Spectral_Efficiency = SpatialBits + log2(M) ;

% Generating all possible nu active antenna combinations
All_Ant_Comb        = flipud(nchoosek(1:Nt,nu)) ;

% Taking a multiple of two number of the possible antenna combinations
Possible_Ant_Comb   = All_Ant_Comb(reshape(repmat( ...
                      1:2^SpatialBits,M,1),2^Spectral_Efficiency,1),:) ;

%% Generating the GSM constellation diagram
```

```
% Creating a matrix containing the indexes of all possible signal
% symbols for the different spatial symbols
PossibleSigSymb = repmat((1:M).',2^SpatialBits,1);

% Creating a matrix containing all possible spatial indexes
PossibleAnt_Ind = zeros(nu,2^Spectral_Efficiency) ;
for i = 1 : nu
    PossibleAnt_Ind(i,:) = sub2ind([Nt 2^Spectral_Efficiency], ...
                    Possible_Ant_Comb(:,i).', 1:2^Spectral_Efficiency) ;
end

% The GSM constellation diagram
GSM_Cons_Diagram                   = zeros(Nt,2^Spectral_Efficiency) ;
GSM_Cons_Diagram(PossibleAnt_Ind) = repmat( ...
                        SigConsDiag(PossibleSigSymb).',nu,1) / sqrt(nu) ;

end
```

A.1.5 QSSK

```
function QSSK_Cons_Diagram = QSSK_Cons_Diagram_Gen(Nt)
% ================================================================
% Book title: Space Modulation Techniques
% Authors:    Raed Mesleh and Abdelhamid Alhassi
% Publisher:  John Wiley & Sons, Ltd
% Date:       2017
% ================================================================
%
% Description
%  This script generates the quadrature space shift keying (QSSK)
%  constellations diagram containing all possible QSSK modulated
%  transmit vectors.
%
% Input
%  Nt                   Number of transmit antennas.
%
% Output
%  QSSK_Cons_Diagram   QSSK constellations diagram.
%
% Usage
%  QSSK_Cons_Diagram = QSSK_Cons_Diagram_Gen(Nt)

%% Generating Possible Antenna bits

% Calculating the spectral efficiency
eta = 2*log2(Nt) ;

% Creating a matrix containing all possible spatial indexes for the
% real part of QSSK
PossAntRe     = reshape(repmat(1:Nt,Nt,1),2^eta,1);
PossAntRe_Ind = sub2ind([Nt 2^eta], PossAntRe.', 1:2^eta) ;
```

```
% Creating a matrix containing all possible spatial indexes for the
% imaginary part of QSSK
PossAntIm     = reshape(repmat(1:Nt,1,Nt),2^eta,1) ;
PossAntIm_Ind = sub2ind([Nt 2^eta], PossAntIm.',  1:2^eta) ;

%% Generating the QSSK constellation diagram

% The real part of QSSK
QSSKCons_MatRe = zeros( Nt , 2^eta ) ;
QSSKCons_MatRe(PossAntRe_Ind) = 1 ;

% The imaginary part of QSSK
QSSKCons_MatIm = zeros( Nt , 2^eta ) ;
QSSKCons_MatIm(PossAntIm_Ind) = 1j ;

% The QSSK constellation diagram
QSSK_Cons_Diagram = QSSKCons_MatRe +  QSSKCons_MatIm ;

end
```

A.1.6 QSM

```
function QSM_Cons_Diagram = QSM_Cons_Diagram_Gen(Nt,M)
% =====================================================================
% Book title: Space Modulation Techniques
% Authors:    Raed Mesleh and Abdelhamid Alhassi
% Publisher:  John Wiley & Sons, Ltd
% Date:       2017
% =====================================================================
%
% Description
%  This script generates the quadrature spatial modulation (QSM)
%  constellations diagram containing all possible QSM modulated
%  transmit vectors.
%
% Inputs
%  Nt                  Number of transmit antennas.
%  M                   Size of the signal constellation diagram.
%
% Output
%  QSM_Cons_Diagram   QSM constellations Diagram
%
% Usage
%  QSM_Cons_Diagram = QSM_Cons_Diagram_Gen(Nt,M)

%% Generating the M-QAM signal constellation symbols

% Constructing an M-QAM modulator
alphabit        = modem.qammod( 'M',  M ) ;
% Generating signal constellation diagram
SigConsDiag     = modulate( alphabit , (0:M-1)).';
% Calculating average power of the signal constellation diagram
SigConsDiag_avg = SigConsDiag'*SigConsDiag / M ;
```

```
% Normalizing the power of the signal constellation diagram
SigConsDiag      = SigConsDiag / sqrt(SigConsDiag_avg);

%% QSM modulation

% Calculating the spectral efficiency
eta    = 2*log2(Nt) + log2(M) ;

% Creating a matrix containing all possible spatial indexes for the
% real part of QSM
PossAntRe      = reshape(repmat(1:Nt,Nt*M,1),2^eta,1);
PossAntRe_Ind = sub2ind([Nt 2^eta], PossAntRe.', 1:2^eta) ;

% Creating a matrix containing all possible spatial indexes for the
% imaginary part of QSM
PossAntIm      = reshape(repmat(1:Nt,M,Nt),2^eta,1) ;
PossAntIm_Ind = sub2ind([Nt 2^eta], PossAntIm.', 1:2^eta) ;

% Creating a matrix containing the indexes of all possible signal
% symbols for the different spatial symbols
PossSigSym = repmat((1:M).',Nt^2,1);

% The real part of the QSM
QSMCons_MatRe                   = zeros( Nt , 2^eta ) ;
QSMCons_MatRe(PossAntRe_Ind) = real(SigConsDiag(PossSigSym)) ;

% The imaginary part of the QSM
QSMCons_MatIm                   = zeros( Nt , 2^eta ) ;
QSMCons_MatIm(PossAntIm_Ind) = 1j * imag(SigConsDiag(PossSigSym)) ;

% The QSM constellation diagram
QSM_Cons_Diagram = QSMCons_MatRe + QSMCons_MatIm ;

end
```

A.1.7 GQSSK

```
function GQSSK_Cons_Diagram = GQSSK_Cons_Diagram_Gen(Nt,nu)
% ==================================================================
% Book title: Space Modulation Techniques
% Authors:    Raed Mesleh and Abdelhamid Alhassi
% Publisher:  John Wiley & Sons, Ltd
% Date:       2017
% ==================================================================
%
% Description
%  This script generates the generalized quadrature space shift keying
%  (GQSSK) constellations diagram containing all possible GQSSK
%  modulated transmit vectors.
%
% Inputs
%  Nt                  Number of transmit antennas.
%  nu                  Number of active transmit antennas.
```

```
%
% Output
%   GQSSK_Cons_Diagram    GQSSK constellations diagram.
%
% Usage
%   GQSSK_Cons_Diagram = GQSSK_Cons_Diagram_Gen(Nt,nu)

%% Generating the antenna combinations

% Number of bits transmitted in the spatial domain
SpatialBits            = floor(log2(nchoosek(Nt,nu))) ;

% Calculating the spectral efficiency
Spectral_Efficiency    = 2*SpatialBits ;

% Generating all possible nu active antenna combinations
All_Ant_Comb           = flipud(nchoosek(1:Nt,nu)) ;

% Taking a multiple of two number of the possible antenna combinations
Possible_Ant_Comb_Re = All_Ant_Comb(reshape(repmat( ...
        1:2^SpatialBits,2^SpatialBits,1),2^Spectral_Efficiency,1),:) ;

% Creating a matrix containing all possible spatial indexes
PossibleAntRe_Ind      = zeros(nu,2^Spectral_Efficiency) ;
for i = 1 : nu
    PossibleAntRe_Ind(i,:) = sub2ind([Nt 2^Spectral_Efficiency], ...
              Possible_Ant_Comb_Re(:,i).', 1:2^Spectral_Efficiency) ;
end

% Taking a multiple of two number of the possible antenna combinations
Possible_Ant_Comb_Im = All_Ant_Comb(reshape(repmat( ...
        1:2^SpatialBits,1,2^SpatialBits),2^Spectral_Efficiency,1),:) ;

% Creating a matrix containing all possible spatial indexes
PossibleAntIm_Ind      = zeros(nu,2^Spectral_Efficiency) ;
for i = 1 : nu
    PossibleAntIm_Ind(i,:) = sub2ind([Nt 2^Spectral_Efficiency], ...
              Possible_Ant_Comb_Im(:,i).', 1:2^Spectral_Efficiency) ;
end

%% GQSSK modulation

% The real part of the QSM
GQSSKCons_Diagram_Re = zeros( Nt , 2^Spectral_Efficiency ) ;
GQSSKCons_Diagram_Re(PossibleAntRe_Ind) = 1/sqrt(nu) ;

% The imaginary part of the QSM
GQSSKCons_Diagram_Im = zeros( Nt , 2^Spectral_Efficiency ) ;
GQSSKCons_Diagram_Im(PossibleAntIm_Ind) = 1j * 1/sqrt(nu) ;

% The GQSSK constellation diagram
GQSSK_Cons_Diagram = GQSSKCons_Diagram_Re + GQSSKCons_Diagram_Im ;

end
```

A.1.8 GQSM

```
function GQSM_Cons_Diagram = GQSM_Cons_Diagram_Gen(Nt,nu,M)
% ====================================================================
% Book title: Space Modulation Techniques
% Authors:    Raed Mesleh and Abdelhamid Alhassi
% Publisher:  John Wiley & Sons, Ltd
% Date:       2017
% ====================================================================
%
% Description
%  This script generates the generalized quadrature spatial modulation
%  (GQSM) constellations diagram containing all possible GQSM
%  modulated transmit vectors.
%
% Inputs
%  Nt                    Number of transmit antennas.
%  nu                    Number of active transmit antennas.
%  M                     Size of the signal constellation diagram.
%
% Output
%  GQSM_Cons_Diagram   GQSM constellations diagram
%
% Usage
%  GQSM_Cons_Diagram = GQSM_Cons_Diagram_Gen(Nt,nu,M)

%% Generating the M-QAM signal constellation symbols

% Constructing an M-QAM modulator
alphabit         = modem.qammod( 'M' ,  M ) ;
% Generating signal constellation diagram
SigConsDiag      = modulate( alphabit , (0:M-1)).';
% Calculating average power of the signal constellation diagram
SigConsDiag_avg = SigConsDiag'*SigConsDiag / M ;
% Normalizing the power of the signal constellation diagram
SigConsDiag      = SigConsDiag / sqrt(SigConsDiag_avg);

%% Generating the antenna combinations

% Number of bits transmitted in the spatial domain
SpatialBits           = floor(log2(nchoosek(Nt,nu))) ;

% Calculating the spectral efficiency
Spectral_Efficiency  = 2 * SpatialBits + log2(M) ;

% Generating all possible nu active antenna combinations
All_Ant_Comb          = flipud(nchoosek(1:Nt,nu)) ;

% Taking a multiple of two number of the possible antenna combinations
Possible_Ant_Comb_Re = All_Ant_Comb(reshape(repmat( ...
       1:2^SpatialBits,2^SpatialBits*M,1),2^Spectral_Efficiency,1),:) ;

% Creating a matrix containing all possible spatial indexes
PossibleAntRe_Ind              = zeros(nu,2^Spectral_Efficiency) ;
```

```
for i = 1 : nu
    PossibleAntRe_Ind(i,:)    = sub2ind([Nt 2^Spectral_Efficiency], ...
                 Possible_Ant_Comb_Re(:,i).', 1:2^Spectral_Efficiency) ;
end

% Taking a multiple of two number of the possible antenna combinations
Possible_Ant_Comb_Im = All_Ant_Comb(reshape(repmat( ...
        1:2^SpatialBits ,M,2^SpatialBits),2^Spectral_Efficiency ,1),:) ;

% Creating a matrix containing all possible spatial indexes
PossibleAntIm_Ind     = zeros(nu,2^Spectral_Efficiency) ;
for i = 1 : nu
    PossibleAntIm_Ind(i,:)    = sub2ind([Nt 2^Spectral_Efficiency], ...
                 Possible_Ant_Comb_Im(:,i).', 1:2^Spectral_Efficiency) ;
end

% Creating a matrix containing the indexes of all possible signal
% symbols for the different spatial symbols
PossibleSignalSymbols        = repmat((1:M).',2^(2*SpatialBits),1);

%% GQSM modulation

% The real part of the GQSM
GQSMCons_Diagram_Re = zeros( Nt , 2^Spectral_Efficiency ) ;
GQSMCons_Diagram_Re(PossibleAntRe_Ind) = repmat(real( ...
            SigConsDiag(PossibleSignalSymbols)).',nu,1) / sqrt(nu) ;

% The imaginary part of the GQSM
GQSMCons_Diagram_Im = zeros( Nt , 2^Spectral_Efficiency ) ;
GQSMCons_Diagram_Im(PossibleAntIm_Ind) = 1j * repmat(imag( ...
            SigConsDiag(PossibleSignalSymbols)).',nu,1) / sqrt(nu) ;

% The GQSM constellation diagram
GQSM_Cons_Diagram = GQSMCons_Diagram_Re + GQSMCons_Diagram_Im ;

end
```

A.1.9 SMTs

```
function SMTConstellation_Mat = SMT_Cons_Diagram_Gen(SMT_SyS,Nt,M)
% ======================================================================
% Book title: Space Modulation Techniques
% Authors:    Raed Mesleh and Abdelhamid Alhassi
% Publisher:  John Wiley & Sons, Ltd
% Date:       2017
% ======================================================================
%
% Description
%  This script uses previous scripts to generate the constellation
%  diagram for any space modulation technique (SMT).
%
% Inputs
%  SMT_SyS            A string indicating the SMT system to calculate
%                     the analytical ABER for:
%                        - 'SSK'   for space shift keying (SSK);
```

```
%                       - 'SM'    for spatial modulation (SM);
%                       - 'QSSK'  for quadrature SSK (QSSK);
%                       - 'QSM'   for quadrature SM (QSM);
%                       - 'GSSK'  for generalized SSK (GSSK);
%                       - 'GSM'   for generalized SM (GSM);
%                       - 'GQSSK' for generalized QSSK (GQSSK);
%                       - 'GQSM'  for generalized QSM (GQSM).
%
%  Nt                 For SMTs: Nt is a single element indicating the
%                               number of transmit antennas.
%                     For GSMTs and GQSMTs: Nt is a two element vector:
%                     - The first element, Nt(1), is the number of
%                       transmit antennas,
%                     - The second element, Nt(2), is the number of
%                       active transmit antennas at a time (nu).
%
%  M                  Size of the signal constellation diagram.
%
% Output
%  SMTConstellation_Mat   SMT constellations Diagram.
%
% Usage
%  SMTConstellation_Mat = SMT_Cons_Diagram_Gen(SMT_SyS,Nt,M)

%%
switch lower(SMT_SyS)

    case 'ssk'
        SMTConstellation_Mat = SSK_Cons_Diagram_Gen(Nt) ;

    case 'sm'
        SMTConstellation_Mat = SM_Cons_Diagram_Gen(Nt,M) ;

    case 'qssk'
        SMTConstellation_Mat = QSSK_Cons_Diagram_Gen(Nt) ;

    case 'qsm'
        SMTConstellation_Mat = QSM_Cons_Diagram_Gen(Nt,M) ;

    case 'gssk'
        SMTConstellation_Mat = GSSK_Cons_Diagram_Gen(Nt(1),Nt(2)) ;

    case 'gsm'
        SMTConstellation_Mat = GSM_Cons_Diagram_Gen(Nt(1),Nt(2),M) ;

    case 'gqssk'
        SMTConstellation_Mat = GQSSK_Cons_Diagram_Gen(Nt(1),Nt(2)) ;

    case 'gqsm'
        SMTConstellation_Mat = GQSM_Cons_Diagram_Gen(Nt(1),Nt(2),M) ;

    otherwise
        error('Unknown system')

end

end
```

A.1.10 DSSK

```
function DSSK_Cons_Diagram = DSSK_Cons_Diagram_Gen(Nt)
% ==================================================================
% Book title: Space Modulation Techniques
% Authors:    Raed Mesleh and Abdelhamid Alhassi
% Publisher:  John Wiley & Sons, Ltd
% Date:       2017
% ==================================================================
%
% Description
%  This script generates the differential space shift keying (DSSK)
%  constellations diagram containing all possible DSSK modulated
%  transmit vectors.
%
% Input
%  Nt                  Number of transmit antennas.
%
% Output
%  DSSK_Cons_Diagram   DSSK constellations diagram.
%
% Usage
%  DSSK_Cons_Diagram = DSSK_Cons_Diagram_Gen(Nt)

%% DSSK modulation

AllPossAnt = perms(Nt:-1:1) ;

AntBits     = floor(log2(factorial(Nt))) ;

eta         = AntBits ;

DSSK_Cons_Diagram   = zeros( Nt, Nt , 2^eta   ) ;

for t = 1 : Nt

    DSMPossAnt  = AllPossAnt(1:2^AntBits,t) ;

    PossAnt_Ind = sub2ind([Nt 2^eta], DSMPossAnt.', 1:2^eta) ;

    DSSK_Cons_Diagram(t, PossAnt_Ind) = ones(size(PossAnt_Ind)) ;

end

DSSK_Cons_Diagram = permute(DSSK_Cons_Diagram,[2 1 3]);

end
```

A.1.11 DSM

```
function DSM_Cons_Diagram = DSM_Cons_Diagram_Gen(Nt,M)
% ==================================================================
% Book title: Space Modulation Techniques
% Authors:    Raed Mesleh and Abdelhamid Alhassi
% Publisher:  John Wiley & Sons, Ltd
% Date:       2017
% ==================================================================
%
% Description
%  This script generates the differential spatial modulation (DSM)
%  constellations diagram containing all possible DSM modulated
%  transmit vectors.
%
% Inputs
%  Nt                  Number of transmit antennas.
%  M                   Size of the signal constellation diagram.
%
% Output
%  DSM_Cons_Diagram    DSM constellations diagram.
%
% Usage
%  DSM_Cons_Diagram = DSM_Cons_Diagram_Gen(Nt,M)

%% Generating the M-QAM signal constellation symbols

% Constructing an M-QAM modulator
alphabit         = modem.pskmod('M', M) ;
% Generating signal constellation diagram
SigConsDiag      = modulate( alphabit , (0:M-1)).';
% Calculating average power of the signal constellation diagram
SigConsDiag_avg  = SigConsDiag'*SigConsDiag / M ;
% Normalizing the power of the signal constellation diagram
SigConsDiag      = SigConsDiag / sqrt(SigConsDiag_avg);

%% DSM modulation

AllPossAnt = perms(Nt:-1:1) ;

AntBt      = floor(log2(factorial(Nt))) ;

% The spectral efficiency
eta        = log2(M)*Nt + AntBt ;

DSM_Cons_Diagram = zeros( Nt, Nt , 2^eta  ) ;
```

```
for t = 1 : Nt

    DSMPossAnt   = AllPossAnt(1:2^AntBt,t) ;
    PossAnt      = reshape(repmat(DSMPossAnt.',M^(Nt),1),[2^eta 1]) ;

    PossAnt_Ind = sub2ind([Nt 2^eta],PossAnt.', 1:2^eta) ;

    DSM_Cons_Diagram(t, PossAnt_Ind) = repmat(reshape(repmat( ...
                SigConsDiag,M^(t-1),M^(Nt-t)).',M^Nt,1),2^AntBt,1) ;

end

DSM_Cons_Diagram = permute(DSM_Cons_Diagram,[2 1 3]);

end
```

A.1.12 DSMTs

```
function DSMTConstellation_Mat = DSMT_Cons_Diagram_Gen(DSMT_SyS,Nt,M)
% =====================================================================
% Book title: Space Modulation Techniques
% Authors:    Raed Mesleh and Abdelhamid Alhassi
% Publisher:  John Wiley & Sons, Ltd
% Date:       2017
% =====================================================================
%
% Description
%  This script uses previous scripts to generate the constellation
%  diagram for differential space shift keying (DSSK) and differential
%  spatial modulation (DSM).
%
% Inputs
%  DSMT_SyS                A string indicating the DSMT system to
%                          calculate the analytical ABER for:
%                          -'DSSK' for differential space shift keying;
%                          - DSM'  for differential spatial modulation.
%
%  Nt                      Nt is the number of transmit antennas.
%
%  M                       Size of the signal constellation diagram.
%                          Note, for DSSK, the value of M does not matter.
%
% Output
%  DSMTConstellation_Mat   DSMT constellations Diagram.
%
% Usage
%  DSMTConstellation_Mat = DSMT_Cons_Diagram_Gen(SMT_SyS,Nt,M)

%%
switch lower(DSMT_SyS)

    case 'dssk'
        DSMTConstellation_Mat = DSSK_Cons_Diagram_Gen(Nt) ;
```

```
    case 'dsm'
        DSMTConstellation_Mat = DSM_Cons_Diagram_Gen(Nt,M) ;

    otherwise
        error('Unknown system')

end

end
```

A.2 Receivers

A.2.1 SMTs ML Receiver

```
function ML_Binary_Results = SMT_ML_Receiver(y,H,SMT_Dg)
% ================================================================
% Book title: Space Modulation Techniques
% Authors:    Raed Mesleh and Abdelhamid Alhassi
% Publisher:  John Wiley & Sons, Ltd
% Date:       2017
% ================================================================
%
% Description
%  This script performs the maximum-likelihood (ML) receiver to
%  retrieve the transmitted binary data using space modulation
%  techniques (SMTs).
%
% Inputs
%  y                  The received vector.
%  H                  The fading channel matrix.
%  SMT_Dg             The used SMT constellations diagram.
%
% Output
%  ML_Binary_Results   The binary data decoded by the ML receiver.
%
% Usage
%  ML_Binary_Results = SMT_ML_receiver(y,H,SMT_Cons_Diagram)

%% The size of the used SMT constellation diagram
CSiz = size(SMT_Dg,2) ;

%% The ML receiver

% Calculating the Euclidean distances of all possible SMT transmitted
% vectors, and then finding the index of the SMT symbol with the
% minimum Euclidean distance, i.e. error.
[ ~ , Idx_min_Error ] = min(sum(abs(repmat(y,1,CSiz)-H*SMT_Dg).^2,1));

% Converting the index to binary, to retrieve the transmitted binary bits
ML_Binary_Results = dec2bin(Idx_min_Error - 1 , log2(CSiz));

end
```

A.2.2 DSMTs ML Receiver

```
function ML_Binary_Results = DSMT_ML_Receiver(Yt,Yt1,DSMConsD)
% ======================================================================
% Book title: Space Modulation Techniques
% Authors:    Raed Mesleh and Abdelhamid Alhassi
% Publisher:  John Wiley & Sons, Ltd
% Date:       2017
% ======================================================================
%
% Description
%  This script performs the maximum likelihood (ML) receiver to
%  retrieve the transmitted binary data using differential space shift
%  keying (DSSK) and  differential spatial modulation (DSM).
%
% Inputs
%  Yt                 The received block at time t.
%  Yt1                The receiver block at time t-1.
%  DSMConsD           The used DSMT constellations diagram.
%
% Output
%  ML_Binary_Results  The binary data decoded by the ML receiver.
%
% Usage
%  ML_Binary_Results = DSMT_ML_Receiver(Yt,Yt1,DSMConstellation_Mat)

%% Setting up the receiver

% Getting the size of the used DSMT constellation diagram
ConsSize = size(DSMConsD,3) ;

% Preallocate variables to store calculated Euclidean distances
Euclidean_Distances = zeros( ConsSize , 1 ) ;

%% DSMT ML Receiver
for i = 1 : ConsSize
    % Calculate the Euclidean distance for each possible transmitted
    % DSMT symbol
    Euclidean_Distances(i) = sum(sum(abs(Yt-Yt1*DSMConsD(:,:,i)).^2));
end

% Finding the index of the possible DSMT transmitted vector with the
% minimum Euclidean distance, i.e., error
[~ , Idx_min_Error] = min(Euclidean_Distances);

% Converting the index to binary, to retrieve the transmitted binary bits
ML_Binary_Results   = dec2bin(Idx_min_Error - 1 , log2(ConsSize));

end
```

A.3 Analytical and Simulated ABER

A.3.1 ABER of SM over Rayleigh Fading Channels with No CSE

```
function ABER_Ana = ABER_ANA_SM_Rayleigh_no_CSE(M,Nt,Nr,SNR_Vector)
% ====================================================================
% Book title: Space Modulation Techniques
% Authors:    Raed Mesleh and Abdelhamid Alhassi
% Publisher:  John Wiley & Sons, Ltd
% Date:       2017
% ====================================================================
%
% Description
%  This script calculates the analytical average bit error ratio (ABER)
%  of spatial modulation (SM) over Rayleigh fading channel with no
%  channel estimation errors (CSEs).
%
% Inputs
% M               Size of the signal constellation diagram.
%
%  Nt             Number of transmit antennas.
%
%  Nr             Number of receive antennas.
%
%  SNR_Vector     Vector containing the signal-to-noise ratio (SNR)
%                 values to calculate the ABER over.
%
% Output
%  ABER_Ana       The calculated analytical ABER of SM.
%
% Usage
%  ABER_Ana = ABER_ANA_SM_Rayleigh_no_CSE(M,Nt,Nr,SNR_Vector)

%% Generating the M-QAM signal constellation symbols

% Constructing an M-QAM modulator
alphabit        = modem.qammod('M', M);
% Generating signal constellation diagram
SigConsDiag     = modulate( alphabit , (0:M-1)).';
% Calculating average power of the signal constellation diagram
SigConsDiag_avg = SigConsDiag'*SigConsDiag / M ;
% Normalizing the power of the signal constellation diagram
SigConsDiag     = SigConsDiag / sqrt(SigConsDiag_avg);

%% Calculating the analytical ABER

% Calculating the spectral efficiency of SM
eta = log2(Nt) + log2(M) ;
```

```
% Preallocate variables to store the calculated ABER
ABER_Ana = zeros(size(SNR_Vector)) ;

for snr = 1 : length(SNR_Vector)

    % Calculating the noise variance sigma_n^2
    % Note, E[|Hx|^2]=1 is assumed
    sigma_n  = (1 / 10^( SNR_Vector(snr) / 10 )) ;

    for ell_t = 1 : Nt

        for ell = 1 : Nt

            for i_t = 1 : M

                for i = 1 : M

                    if ell_t == ell
                        Psi = abs(SigConsDiag(i_t)-SigConsDiag(i))^2;
                        BarGamma_SM = (1/(2*sigma_n)) * Psi ;
                    else
                        Psi = abs(SigConsDiag(i_t))^2 + ...
                                                  abs(SigConsDiag(i))^2;
                        BarGamma_SM = (1/(2*sigma_n)) * Psi ;
                    end

                    alpha_a  =(1/2)*(1-sqrt((BarGamma_SM/2)/ ...
                                                  (1+(BarGamma_SM/2))));

                    % Calculating the PEP for this event
                    PEP = 0 ;
                    for nr = 0 : (Nr-1)
                        PEP = PEP+nchoosek(Nr-1+nr,nr)*(1-alpha_a)^nr;
                    end
                    PEP =  alpha_a^Nr  * PEP ;

                    % Number of bits in error for this event
                    bitError = (biterr(ell_t-1,ell-1,log2(Nt)) ...
                               +biterr(i_t-1,i-1,log2(M))) / eta ;

                    % Calculating the ABER
                    ABER_Ana(snr) = ABER_Ana(snr) + bitError * PEP ;

                end

            end

        end

    end

end

ABER_Ana = ABER_Ana / 2^eta ;
```

A.3.2 ABER of SM over Rayleigh Fading Channels with CSE

```
function ABER_Ana = ABER_ANA_SM_Rayleigh_CSE(varargin)
% ==================================================================
% Book title: Space Modulation Techniques
% Authors:    Raed Mesleh and Abdelhamid Alhassi
% Publisher:  John Wiley & Sons, Ltd
% Date:       2017
% ==================================================================
%
% Description
%  This script calculates the analytical average bit error ratio (ABER)
%  of spatial modulation (SM) over Rayleigh fading channel with
%  channel estimation errors (CSEs).
%
% Inputs
%  M                 Size of the signal constellation diagram.
%
%  Nt                Number of transmit antennas.
%
%  Nr                Number of receive antennas.
%
%  SNR_Vec           Vector containing the signal-to-noise ratio (SNR)
%                    values to calculate the ABER over.
%
%  sigma_e_Vec       - a single double containing sigma_e.
%                    - a vector containing a variable sigma_e.
%                      Note, the vector has to have the same length as
%                      the range of the SNR defined next.
%
% Output
%  ABER_Ana          The calculated analytical ABER of SM.
%
% Usage
%  For sigma_e = sigma_n
%  ABER_Ana = ABER_ANA_SM_Rayleigh_CSE(M,M,Nr,SNR_Vec)
%  For a given sigma_e
%  ABER_Ana = ABER_ANA_SM_Rayleigh_CSE(M,M,Nr,SNR_Vec,sigma_e_Vec)

%% Defining input parameters
switch nargin
    case 4 % For sigma_e = sigma_n
        M           = varargin{1} ;
        Nt          = varargin{2} ;
        Nr          = varargin{3} ;
        SNR_Vec     = varargin{4} ;
        sigma_e_Vec = (1 ./ 10.^( SNR_Vec / 10 )) ;

    case 5 % For a given sigma_e
        M           = varargin{1} ;
        Nt          = varargin{2} ;
        Nr          = varargin{3} ;
        SNR_Vec     = varargin{4} ;
        if length(varargin{5}) == 1
            sigma_e_Vec = varargin{5} * ones(size(SNR_Vec)) ;
```

```
        elseif length(varargin{5}) == length(SNR_Vec)
            sigma_e_Vec = varargin{5} ;
        else
            error('sigma_e and SNR_Vector has to be the same length')
        end

end

%% Generating the M-QAM signal constellation symbols

% Constructing an M-QAM modulator
alphabit          = modem.qammod( 'M',  M ) ;
% Generating signal constellation diagram
SigConsDiag       = modulate( alphabit , (0:M-1)).';
% Calculating average power of the signal constellation diagram
SigConsDiag_avg = SigConsDiag'*SigConsDiag / M ;
% Normalizing the power of the signal constellation diagram
SigConsDiag       = SigConsDiag / sqrt(SigConsDiag_avg);

%% Calculating Analytical ABER

% Calculating the spectral efficiency of SM
eta = log2(Nt) + log2(M) ;

% Preallocate variables to store the calculated ABER
ABER_Ana = zeros(size(SNR_Vec)) ;

for snr = 1 : length(SNR_Vec)

    % Calculating the noise variance sigma_n^2
    % Note, E[|Hx|^2]=1 is assumed
    sigma_n  = (1 / 10^( SNR_Vec(snr) / 10 )) ;

    sigma_e  = sigma_e_Vec(snr) ;

    for ell_t = 1 : Nt

        for ell = 1 : Nt

            for i_t = 1 : M

                for i = 1 : M

                    varphi = ( 1 + sigma_e ) / ...
                        (2*(sigma_e*abs(SigConsDiag(i_t))^2+sigma_n));

                    if ell_t == ell
                       Psi = abs(SigConsDiag(i_t) - SigConsDiag(i))^2;
                       BarGamma_SM = varphi * Psi ;
                    else
                       Psi = abs(SigConsDiag(i_t))^2 + ...
                                                abs(SigConsDiag(i))^2;
                       BarGamma_SM = varphi * Psi;
                    end
```

```
                    alpha_a   =(1/2)*(1 - sqrt((BarGamma_SM/2)/ ...
                                        (1+(BarGamma_SM/2))));

                    % Calculating the PEP for this event
                    PEP = 0 ;
                    for nr = 0 : (Nr-1)
                        PEP  =PEP+nchoosek(Nr-1+nr,nr)*(1-alpha_a)^nr;
                    end
                    PEP =  alpha_a^Nr  * PEP ;

                    % Number of bits in error for this event
                    bitError = (biterr(ell_t -1,ell -1,log2(Nt)) ...
                              +biterr(i_t -1,i -1,log2(M))) / eta ;

                    % Calculating the ABER
                    ABER_Ana(snr) = ABER_Ana(snr) + bitError * PEP ;

                end

            end

        end

    end

end

ABER_Ana = ABER_Ana / 2^eta ;
```

A.3.3 ABER of QSM over Rayleigh Fading Channels with No CSE

```
function ABER_Ana = ABER_ANA_QSM_Rayleigh_no_CSE(M,Nt,Nr,SNR_Vector)
% ==================================================================
% Book title: Space Modulation Techniques
% Authors:    Raed Mesleh and Abdelhamid Alhassi
% Publisher:  John Wiley & Sons, Ltd
% Date:       2017
% ==================================================================
%
% Description
%  This script calculates the analytical average bit error ratio (ABER)
%  of quadrature spatial modulation (QSM) over Rayleigh fading channel
%  with no channel estimation errors (CSEs).
%
% Inputs
%  M                Size of the signal constellation diagram.
%
%  Nt               Number of transmit antennas.
%
%  Nr               Number of receive antennas.
%
%  SNR_Vector       Vector containing the signal-to-noise ratio (SNR)
```

```
%                       values to calculate the ABER over.
%
% Output
%  ABER_Ana             The calculated analytical ABER of QAM.
%
% Usage
%  ABER_Ana = ABER_ANA_QSM_Rayleigh_no_CSE(M,Nt,Nr,SNR_Vector)

%% Generating the M-QAM signal constellation symbols

% Constructing an M-QAM modulator
alphabit = modem.qammod('M', M ) ;
% Generating signal constellation diagram
S        = modulate( alphabit , (0:M-1)).';
% Calculating average power of the signal constellation diagram
S_avg    = S'*S / M ;
% Normalizing the power of the signal constellation diagram
S        = S / sqrt(S_avg);

%% Calculating Analytical ABER

% Calculating the spectral efficiency of SM
eta = 2*log2(Nt) + log2(M) ;

% Preallocate variables to store the calculated ABER
ABER_Ana = zeros(size(SNR_Vector)) ;

for snr = 1 : length(SNR_Vector)

    % Calculating the noise variance sigma_n^2
    % Note, E[|Hx|^2]=1 is assumed
    sigma_n  = (1 / 10^( SNR_Vector(snr) / 10 )) ;

    for ellRe_t = 1 : Nt

        for ellIm_t = 1 : Nt

            for ellRe = 1 : Nt

                for ellIm = 1 : Nt

                    for i_t = 1 : M

                        for i = 1 : M

                            if ellRe_t == ellRe && ellIm_t == ellIm

                                Psi=abs(real(S(i_t))-real(S(i)))^2 ...
                                    +abs(imag(S(i_t))-imag(S(i)))^2 ;

                                BarGamma_QSM = (1/(2*sigma_n)) * Psi ;

                            elseif ellRe_t ~= ellRe && ellIm_t == ellIm
```

```
                        Psi = abs(real(S(i_t)))^2 + ...
                              abs(real(S(i)))^2   + ...
                              abs(imag(S(i_t))-imag(S(i)))^2;

                        BarGamma_QSM = (1/(2*sigma_n)) * Psi ;
                    elseif ellRe_t == ellRe && ellIm_t ~= ellIm

                        Psi=abs(real(S(i_t))-real(S(i)))^2 ...
                            + abs(imag(S(i_t)))^2 ...
                            + abs(imag(S(i)))^2 ;

                        BarGamma_QSM = (1/(2*sigma_n)) * Psi ;

                    elseif ellRe_t ~= ellRe && ellIm_t ~= ellIm

                        Psi = abs(S(i_t))^2 + abs(S(i_t))^2 ;

                        BarGamma_QSM = (1/(2*sigma_n)) * Psi ;

                    end

                    alpha_a=(1/2)*(1-sqrt((BarGamma_QSM/2)/...
                                    (1+(BarGamma_QSM/2))));

                    % Calculating the PEP for this event
                    PEP = 0 ;
                    for nr = 0 : (Nr-1)
                        PEP = PEP + ...
                          nchoosek(Nr-1+nr,nr)*(1-alpha_a)^nr;
                    end
                    PEP =  alpha_a^Nr  * PEP ;

                    % Number of bits in error for this event
                    btE=(biterr(ellRe_t-1,ellRe-1,log2(Nt))...
                        + biterr(ellIm_t-1,ellIm-1,log2(Nt))...
                        + biterr(i_t-1,i-1,log2(M))) / eta ;

                    ABER_Ana(snr) = ABER_Ana(snr) + btE*PEP ;

                end

            end

        end

    end

  end

end

end

ABER_Ana = ABER_Ana / 2^eta ;
```

A.3.4 ABER of QSM over Rayleigh Fading Channels with CSE

```
function ABER_Ana = ABER_ANA_QSM_Rayleigh_CSE(varargin)
% ======================================================================
% Book title: Space Modulation Techniques
% Authors:    Raed Mesleh and Abdelhamid Alhassi
% Publisher:  John Wiley & Sons, Ltd
% Date:       2017
% ======================================================================
%
% Desctription
%  This script calculates the analytical average bit error ratio (ABER)
%  of quadrature spatial modulation (QSM) over Rayleigh fading channel
%  with channel estimation errors (CSE).
%
% Inputs
%  M                  Size of the signal constellation diagram.
%
%  Nt                 Number of transmit antennas.
%
%  Nr                 Number of receive antennas.
%
%  SNR_Vector         Vector containing the signal to noise ratio (SNR)
%                     values to calculate the ABER over.
%
%  sigma_e_Vec        - a single double containig sigma_e.
%                     - a vector containing a variable sigma_e.
%                       Note, the vector has to have the same lenght as
%                       the range of the SNR defined next.
%
% Output
%  ABER_Ana           The calculated analytical ABER of QAM.
%
% Usage
%  For sigma_e = sigma_n
%  ABER_Ana = ABER_ANA_QSM_Rayleigh_CSE(M,M,Nr,SNR_Vec)
%
%  For a given sigma_e
%  ABER_Ana = ABER_ANA_QSM_Rayleigh_CSE(M,M,Nr,SNR_Vec,sigma_e_Vec)

%% Defining input parameters
switch nargin
    case 4 % For sigma_e = sigma_n
        M           = varargin{1} ;
        Nt          = varargin{2} ;
        Nr          = varargin{3} ;
        SNR_Vec     = varargin{4} ;
        sigma_e_Vec = (1 ./ 10.^( SNR_Vec / 10 )) ;

    case 5 % For a given sigma_e
        M           = varargin{1} ;
        Nt          = varargin{2} ;
        Nr          = varargin{3} ;
        SNR_Vec     = varargin{4} ;
        if length(varargin{5}) == 1
            sigma_e_Vec = varargin{5} * ones(size(SNR_Vec)) ;
```

```
        elseif length(varargin{5}) == length(SNR_Vec)
            sigma_e_Vec = varargin{5} ;
        else
            error('sigma_e and SNR_Vector has to be the same length')
        end

end

%% Generating the M-QAM signal constellation symbols

% Constructing an M-QAM modulator
alphabit = modem.qammod( 'M', M ) ;
% Generating signal constellation diagram
S        = modulate( alphabit , (0:M-1)).';
% Calculating average power of the signal constellation diagram
S_avg    = S'*S / M ;
% Normalizing the power of the signal constellation diagram
S        = S / sqrt(S_avg);

%% Calaculating Analaytical ABER

% Calulating the spectral efficiency of SM
eta = 2*log2(Nt) + log2(M) ;

% Pre-allocate variables to store the calculated ABER
ABER_Ana = zeros(size(SNR_Vec)) ;

for snr = 1 : length(SNR_Vec)

    % Calculating the noise variance sigma_n^2
    % Note, E[|Hx|^2]=1 is assumed
    sigma_n  = (1 / 10^( SNR_Vec(snr) / 10 )) ;

    sigma_e  = sigma_e_Vec(snr) ;

    for ellRe_t = 1 : Nt

        for ellIm_t = 1 : Nt

            for ellRe = 1 : Nt

                for ellIm = 1 : Nt

                    for i_t = 1 : M

                        for i = 1 : M

                            varphi = ( 1 + sigma_e ) / ...
                                (2*(sigma_e*abs(S(i_t))^2+sigma_n));

                            if ellRe_t == ellRe && ellIm_t == ellIm


                                Psi=abs(real(S(i_t))-real(S(i)))^2 ...
                                    +abs(imag(S(i_t))-imag(S(i)))^2 ;

                                BarGamma_QSM = varphi * Psi ;
```

```
                        elseif ellRe_t ~= ellRe && ellIm_t == ellIm

                            Psi = abs(real(S(i_t)))^2 + ...
                                  abs(real(S(i)))^2   + ...
                                  abs(imag(S(i_t))-imag(S(i)))^2;

                            BarGamma_QSM = varphi * Psi ;

                        elseif ellRe_t == ellRe && ellIm_t ~= ellIm

                            Psi=abs(real(S(i_t))-real(S(i)))^2 ...
                                + abs(imag(S(i_t)))^2 ...
                                + abs(imag(S(i)))^2 ;

                            BarGamma_QSM = varphi * Psi ;

                        elseif ellRe_t ~= ellRe && ellIm_t ~= ellIm

                            Psi = abs(S(i_t))^2 + abs(S(i_t))^2 ;

                            BarGamma_QSM = varphi * Psi ;

                        end

                        alpha_a=(1/2)*(1-sqrt((BarGamma_QSM/2)/...
                                       (1+(BarGamma_QSM/2))));

                        % Calculating the PEP fot this event
                        PEP = 0 ;
                        for nr = 0 : (Nr-1)
                            PEP  = PEP + ...
                               nchoosek(Nr-1+nr,nr)*(1-alpha_a)^nr;
                        end
                        PEP =  alpha_a^Nr  * PEP ;

                        % Number of bits in error for this event
                        btE=(biterr(ellRe_t -1,ellRe -1,log2(Nt)) ...
                            + biterr(ellIm_t -1,ellIm -1,log2(Nt))...
                            + biterr(i_t -1,i -1,log2(M))) / eta ;

                        ABER_Ana(snr) = ABER_Ana(snr) + btE*PEP ;

                    end

                end

            end

        end

    end

  end

end

ABER_Ana = ABER_Ana / 2^eta ;
```

A.3.5 Analytical ABER of SMTs over Generalized Fading Channels and with CSE and SC

```
function ABER_Ana = ABER_SMT_Analytical_General(varargin)
% ==================================================================
% Book title: Space Modulation Techniques
% Authors:    Raed Mesleh and Abdelhamid Alhassi
% Publisher:  John Wiley & Sons, Ltd
% Date:       2017
% ==================================================================
%
% Description
%  This script calculates the analytical average bit error ratio (ABER)
%  of space modulation techniques (SMTs) over generalized fading channel
%  and in the presence of spatial correlation (SC) and channel estimation
%  errors (CSEs).
%
% Inputs
%  SMT_SyS           A string indicating the SMT system to calculate
%                    the analytical ABER for:
%                       - 'SSK'   for space shift keying (SSK);
%                       - 'SM'    for spatial modulation (SM);
%                       - 'QSSK'  for quadrature SSK (QSSK);
%                       - 'QSM'   for quadrature SM (QSM);
%                       - 'GSSK'  for generalized SSK (GSSK);
%                       - 'GSM'   for generalized SM (GSM);
%                       - 'GQSSK' for generalized QSSK (GQSSK);
%                       - 'GQSM'  for generalized QSM (GQSM).
%
%  M                 Size of the signal constellation diagram
%                    Note, for SSK, the value of M does not matter.
%
%  Nt                For SMTs: Nt is a single element indicating the
%                              number of transmit antennas.
%                    For GSMTs and GQSMTs: Nt is a two element vector:
%                    - The first element, Nt(1), is the number of
%                      transmit antennas;
%                    - The second element, Nt(2), is the number of
%                      active transmit antennas at a time (nu).
%
%  Nr                Number of receive antennas
%
%  Ch_Type           A string indicating the type of channel to
%                    calculate the ABER over:
%                       - 'rayleigh' or 'ray'  for Rayleigh;
%                       - 'rician'   or 'rice' for Rician;
%                       - 'nakagami' or 'nak'  for Nakagami.
%  K                 Rician K factor in dB.
%  m                 Shape parameter of the Nakagami-m fading channel.
%
%  Ch_Err            - Logical False for no CSE.
%                    - Logical True for CSE and sigma_e = sigma_n.
%                    - a single double containing sigma_e.
%                    - a vector containing a variable sigma_e.
```

```
%                          Note, the vector has to have the same length as
%                          the range of the SNR defined next.
%
%  SNR_Vector            Vector containing the signal-to-noise ratio (SNR)
%                        values to calculate the ABER over.
%
%  R_tx                  The transmitter correlation matrix.
%  R_rx                  The receiver correlation matrix.
%
% Output
%  ABER_Ana              The calculated analytical ABER of the chosen SMT.
%
% Usage
%
%  For Rayleigh with no Correlation
%  ABER_Ana = ABER_Analytical_General(SMT_SyS,M,Nt,Nr, ...
%                                                  Ch_Err,SNR_Vector)
%
%  For Rayleigh with Correlation
%  ABER_Ana = ABER_Analytical_General(SMT_SyS,M,Nt,Nr, ...
%                                        Ch_Err,R_tx,R_rx,SNR_Vector)
%
%  For Rician with K factor and no Correlation
%  ABER_Ana = ABER_Analytical_General(SMT_SyS,M,Nt,Nr,Ch_Type,K, ...
%                                                  Ch_Err,SNR_Vector)
%
%  For Nakagami-m with m shaping parameter and no Correlation
%  ABER_Ana = ABER_Analytical_General(SMT_SyS,M,Nt,Nr,Ch_Type,m, ...
%                                                  Ch_Err,SNR_Vector)
%
%  For Rician with K factor and Correlation
%  ABER_Ana = ABER_Analytical_General(SMT_SyS,M,Nt,Nr,Ch_Type,K, ...
%                                        Ch_Err,R_tx,R_rx,SNR_Vector)
%
%  For Nakagami-m with m shaping parameter and Correlation
%  ABER_Ana = ABER_Analytical_General(SMT_SyS,M,Nt,Nr,Ch_Type,m, ...
%                                        Ch_Err,R_tx,R_rx,SNR_Vector)

%% Defining input parameters
switch nargin

    case 6      % No correlation Rayleigh fading
        SMT_SyS = lower(varargin{1}) ;

        M       = varargin{2} ;

        Nt      = varargin{3} ;
        if ~strcmp(SMT_SyS(1),'g') && length(Nt)>1
            error(' For SMTs Nt should be a single element')
        elseif strcmp(SMT_SyS(1),'g') && length(Nt)==1
            error(' For GSMTs Nt should be a two element array')
        end

        Nr      = varargin{4} ;
```

```
        Ch_Type = 'rayleigh' ;

        Ch_Err  = varargin{5} ;

        SNR_Vector = varargin{6} ;

        R_tx = eye(Nt(1)) ;
        R_rx = eye(Nr) ;

        R_s  = kron(R_rx,R_tx)  ;

    case 8      % No correlation Rician or Nakagami-m

        SMT_SyS = lower(varargin{1}) ;

        M       = varargin{2} ;

        Nt      = varargin{3} ;
        if ~strcmp(SMT_SyS(1),'g') && length(Nt)>1
            error(' For SMTs Nt should be a single element')
        elseif strcmp(SMT_SyS(1),'g') && length(Nt)==1
            error(' For GSMTs Nt should be a two element array')
        end

        Nr      = varargin{4} ;

        if ischar(varargin{5}) % for Rician or Nakagami-m
            % with no Correlation

            Ch_Type = lower(varargin{5}) ;

            if strcmp(Ch_Type,'rician') || strcmp(Ch_Type,'rice')
                KdB = varargin{6} ;
                K   = 10^(KdB/10) ;
            elseif strcmp(Ch_Type,'nak') || strcmp(Ch_Type,'nakagami')
                m   = varargin{6} ;
            else
                error('Unknown Channel')
            end

            Ch_Err  = varargin{7} ;

            R_tx = eye(Nt(1)) ;
            R_rx = eye(Nr) ;

            R_s  = kron(R_rx,R_tx)  ;

        else                    % for Rayleigh with correlation
            Ch_Type = 'rayleigh' ;
            Ch_Err  = varargin{5} ;

            R_tx    = varargin{6};
            R_rx    = varargin{7} ;
```

```
            R_s = kron(R_rx,R_tx)    ;
        end

        SNR_Vector = varargin{8} ;

    case 10     % for Rician or Nakagami-m with correlation

        SMT_SyS = lower(varargin{1}) ;

        M       = varargin{2} ;

        Nt      = varargin{3} ;
        if ~strcmp(SMT_SyS(1),'g') && length(Nt)>1
            error(' For SMTs Nt should be a single element')
        elseif strcmp(SMT_SyS(1),'g') && length(Nt)==1
            error(' For GSMTs Nt should be a two element array')
        end

        Nr      = varargin{4} ;

        Ch_Type = lower(varargin{5}) ;

        if strcmp(Ch_Type,'rician') || strcmp(Ch_Type,'rice')
            KdB = varargin{6} ;
            K   = 10^(KdB/10) ;
        elseif strcmp(Ch_Type,'nak') || strcmp(Ch_Type,'nakagami')
            m   = varargin{6} ;
        else
            error('Unknown Channel')
        end

        Ch_Err  = varargin{7} ;

        R_tx    = varargin{8};
        R_rx    = varargin{9} ;

        R_s = kron(R_rx,R_tx)   ;

        SNR_Vector = varargin{10} ;

    otherwise

        error('Not enough or too many inputs')

end

%% Generating the SMT constellation matrix
SMTConstellation_Mat = SMT_Cons_Diagram_Gen(SMT_SyS,Nt,M) ;

% The spectral efficiency
eta = log2(size(SMTConstellation_Mat,2)) ;

% In the case of GSMTs, Nu is not needed anymore
Nt  = Nt(1) ;
```

```
%% Calculating the mean and the variance of the chosen channel
switch Ch_Type

    case {'ray','rayleigh'} % Rayleigh
        % Covariance matrix of the Rayleigh fading channel
        LH        = R_s ;

    case {'rice','rician'} % Rician
        % The mean vector of the Rician fading channel including SC
        u         = R_s^(1/2) *  sqrt(K/(1+K)) * ones(Nr*Nt,1)   ;

        % The covariance matrix of the Rician fading channel
        % including SC
        LH        = (1/(1+K)) * R_s ;

    case {'nak','nakagami'} % Nakagami-m
        % The mean vector of the Nakagami-m fading channel
        u_H   = (gamma((m/2)+0.5)/gamma((m/2))) * sqrt(1/(2*(m/2))) ...
           + 1j * (gamma((m/2)+0.5)/gamma((m/2))) * sqrt(1/(2*(m/2))) ;

        % The mean vector of the Nakagami-m fading channel including SC
        u         = R_s^(1/2) *  u_H  * ones(Nr*Nt,1)   ;

        % The variance of the Nakagami-m fading channel
        variNak = ( 1 - 2 * ((gamma((m/2)+0.5)/gamma((m/2))) ...
                                          * sqrt(1/(2*(m/2))))^2  ) ;

        % The covariance matrix of the Nakagami-m fading channel
        % including SC
        LH        = variNak * R_s ;

end

%% Calculating the analytical ABER
% Preallocate variables to store the calculated ABER
ABER_Ana = zeros(size(SNR_Vector)) ;

for snr = 1: length(SNR_Vector)
    % Calculating the noise variance sigma_n^2
    % Note, E[|Hx|^2]=1 is assumed
    sigma_n = 1 / 10^( SNR_Vector(snr) / 10 ) ;

    if islogical(Ch_Err)
        if Ch_Err % CSE with sigma_e = sigma_n
            sigma_e = sigma_n ;
            L       = LH + sigma_e*eye(size(LH)) ;
        else      % no CSE
            sigma_e = 0 ;
            L       = LH ;
        end
    elseif length(Ch_Err)==1 % CSE with constant sigma_e
        sigma_e       = Ch_Err ;
        L             = LH + sigma_e*eye(size(LH)) ;
    else                          % CSE with sigma_e given as a matrix
```

```
        if length(Ch_Err) == length(SNR_Vector)
            sigma_e      = Ch_Err(snr) ;
            L            = LH + sigma_e*eye(size(LH)) ;
        else
            error('sigma_e and SNR_Vector has to be the same length')
        end
    end

    ABER = 0 ;

    for i_t = 1 : 2^eta

        x_t        = SMTConstellation_Mat(:,i_t) ;

        % Note, for no CSE sigma_e = 0. Hence, sigma_T = sigma_n
        sigma_T    = sigma_n + sigma_e * sum(abs(x_t).^2) ;

        gamma_bar =  1 / (2 * sigma_T) ;

        for i = 1 : 2^eta

            x       = SMTConstellation_Mat(:,i) ;

            Psi     = x_t - x ;

            Lambda = kron(eye(Nr),Psi*Psi');

            % Calculating the PEP for this event
            switch Ch_Type

                case {'ray','rayleigh'}%Note, Rayleigh has a zero mean
                    PEP = (1/2) * 1 / det( eye(Nt*Nr) ...
                            + ( gamma_bar / sqrt(2) ) * L * Lambda ) ;

                otherwise
                    PEP = (1/2)*exp(-1 * gamma_bar/2 * u' * Lambda ...
                *(eye(Nr*Nt)+(gamma_bar/sqrt(2))*L*Lambda)^(-1)*u) ...
                  / det(eye(Nt*Nr)+(gamma_bar/sqrt(2)) * L * Lambda) ;

            end

            ABER = ABER + (biterr(i_t-1,i-1,eta)/eta) *  PEP ;

        end

    end

    % Averaging to get the ABER
    ABER_Ana(snr) = ABER / (2^eta) ;

end

end
```

A.3.6 Simulated ABER of SMTs Using Monte Carlo Simulation over Generalized Fading Channels and with CSE and SC

```
function ABER_Sim = ABER_SMT_Monte_Carlo_Simulation(varargin)
% ====================================================================
% Book title: Space Modulation Techniques
% Authors:    Raed Mesleh and Abdelhamid Alhassi
% Publisher:  John Wiley & Sons, Ltd
% Date:       2017
% ====================================================================
%
% Description
%  This script calculates the average bit error ratio (ABER) of space
%  modulation techniques (SMTs) using Monte Carlo simulations, over
%  generalized fading channel and in the presence of spatial
%  correlation (SC) and channel estimation errors (CSEs).
%
% Inputs
%  SMT_SyS            A string indicating the SMT system to calculate
%                     the analytical ABER for,
%                        - 'SSK'    for space shift keying (SSK);
%                        - 'SM'     for spatial modulation (SM);
%                        - 'QSSK'   for quadrature SSK (QSSK);
%                        - 'QSM'    for quadrature SM (QSM);
%                        - 'GSSK'   for generalized SSK (GSSK);
%                        - 'GSM'    for generalized SM (GSM);
%                        - 'GQSSK'  for generalized QSSK (GQSSK);
%                        - 'GQSM'   for generalized QSM (GQSM).
%
%  Nt                 For SMTs: Nt is a single element indicating the
%                               number of transmit antennas.
%                     For GSMTs and GQSMTs: Nt is a two element vector:
%                     - The first element, Nt(1), is the number of
%                       transmit antennas;
%                     - The second element, Nt(2), is the number of
%                       active transmit antennas at a time (nu).
%
%  M                  Size of the signal constellation diagram.
%                     Note, for SSK the value of M does not matter.
%
%  Nr                 Number of receive antennas
%
%  Ch_Type            A string indicating the type of channel to
%                     calculate the ABER over:
%                        - 'rayleigh' or 'ray'  for Rayleigh;
%                        - 'rician'   or 'rice' for Rician;
%                        - 'nakagami' or 'nak'  for Nakagami.
%  K                  Rician K factor in dB.
%  m                  Shape parameter of the Nakagami-m fading channel.
%
%  Ch_Err             - Logical False for no CSE.
%                     - Logical True for CSE and sigma_e = sigma_n.
%                     - a single double containing sigma_e.
```

```
%                        - a vector containing a variable sigma_e.
%                          Note, the vector has to have the same length as
%                          the range of the SNR defined next.
%
%   SNR_Vector           Vector containing the signal-to-noise ratio (SNR)
%                        values to calculate the ABER over.
%   R_tx                 The transmitter correlation matrix.
%   R_rx                 The receiver correlation matrix.
%
%   NumIterations        Number of simulation iterations.
%
% Output
%   ABER_Sim             The calculated ABER of the chosen SMT.
%
% Usage
%
%   For Rayleigh with no Correlation
%   ABER_Sim = SMT_ABER_Monte_Carlo_Simulation(SMT_SyS,M,Nt,Nr, ...
%                                    Ch_Err,SNR_Vector,NumIterations)
%
%   For Rayleigh with Correlation
%   ABER_Sim = SMT_ABER_Monte_Carlo_Simulation(SMT_SyS,M,Nt,Nr, ...
%                          Ch_Err,R_tx,R_rx,SNR_Vector,NumIterations)
%
%   For Rician with K factor and no Correlation
%   ABER_Sim = SMT_ABER_Monte_Carlo_Simulation(SMT_SyS,M,Nt,Nr, ...
%                          Ch_Type,K,Ch_Err,SNR_Vector,NumIterations)
%
%   For Nakagami-m with m shaping parameter and no Correlation
%   ABER_Sim = SMT_ABER_Monte_Carlo_Simulation(SMT_SyS,M,Nt,Nr, ...
%                          Ch_Type,m,Ch_Err,SNR_Vector,NumIterations)
%
%   For Rician with K factor and Correlation
%   ABER_Sim = SMT_ABER_Monte_Carlo_Simulation(SMT_SyS,M,Nt,Nr, ...
%                 Ch_Type,K,Ch_Err,R_tx,R_rx,SNR_Vector,NumIterations)
%
%   For Nakagami-m with m shaping parameter and  Correlation
%   ABER_Sim = SMT_ABER_Monte_Carlo_Simulation(SMT_SyS,M,Nt,Nr, ...
%                 Ch_Type,m,Ch_Err,R_tx,R_rx,SNR_Vector,NumIterations)

%% Defining input parameters
switch nargin

    case 7     % No correlation Rayleigh fading
        SMT_SyS = lower(varargin{1}) ;

        M       = varargin{2} ;

        Nt      = varargin{3} ;
        if ~strcmp(SMT_SyS(1),'g') && length(Nt)>1
            error(' For SMTs Nt should be a single element')
        elseif strcmp(SMT_SyS(1),'g') && length(Nt)==1
            error(' For GSMTs Nt should be a two element array')
        end
```

```
    Nr        = varargin{4} ;

    Ch_Type = 'rayleigh' ;

    Ch_Err   = varargin{5} ;

    SNR_Vector = varargin{6} ;

    R_tx = eye(Nt(1)) ;
    R_rx = eye(Nr) ;

    NumIterations = varargin{7} ;

case 9       % No correlation Rician or Nakagami-m
    SMT_SyS = lower(varargin{1}) ;

    M        = varargin{2} ;

    Nt       = varargin{3} ;
    if ~strcmp(SMT_SyS(1),'g') && length(Nt)>1
        error(' For SMTs Nt should be a single element')
    elseif strcmp(SMT_SyS(1),'g') && length(Nt)==1
        error(' For GSMTs Nt should be a two element array')
    end

    Nr       = varargin{4} ;

    if ischar(varargin{5}) % for Rician or Nakagami-m
                           % with no Correlation

        Ch_Type = lower(varargin{5}) ;

        if strcmp(Ch_Type,'rician') || strcmp(Ch_Type,'rice')
            KdB = varargin{6} ;
            K   = 10^(KdB/10) ;
        elseif strcmp(Ch_Type,'nak') || strcmp(Ch_Type,'nakagami')
            m   = varargin{6} ;
        else
            error('Unknown Channel')
        end

        Ch_Err   = varargin{7} ;

        R_tx = eye(Nt(1)) ;
        R_rx = eye(Nr) ;

    else                         % for Rayleigh with correlation

        Ch_Type = 'rayleigh' ;
        Ch_Err  = varargin{5} ;

        R_tx     = varargin{6};
        R_rx     = varargin{7} ;
```

```
        end

        SNR_Vector     = varargin{8} ;

        NumIterations = varargin{9} ;

    case 11     % for Rician or Nakagami-m with correlation
        SMT_SyS = lower(varargin{1}) ;

        M       = varargin{2} ;

        Nt      = varargin{3} ;
        if ~strcmp(SMT_SyS(1),'g') && length(Nt)>1
            error(' For SMTs Nt should be a single element')
        elseif strcmp(SMT_SyS(1),'g') && length(Nt)==1
            error(' For GSMTs Nt should be a two element array')
        end

        Nr      = varargin{4} ;

        Ch_Type = lower(varargin{5}) ;

        if strcmp(Ch_Type,'rician') || strcmp(Ch_Type,'rice')
            KdB = varargin{6} ;
            K   = 10^(KdB/10) ;
        elseif strcmp(Ch_Type,'nak') || strcmp(Ch_Type,'nakagami')
            m   = varargin{6} ;
        else
            error('Unknown Channel')
        end

        Ch_Err  = varargin{7} ;

        R_tx    = varargin{8};
        R_rx    = varargin{9} ;

        SNR_Vector     = varargin{10} ;

        NumIterations = varargin{11} ;

    otherwise

        error('Not enough or too many inputs')

end

%% Generating the SMT constellation matrix
SMT_Cons_Diagram = SMT_Cons_Diagram_Gen(SMT_SyS,Nt,M) ;

% In case of GSMTs Nu is not needed anymore
Nt = Nt(1) ;

% The spectral efficiency
```

```
eta = log2(size(SMT_Cons_Diagram,2)) ;

%% Monte Carlo ABER Simulation

% Preallocate variables to store the calculated ABER
ABER_Sim    = zeros( size(SNR_Vector) ) ;

% Starting the simulation
for snr = 1 : length(SNR_Vector)

    for itr = 1 : NumIterations

        % Generate the data symbol to be transmitted
        PreGenData = randi([0 (2^eta -1)]);

        % SMT modulation
        xt          = SMT_Cons_Diagram(:,PreGenData+1) ;

        % Passing through channel
        switch Ch_Type
            case {'ray','rayleigh'}
                H = ( randn(Nr,Nt) + 1j.*randn(Nr,Nt) ) / sqrt(2) ;

            case {'rice','rician'}
                H = ( randn(Nr,Nt) + 1j.*randn(Nr,Nt) ) / sqrt(2) ;
                H = sqrt(K/(1+K))*ones(size(H)) + sqrt(1/(1+K))*H ;

            case {'nak','nakagami'}
                H = sqrt(sum(abs(randn(Nr,Nt,m)/sqrt(2*m)).^2,3)) ...
                    +1j.*sqrt(sum(abs(randn(Nr,Nt,m)/sqrt(2*m)).^2,3));
        end

        % Applying SC
        % Note, for no correlation , R_tx and R_rx are all ones diagonal
        % matrices , and thus , H_SC = H
        H_SC     = R_rx^(1/2) * H * R_tx^(1/2) ;

        % Generating the noise
        sigma_n = 1 / 10^( SNR_Vector(snr) / 10 ) ;

        noise    = sqrt(sigma_n)*(randn(Nr,1)+1j*randn(Nr,1))/sqrt(2) ;

        % The received signal
        y        = H_SC * xt + noise ;

        % CSE
        if islogical(Ch_Err)
            if Ch_Err % CSE with sigma_e = sigma_n
                sigma_e = sigma_n ;
            else      % no CSE
                sigma_e = 0 ;
            end
        elseif length(Ch_Err)==1 % CSE with constant sigma_e
            sigma_e     = Ch_Err ;
```

```
        else                          % CSE with sigma_e given as a matrix
            if length(Ch_Err) == length(SNR_Vector)
                sigma_e = Ch_Err(snr) ;
            else
             error('sigma_e and SNR_Vector has to be the same length')
            end
        end

        % Generating CSE noise
        % Note, for no CSE sigma_e=0, and consequently, e will be an
        % Nr*Nt square matrix
        e = sqrt(sigma_e).*(randn(Nr,Nt)+1j*randn(Nr,Nt))./sqrt(2) ;

        % Channel with CSE
        % Note, in the case of no CSE, e is an all zeros matrix.
        % Hence, H_CSE_SC = H_SC
        H_CSE_SC = H_SC - e ;

        % Performing the ML receiver
        SMT_ML_Bin_Res = SMT_ML_Receiver(y,H_CSE_SC,SMT_Cons_Diagram);

        % Calculating the BER
        BER_SMT_ML = sum(dec2bin(PreGenData,eta)~=SMT_ML_Bin_Res)/eta;

        ABER_Sim(snr) = ABER_Sim(snr) + BER_SMT_ML ;

    end

end

% Averaging to get the ABER
ABER_Sim = ABER_Sim / NumIterations ;

end
```

A.3.7 Analytical ABER of DSMTs over Generalized Fading Channels

```
function ABER_Sim = ABER_DSMT_Monte_Carlo_Simulation(varargin)
% =================================================================
% Book title: Space Modulation Techniques
% Authors:    Raed Mesleh and Abdelhamid Alhassi
% Publisher:  John Wiley & Sons, Ltd
% Date:       2017
% =================================================================
%
% Description
%  This script calculates the average bit error ratio (ABER) of
%  differential space modulation techniques (DSMTs) using Monte Carlo
%  simulations, over generalized fading channel and in the presence of
%  channel estimation errors (CSEs).
%
% Inputs
```

```
%  DSMT_SyS            A string indicating the SMT system to calculate
%                      the analytical ABER for:
%                         - 'DSSK' for differential space shift keying;
%                         - 'DSM'  for differential spatial modulation.
%
%  Nt                  Nt is the number of transmit antennas.
%
%  M                   Size of the signal constellation diagram.
%                      Note, for DSSK the value of M does not matter.
%
%  Nr                  Number of receive antennas.
%
%  Ch_Type             A string indicating the type of channel to
%                      calculate the ABER over:
%                         - 'rayleigh' or 'ray'  for Rayleigh;
%                         - 'rician'   or 'rice' for Rician;
%                         - 'nakagami' or 'nak'  for Nakagami.
%  K                   Rician K factor in dB.
%  m                   Shape parameter of the Nakagami-m fading channel.
%
%
%  SNR_Vector          Vector containing the signal-to-noise ratio (SNR)
%                      values to calculate the ABER over.
%
%  NumIterations       Number of simulation iterations.
%
% Output
%  ABER_Sim            The calculated ABER of the chosen DSMT.
%
% Usage
%
%  For Rayleigh
%  ABER_Sim = DSMT_ABER_Monte_Carlo_Simulation(SMT_SyS,M,Nt,Nr, ...
%                                         SNR_Vector, NumIterations)
%
%  For Rician with K factor
%  ABER_Sim = DSMT_ABER_Monte_Carlo_Simulation(SMT_SyS,M,Nt,Nr, ...
%                                 Ch_Type,K,SNR_Vector,NumIterations)
%
%  For Nakagami-m with m shaping parameter
%  ABER_Sim = DSMT_ABER_Monte_Carlo_Simulation(SMT_SyS,M,Nt,Nr, ...
%                                 Ch_Type,m,SNR_Vector,NumIterations)

%% Defining input parameters
switch nargin

    case 6     % No correlation Rayleigh fading
        DSMT_SyS = lower(varargin{1}) ;

        M        = varargin{2} ;

        Nt       = varargin{3} ;

        Nr       = varargin{4} ;
```

```
        Ch_Type = 'rayleigh' ;

        SNR_Vector = varargin{5} ;

        NumIterations = varargin{6} ;

    case 8      % No correlation Rician or Nakagami-m

        DSMT_SyS = lower(varargin{1}) ;

        M        = varargin{2} ;

        Nt       = varargin{3} ;

        Nr       = varargin{4} ;

        Ch_Type = lower(varargin{5}) ;

        if strcmp(Ch_Type,'rician') || strcmp(Ch_Type,'rice')
            KdB = varargin{6} ;
            K   = 10^(KdB/10) ;
        elseif strcmp(Ch_Type,'nak') || strcmp(Ch_Type,'nakagami')
            m   = varargin{6} ;
        else
            error('Unknown Channel')
        end

        SNR_Vector     = varargin{7} ;

        NumIterations = varargin{8} ;

    otherwise

        error('Not enough or too many inputs')

end

%% Generating the SMT constellation matrix
DSMT_Cons_Diagram = DSMT_Cons_Diagram_Gen(DSMT_SyS,Nt,M) ;

% The spectral efficiency
eta = log2(size(DSMT_Cons_Diagram,3)) ;

%% Monte Carlo ABER Simulation

% Preallocate variables to store the calculated ABER
ABER_Sim = zeros( length(SNR_Vector) , 1 ) ;

% The coherence time
CohTime  = Nt ;

for snr = 1 : length(SNR_Vector)
```

```
for itr = 1 : NumIterations

    % Generating the channel
    switch Ch_Type
        case {'ray','rayleigh'}
            H = ( randn(Nr,Nt) + 1j.*randn(Nr,Nt) ) / sqrt(2) ;

        case {'rice','rician'}
            H = ( randn(Nr,Nt) + 1j.*randn(Nr,Nt) ) / sqrt(2) ;
            H = sqrt(K/(1+K))*ones(size(H)) + sqrt(1/(1+K))*H ;

        case {'nak','nakagami'}
            H = sqrt(sum(abs(randn(Nr,Nt,m)/sqrt(2*m)).^2,3)) ...
             + 1j .* sqrt(sum(abs(randn(Nr,Nt,m)/sqrt(2*m)).^2,3));
    end

    for timeSlot = 0 : CohTime

        % Bit stream to be transmitted already separated in bit
        % blocks to be used for determining the antenna index and
        % the rest to be modulated.

        % At the first time slot no information is transmitted,
        % where the information is transmitted in the next Nt slots
        if timeSlot == 0
            Xt = eye(Nt) ;

        else
            % Generate the data symbol to be transmitted
            PreGenData = randi([0 (2^eta -1)]);

            % DSM modulation
            Xt = Xt1*squeeze(DSMT_Cons_Diagram(:,:,PreGenData+1));

        end

        % Generating the noise block
        sn       = 1 / 10^( SNR_Vector(snr) / 10 ) ;

        noise    = sqrt(sn)*(randn(Nr,Nt)+1j*randn(Nr,Nt))/sqrt(2);

        % The received signal block
        Yt       = H * Xt + noise ;

        % At the first time slot no information is transmitted,
        % where the information is transmitted in the next Nt slots
        if timeSlot ~= 0
            % Performing the ML receiver
            ML_Bin = DSMT_ML_Receiver(Yt,Yt1,DSMT_Cons_Diagram);

            % Calculating the BER
            BER_ML = sum(dec2bin(PreGenData,eta) ~= ML_Bin) / eta;

            ABER_Sim(snr) = ABER_Sim(snr) + BER_ML ;
```

```
            end

            Xt1 = Xt ;
            Yt1 = Yt ;

        end

    end

end

% Averaging to get the ABER
ABER_Sim = ABER_Sim / (CohTime * NumIterations );

end
```

A.3.8 Simulated ABER of DSMTs Using Monte Carlo Simulation over Generalized Fading Channels

```
function ABER_Ana = ABER_DSMT_Analytical_General(varargin)
% ================================================================
% Book title: Space Modulation Techniques
% Authors:    Raed Mesleh and Abdelhamid Alhassi
% Publisher:  John Wiley & Sons, Ltd
% Date:       2017
% ================================================================
%
% Description
%  This script calculates the average bit error ratio (ABER) of
%  differential space modulation techniques (DSMTs) using Monte Carlo
%  simulations, over generalized fading channel and in the presence of
%  channel estimation errors (CSEs).
%
% Inputs
%  DSMT_SyS          A string indicating the SMT system to calculate
%                    the analytical ABER for:
%                       - 'DSSK' for differential space shift keying;
%                       - 'DSM'  for differential spatial modulation.
%
%  Nt                Nt is the number of transmit antennas.
%
%  M                 Size of the signal constellation diagram.
%                    Note, for DSSK, the value of M does not matter.
%
%  Nr                Number of receive antennas.
%
%  Ch_Type           A string indicating the type of channel to
%                    calculate the ABER over:
%                       - 'rayleigh' or 'ray'  for Rayleigh;
%                       - 'rician'   or 'rice' for Rician;
%                       - 'nakagami' or 'nak'  for Nakagami.
```

```
%  K                   Rician K factor in dB.
%  m                   Shape parameter of the Nakagami-m fading channel.
%
%
%  SNR_Vector          Vector containing the signal-to-noise ratio (SNR)
%                      values to calculate the ABER over.
%
%
% Output
%  ABER_Sim            The calculated ABER of the chosen DSMT.
%
% Usage
%
%  For Rayleigh
%  ABER_Sim = DSMT_ABER_Monte_Carlo_Simulation(SMT_SyS,M,Nt,Nr,  ...
%                                        SNR_Vector,NumIterations)
%
%  For Rician with K factor
%  ABER_Sim = DSMT_ABER_Monte_Carlo_Simulation(SMT_SyS,M,Nt,Nr,  ...
%                                 Ch_Type,K,SNR_Vector,NumIterations)
%
%  For Nakagami-m with m shaping parameter
%  ABER_Sim = DSMT_ABER_Monte_Carlo_Simulation(SMT_SyS,M,Nt,Nr,  ...
%                                 Ch_Type,m,SNR_Vector,NumIterations)

%% Defining input parameters
switch nargin

    case 5       % No correlation Rayleigh fading
        DSMT_SyS = lower(varargin{1}) ;

        M        = varargin{2} ;

        Nt       = varargin{3} ;

        Nr       = varargin{4} ;

        Ch_Type = 'rayleigh' ;

        SNR_Vector = varargin{5} ;


    case 7       % No correlation Rician or Nakagami-m

        DSMT_SyS = lower(varargin{1}) ;

        M        = varargin{2} ;

        Nt       = varargin{3} ;

        Nr       = varargin{4} ;

        Ch_Type = lower(varargin{5}) ;
```

```
        if strcmp(Ch_Type,'rician') || strcmp(Ch_Type,'rice')
            KdB = varargin{6} ;
            K   = 10^(KdB/10) ;
        elseif strcmp(Ch_Type,'nak') || strcmp(Ch_Type,'nakagami')
            m   = varargin{6} ;
        else
            error('Unknown Channel')
        end

        SNR_Vector      = varargin{7} ;


    otherwise

        error('Not enough or too many inputs')

end

%% Generating the SMT constellation matrix
DSMT_Cons_Diagram = DSMT_Cons_Diagram_Gen(DSMT_SyS,Nt,M) ;

%% Calculating the mean and the variance of the chosen channel
switch Ch_Type

    case {'rice','rician'} % Rician
        % The mean vector of the Rician fading channel including SC
        u  =  eye(Nr*Nt*Nt) * sqrt(K/(1+K)) * ones(Nr*Nt*Nt,1)   ;

        % The covariance matrix of the Rician fading channel
        % including SC
        LH = (1/(1+K)) * eye(Nr*Nt*Nt) ;

    case {'nak','nakagami'} % Nakagami-m
        % The mean vector of the Nakagami-m fading channel
        u_H  =  (gamma((m/2)+0.5)/gamma((m/2))) * sqrt(1/(2*(m/2)))...
             +1j*(gamma((m/2)+0.5)/gamma((m/2))) * sqrt(1/(2*(m/2)));

        % The mean vector of the Nakagami-m fading channel including SC
        u     = eye(Nr*Nt*Nt) *  u_H  * ones(Nr*Nt*Nt,1)   ;

        % The variance of the Nakagami-m fading channel
        variNak = ( 1 - 2 * ((gamma((m/2)+0.5)/gamma((m/2))) ...
                                        * sqrt(1/(2*(m/2))))^2 ) ;

        % The covariance matrix of the Nakagami-m fading channel
        LH       = variNak * eye(Nr*Nt*Nt) ;

end

%% Monte Carlo ABER Simulation
```

```
% Preallocate variables to store the calculated ABER
ABER_Ana = zeros( length(SNR_Vector) , 1 ) ;

% The spectral efficiency
eta = log2(size(DSMT_Cons_Diagram,3)) ;

for snr = 1 : length(SNR_Vector)

    ABER = 0 ;

    for i_t = 1 : 2^eta

        x_t         = DSMT_Cons_Diagram(:,:,i_t) ;

        sigma_n     = (1 / 10^( SNR_Vector(snr) / 10 )) ;

        varphi      =  1 / (sigma_n*2*sqrt(2)) ;

        for i = 1 : 2^eta

            x       = DSMT_Cons_Diagram(:,:,i) ;

            Psi     = x_t - x ;

            Lambda = kron(eye(Nr),Psi*Psi');

            % Calculating the PEP for this event
            switch Ch_Type
                case {'ray','rayleigh'}%Note, Rayleigh has a zero mean
                    PEP = (1/2)*1/det(eye(Nt*Nr)+(varphi/(2))*Lambda);

                otherwise
                    PEP = (1/2)*exp( -1 * (varphi/2) * u' * Lambda ...
                  *(eye(Nr*Nt)+(varphi/sqrt(2))*LH* Lambda)^(-1)*u)...
                        / det(eye(Nt*Nr)+(varphi/sqrt(2))*LH*Lambda) ;
            end

            ABER = ABER + (biterr(i_t-1,i-1,eta)/eta) *  PEP ;

        end

    end

    % Averaging to get the ABER
    ABER_Ana(snr) = ABER / (2^eta) ;

end

end
```

A.4 Mutual Information and Capacity

A.4.1 SMTs Simulated Mutual Information over Generalized Fading Channels and with CSE

```
function I = I_SMT(varargin)
% ======================================================================
% Book title: Space Modulation Techniques
% Authors:    Raed Mesleh and Abdelhamid Alhassi
% Publisher:  John Wiley & Sons, Ltd
% Date:       2017
% ======================================================================
%
% Description
%  This script calculates the simulated mutual information of space
%  modulation techniques (SMTs) over generalized fading channel and in
%  the presence of channel estimation errors (CSEs)
%
%
% Inputs
%  SMT_SyS             A string indicating the SMT system to calculate
%                      the analytical ABER for:
%                         - 'SSK'    for space shift keying (SSK);
%                         - 'SM'     for spatial modulation (SM);
%                         - 'QSSK'   for quadrature SSK (QSSK);
%                         - 'QSM'    for quadrature SM (QSM);
%                         - 'GSSK'   for generalized SSK (GSSK);
%                         - 'GSM'    for generalized SM (GSM);
%                         - 'GQSSK'  for generalized QSSK (GQSSK);
%                         - 'GQSM'   for generalized QSM (GQSM).
%
%  M                   Size of the signal constellation diagram.
%                      Note, for SSK the value of M does not matter.
%
%  Nt                  For SMTs: Nt is a single element indicating the
%                                number of transmit antennas.
%                      For GSMTs and GQSMTs: Nt is a two element vector:
%                      - The first element, Nt(1), is the number of
%                        transmit antennas;
%                      - The second element, Nt(2), is the number of
%                        active transmit antennas at a time (nu)
%
%  Nr                  Number of receive antennas.
%
%  Ch_Type             A string indicating the type of channel to
%                      calculate the ABER over:
%                         - 'rayleigh' or 'ray'  for Rayleigh;
%                         - 'rician'   or 'rice' for Rician;
%                         - 'nakagami' or 'nak'  for Nakagami.
%  K                   Rician K factor in dB.
%  m                   Shape parameter of the Nakagami-m fading channel.
%
%  Ch_Err              - Logical False for no CSE.
```

```
%                      - Logical True for CSE and sigma_e = sigma_n.
%                      - a single double containing sigma_e.
%                      - a vector containing a variable sigma_e.
%                        Note, the vector has to have the same length as
%                        the range of the SNR defined next.
%
%   SNR_Vector         Vector containing the signal-to-noise ratio (SNR)
%                      values to calculate the ABER over.
%
%   NumItr             Number of simulation iterations.
%
% Output
%   I                  The simulated mutual information.
%
% Usage
%
%   For Rayleigh
%   I = I_SMT(SMT_SyS,M,Nt,Nr,Ch_Err,SNR_Vector,NumItr)
%
%
%   For Rician with K factor
%   I = I_SMT(SMT_SyS,M,Nt,Nr,Ch_Type,K,Ch_Err,SNR_Vector,NumItr)
%
%
%   For Nakagami-m with m shaping parameter
%   I = I_SMT(SMT_SyS,M,Nt,Nr,Ch_Type,m,Ch_Err,SNR_Vector,NumItr)

%% Defining input parameters
switch nargin

    case 7      % Rayleigh fading
        SMT_SyS = lower(varargin{1}) ;

        M       = varargin{2} ;

        Nt      = varargin{3} ;
        if ~strcmp(SMT_SyS(1),'g') && length(Nt)>1
            error(' For SMTs Nt should be a single element')
        end

        Nr      = varargin{4} ;

        Ch_Type = 'rayleigh' ;

        Ch_Err  = varargin{5} ;

        SNR_Vector = varargin{6} ;

        NumItr = varargin{7} ;

    case 8      % Rayleigh fading
        SMT_SyS = lower(varargin{1}) ;

        M       = varargin{2} ;
```

```
        Nt       = varargin{3} ;
        if ~strcmp(SMT_SyS(1),'g') && length(Nt)>1
            error(' For SMTs Nt should be a single element')
        end

        Nr       = varargin{4} ;

        Ch_Type = lower(varargin{5}) ;
        if ~strcmp(Ch_Type,'rayleigh') || ~strcmp(Ch_Type,'ray')
            error('Unknown Channel or missing channel parameter')
        end

        Ch_Err   = varargin{6} ;

        SNR_Vector = varargin{7} ;

        NumItr = varargin{8} ;


    case 9      % Rician or Nakagami-m
        SMT_SyS = lower(varargin{1}) ;

        M        = varargin{2} ;

        Nt       = varargin{3} ;
        if ~strcmp(SMT_SyS(1),'g') && length(Nt)>1
            error(' For SMTs Nt should be a single element')
        end

        Nr       = varargin{4} ;

        Ch_Type = lower(varargin{5}) ;

        if strcmp(Ch_Type,'rician') || strcmp(Ch_Type,'rice')
            KdB = varargin{6} ;
            K   = 10^(KdB/10) ;
        elseif strcmp(Ch_Type,'nak') || strcmp(Ch_Type,'nakagami')
            m   = varargin{6} ;
        else
            error('Unknown Channel')
        end

        Ch_Err   = varargin{7} ;

        SNR_Vector = varargin{8} ;

        NumItr = varargin{9} ;

    otherwise

        error('Not enough or too many inputs')

end
```

```
%% Generating the SMT constellation matrix
SMT_CD = SMT_Cons_Diagram_Gen(SMT_SyS,Nt,M) ;

% The spectral efficiency
eta      = log2(size(SMT_CD,2)) ;

% In case of GSMTs Nu is not needed anymore
Nt       = Nt(1) ;

%% Simulation to calculate the Mutual Information

% Preallocate variables to store the calculated ABER
I       = zeros(size(SNR_Vector)) ;

% The range of the noise variance
sig_n = 1 ./ 10.^( SNR_Vector / 10 ) ;

% CSE
if islogical(Ch_Err)
    if Ch_Err % CSE with sigma_e = sigma_n
        sig_e = sig_n ;
    else      % no CSE
        sig_e = zeros(size(sig_n)) ;
    end
elseif length(Ch_Err)==1 % CSE with constant sigma_e
    sig_e      = Ch_Err * ones(size(sig_n)) ;
else                     % CSE with sigma_e given as a matrix
    if length(Ch_Err) == length(SNR_Vector)
        sig_e = Ch_Err ;
    else
        error('sigma_e and SNR_Vector has to be the same length')
    end
end

for snr = 1 : length(SNR_Vector)

    for itr = 1 : NumItr

        % Generate the data symbol to be transmitted
        PreGenData = randi([0 (2^eta -1)]);

        % SMT modulation
        xt          = SMT_CD(:,PreGenData+1) ;

        % Passing through channel
        switch Ch_Type
          case {'ray','rayleigh'}
          H_CSE = ( randn(Nr,Nt) + 1j.*randn(Nr,Nt) ) / sqrt(2) ;

          case {'rice','rician'}
          H_CSE = ( randn(Nr,Nt) + 1j.*randn(Nr,Nt) ) / sqrt(2) ;
          H_CSE = sqrt(K/(1+K))*ones(size(H_CSE))+sqrt(1/(1+K))*H_CSE;
```

```
            case {'nak','nakagami'}
            H_CSE = sqrt(sum(abs(randn(Nr,Nt,m)/sqrt(2*m)).^2,3)) ...
                    +1j.*sqrt(sum(abs(randn(Nr,Nt,m)/sqrt(2*m)).^2,3));
        end

        % Generating CSE noise
        % Note, for no CSE sigma_e=0, and consequently, e will be an
        % Nr*Nt square matrix
        e    = sqrt(sig_e(snr))*(randn(Nr,Nt)+1j*randn(Nr,Nt))/sqrt(2);

        % Channel without CSE
        % Note, in the case of no CSE, e is an all zeros matrix.
        % Hence, H = H_CSE
        H       = H_CSE + e ;

        % Generating the noise
        noise = sqrt(sig_n(snr))*(randn(Nr,1)+1j*randn(Nr,1))/sqrt(2);

        % The received signal
        y       = H * xt + noise ;

        % Calculating the Mutual Information

        sg_y    = (sig_n(snr)+sig_e(snr)*sum(abs(SMT_CD).^2)) ;

        EucDis = sum(abs(repmat(y,1,2^eta)-H_CSE*SMT_CD).^2,1) ;
        I(snr) = I(snr) - log2(mean(exp(-EucDis./sg_y)./sg_y.^Nr)) ...
                        - mean(Nr*log2(sg_y.*exp(1))) ;

    end

end

I = I / NumItr ;

end
```

A.4.2 SMTs Capacity

```
function Capacity = Capacity_SMT(varargin)
% ================================================================
% Book title: Space Modulation Techniques
% Authors:    Raed Mesleh and Abdelhamid Alhassi
% Publisher:  John Wiley & Sons, Ltd
% Date:       2017
% ================================================================
%
% Description
%  This script calculates the capacity for any space modulation
%  technique (SMT)
%
% Inputs
```

```
%   SMT_SyS             A string indicating the SMT system to calculate
%                       the analytical ABER for:
%                          - 'SSK'    for space shift keying (SSK);
%                          - 'SM'     for spatial modulation (SM);
%                          - 'QSSK'   for quadrature SSK (QSSK);
%                          - 'QSM'    for quadrature SM (QSM);
%                          - 'GSSK'   for generalized SSK (GSSK);
%                          - 'GSM'    for generalized SM (GSM);
%                          - 'GQSSK'  for generalized QSSK (GQSSK);
%                          - 'GQSM'   for generalized QSM (GQSM).
%
%   Nr                  Number of receive antennas.
%
%   Ch_Err              - Logical False for no CSE.
%                       - Logical True for CSE and sigma_e = sigma_n.
%                       - a single double containing sigma_e.
%                       - a vector containing a variable sigma_e.
%                         Note, the vector has to have the same length as
%                         the range of the SNR defined next.
%
%   SNR_Vector          Vector containing the signal-to-noise ratio (SNR)
%                       values to calculate the ABER over.
%
% Output
%   Capacity            SMT capacity.
%
% Usage
%   Capacity of SMT with no CSE
%   Capacity = Capacity_SMT(Nr,SNR_Vector)
%
%   Capacity of SSK with CSE
%   Capacity = Capacity_SMT(Nr,Ch_Err,SNR_Vector)
%
%   Capacity of SMT with CSE
%   Capacity = Capacity_SMT(SMT_sys,Nr,Ch_Err,SNR_Vector)

%%
switch nargin
    case 2 % SMT no CSE
        Nr         = varargin{1} ;
        SNR_Vector = varargin{2} ;
        Ch_Err     = false       ;
    case 3 % SSK with CSE
        Nr         = varargin{1} ;
        Ch_Err     = varargin{2} ;
        SNR_Vector = varargin{3} ;
        SMT_sys    = 'ssk' ;
    case 4 % SMT with CSE
        SMT_sys    = lower(varargin{1}) ;
        Nr         = varargin{2} ;
        Ch_Err     = varargin{3} ;
        SNR_Vector = varargin{4} ;
end
```

```
%%

sigma_n = 1 ./ 10.^(SNR_Vector/10) ;

if islogical(Ch_Err)
    if Ch_Err % CSE with sigma_e = sigma_n
        sigma_e = sigma_n ;
    else      % no CSE
        sigma_e = 0 ;
    end
elseif length(Ch_Err)==1 % CSE with constant sigma_e
    sigma_e     = Ch_Err * ones(size(sigma_n)) ;
else                     % CSE with sigma_e given as a matrix
    if length(Ch_Err) == length(SNR_Vector)
        sigma_e = Ch_Err ;
    else
        error('sigma_e and SNR_Vector has to be the same length')
    end
end

if sigma_e == 0

    Capacity   = Nr * log2( 1 +  10.^(SNR_Vector/10)) ;

else

    switch SMT_sys

        case 'ssk'
            S = 1 ;

        otherwise
            % Generating the M-QAM signal constellation symbols

            % Constructing an M-QAM modulator
            M = 2^17;
            alphabit = modem.qammod('M', M ) ;
            % Generating signal constellation diagram
            S        = modulate( alphabit , (0:M-1)).';
            % Calculating average power of the signal  diagram
            S_avg    = S'*S / M ;
            % Normalizing the power of the signal diagram
            S        = S / sqrt(S_avg);

    end

    Capacity = zeros(size(SNR_Vector)) ;
    for snr = 1 : length(SNR_Vector)
     Capacity(snr) = mean(Nr * log2((1+sigma_e(snr)+sigma_n(snr)) ...
                     ./(sigma_n(snr)+sigma_e(snr)*(abs(S).^2)))) ;
    end

end
```

References

1 Marconi, G. (1897). Improvements in transmitting electrical impulses and signals, and in apparatus therefor. British Patent No. 12039. Date of application, 2 June 1896; Complete specification, 2 March 1897.

2 Barrett, R. (1997). Popov versus Marconi: the century of radio. *GEC Review* **12** (2): 107–116.

3 Shannon, C. (1948). A mathematical theory of communication. *Bell System Technical Journal* **27**: 379–423 and 623–656.

4 Tse, D. and Viswanath, P. (2005). *Fundamentals of Wireless Communication*. Cambridge University Press.

5 Billström, O., Cederquist, L., Ewerbring, M. et al.(2006). Fifty years with mobile phones – from novelty to no. 1 consumer product. *Ericsson Review* (3): 101–106.

6 Padgett, J.E., Gunther, C.G., and Hattori, T. (1995). Overview of wireless personal communications. *IEEE Communications Magazine* **33** (1): 28–41. doi: 10.1109/35.339877.

7 Erdman, W. (1993). Wireless communications: a decade of progress. *IEEE Communications Magazine* **31** (12): 48–51. doi: 10.1109/35.247957.

8 Steele, R. and Hanzo, L. (1999). *Mobile Radio Communications: Second and Third Generation Cellular and WATM Systems*. 2e. IEEE Press - Wiley.

9 Sarkar, T.K., Mailloux, R.J., Oliner, A.A. et al. (2006). *History of Wireless*. Wiley and Hoboken, NJ.

10 Halonen, T., Romero, J., and Melero, J. (2003). *GSM, GPRS and EDGE Performance*, 2e. Wiley.

11 Cox, D. (1995). Wireless personal communications: what is it? *IEEE Personal Communications* **2** (2): 20–35.

12 Rizzo, J. and Sollenberger, N. (1995). Multitier wireless access. *IEEE Personal Communications* **2** (3): 18–30.

13 ITU-R (2003). Framework and Overall Objectives of the Future Development of IMT-2000 and Systems Beyond IMT-2000. Tech. Rep. ITU-R M.1645. ITU.

Space Modulation Techniques, First Edition. Raed Mesleh and Abdelhamid Alhassi.

14 Schulze, H. and Lüders, C. (2005). *Theory and Applications of OFDM and CDMA*. Wiley.
15 Viterbi, A.J. and Padovani, R. (1992). Implications of mobile cellular CDMA. *IEEE Communications Magazine* **30** (12): 38–41. doi: 10.1109/35.210354.
16 Correia, L. (2006). *Mobile Broadband Multimedia Networks: Techniques, Models and Tools for 4G*. Elsevier Ltd.
17 Agilent Technologies (2009). *LTE and the Evolution to 4G Wireless: Design and Measurement Challenges*. Agilent Technologies.
18 ITU-R (2008). Requirements Related to Technical Performance for IMT-Advanced Radio Interface(s). Tech. Rep. ITU-R M.2134. ITU.
19 CISCO (2016). Cisco visual networking index: global mobile data traffic forecast update, 2015–2020, White paper, CISCO.
20 ITU-R (2015). Framework and Overall Objectives of the Future Development of IMT for 2020 and Beyond. Tech. Rep. ITU-R M.2083. ITU.
21 ITU-R (2014). Future Technology Trends of Terrestrial IMT Systems. Tech. Rep. *ITU-R M.2320*. ITU.
22 Winters, J.H. (1984). Optimum combining in digital mobile radio with cochannel interference. *IEEE Transactions on Vehicular Technology* **33** (3): 144–155. doi: 10.1109/T-VT.1984.24001.
23 Salz, J. (1985). Digital transmission over cross-coupled linear channels. *AT & T Technical Journal* **64**: 1147–1159.
24 Alamouti, S.M. (1998). A simple transmit diversity technique for wireless communications. *IEEE Journal on Selected Areas in Communications* **16** (8): 1451–1458. doi: 10.1109/49.730453.
25 Telatar, E. (1999). Capacity of multi-antenna Gaussian channels. *European Transactions on Telecommunications* **10** (6): 585–595.
26 Foschini, G.J. and Gans, M.J. (1998). On limits of wireless communications in a fading environment when using multiple antennas. *Wireless Personal Communications,* **1.6**, 311–335.
27 Foschini, G.J. (1996). Layered space-time architecture for wireless communication in a fading environment when using multi-element antennas. *Bell Labs Technical Journal* **1** (2): 41–59.
28 Zhang, Y.J. and Letaief, K. (2005). Adaptive resource allocation for multiaccess MIMO/OFDM systems with matched filtering. *IEEE Transactions on Communications* **53** (11): 1810–1816.
29 Winters, J. (1987). On the capacity of radio communication systems with diversity in a Rayleigh fading environment. *IEEE Journal on Selected Areas in Communication* **5** (5): 871–878.
30 Wittneben, A. (1991). Basestation modulation diversity for digital SIMULCAST. IEEE 41st Vehicular Technology Conference, 1991. 'Gateway to the Future Technology in Motion', pp. 848–853.

31 Winters, J. (1998). The diversity gain of transmit diversity in wireless systems with Rayleigh fading. *IEEE Transactions on Vehicular Technology* **47** (1): 119–123.

32 Tarokh, V., Seshadri, N., and Calderbank, A. (1998). Space-time codes for high data rate wireless communication: performance criterion and code construction. *IEEE Transactions on Information Theory* **44** (2): 744–765.

33 Goldsmith, A., Jafar, S., Jindal, N., and Vishwanath, S. (2003). Capacity limits of MIMO channels. *IEEE Journal on Selected Areas in Communication* **21** (5): 684–702.

34 Damen, M., Abdi, A., and Kaveh, M. (2001). On the effect of correlated fading on several space-time coding and detection schemes. Proceedings of the 2001 IEEE 54th Vehicular Technology Conference, vol. 1, Atlantic City, NJ, pp. 13–16.

35 Haas, H., Costa, E., and Schulz, E. (2002). Increasing spectral efficiency by data multiplexing using antenna arrays. Proceedings of the 13th IEEE International Symposium on Personal, Indoor and Mobile Radio Communications (PIMRC), vol. 2, pp. 610–613.

36 Mesleh, R., Haas, H., Ahn, C.W., and Yun, S. (2006). Spatial modulation – a new low complexity spectral efficiency enhancing technique. IEEE International Conference on Communication and Networking in China (CHINACOM), Beijing, China, pp. 1–5.

37 Mesleh, R., Haas, H., Sinanović, S. et al. (2008). Spatial modulation. *IEEE Transactions on Vehicular Technology* **57** (4): 2228–2241.

38 Younis, A., Sinanovic, S., Di Renzo, M. et al. (2013). Generalised sphere decoding for spatial modulation. *IEEE Transactions on Communications* **61** (7): 2805–2815. doi: 10.1109/TCOMM.2013.061013.120547.

39 Serafimovski, N., Younis, A., Mesleh, R. et al. (2013). Practical implementation of spatial modulation. *IEEE Transactions on Vehicular Technology* **62** (9): 4511–4523. doi: 10.1109/TVT.2013.2266619.

40 Mesleh, R., Hiari, O., Younis, A., and Alouneh, S. (2017). Transmitter design and hardware considerations for different space modulation techniques. *IEEE Transactions on Wireless Communications* **16** (11): 7512–7522.

41 Mesleh, R., Haas, H., Lee, Y., and Yun, S. (2005). Interchannel interference avoidance in MIMO transmission by exploiting spatial information. Proceedings of the 16th IEEE International Symposium on Personal, Indoor and Mobile Radio Communications (PIMRC), vol. 1, Berlin, Germany, pp. 141–145. doi: 10.1109/PIMRC.2005.1651415.

42 Mesleh, R., Haas, H., Ahn, C.W., and Yun, S. (2006). Spatial modulation – OFDM. Proceedings of the International OFDM Workshop, Hamburg, Germany.

43 Mesleh, R., Gansean, S., and Haas, H. (2007). Impact of channel imperfections on spatial modulation OFDM. IEEE 18th International Symposium

on Personal, Indoor and Mobile Radio Communications (PIMRC), Athens, Greece, pp. 1–5.
44 Mesleh, R., Engelken, S., Sinanović, S., and Haas, H. (2008). Analytical SER calculation of spatial modulation. IEEE International Symposium on Spread Spectrum Techniques and Applications (ISSSTA), Bologna, Italy.
45 Younis, A., Thompson, W., Renzo, M.D. et al. (2013). Performance of spatial modulation using measured real-world channels. Proceedings of the 78th IEEE Vehicular Technology Conference (VTC), Las Vegas, NV.
46 Younis, A., Basnayaka, D.A., and Haas, H. (2014). Performance analysis for generalised spatial modulation. Proceedings of European Wireless Conference (EW 2014), Barcelona, Spain, pp. 207–212.
47 Serafimovski, N., Sinanovic, S., Younis, A. et al. (2011). 2-user multiple access spatial modulation. IEEE GLOBECOM Workshops (GC Wkshps), Houston, TX, pp. 343–347. doi: 10.1109/GLOCOMW.2011.6162467.
48 Younis, A., Mesleh, R., and Haas, H. (2015). Quadrature spatial modulation performance over Nakagami–m fading channels. *IEEE Transactions on Vehicular Technology* **65** (12): 10227–10231. doi: 10.1109/TVT.2015.2478841.
49 Thompson, W., Beach, M., McGeehan, J. et al. (2011). Spatial modulation explained and routes for practical evaluation. European Cooperation in the Field of Scientific and Technical Research (COST), Lisbon, Portugal.
50 Thompson, W., Younis, A., Beach, M. et al. (2012). Initial investigations into the sensitivity of spatial modulation systems on subchannel correlation and power balances. IC 1004 TD(12)03046 Cost Meeting, EURO-COST, Barcelona, Spain, p. 9 pages.
51 Mesleh, R. and Younis, A. (2017). Capacity analysis for LOS millimeter–wave quadrature spatial modulation. *Wireless Networks*. doi: 10.1007/s11276-017-1444-y.
52 Younis, A., Mesleh, R., Renzo, M.D., and Haas, H. (2014). Generalised spatial modulation for large–scale MIMO. Proceedings of the 22nd European Signal Processing Conference (EUSIPCO), Lisbon, Portugal.
53 Elkawafi, S., Younis, A., Mesleh, R. et al. (2017). Spatial modulation and spatial multiplexing capacity analysis over 3D mmwave communications. 23th European Wireless Conference European Wireless, Dresden, Germany.
54 Elkawafi, S., Younis, A., Mesleh, R. et al. (2017). Spatial Modulation and Spatial Multiplexing Performance Comparison over 3D mmWave Communications. International Conference on Wireless Communications, Signal Processing and Networking (WiSPNET), Chennai, India.
55 Younis, A., Abuzgaia, N., Mesleh, R., and Haas, H. (2017). Quadrature spatial modulation for 5G outdoor millimeter–wave communications: capacity analysis. *IEEE Transactions on Wireless Communications* **16** (5): 2882–2890.

56 Mesleh, R. and Younis, A. (2016). LOS millimeter-wave communication with quadrature spatial modulation. IEEE International Symposium on Signal Processing and Information Technology (ISSPIT), pp. 109–113. doi: 10.1109/ISSPIT.2016.7886018.

57 Ganesan, S., Mesleh, R., Haas, H. et al. (2006). On the performance of spatial modulation OFDM. Asilomar Conference on Signals, Systems, and Computers, Pacific Grove, CA, pp. 1825–1829.

58 Serafimovski, N., Di Renzo, M., Sinanović, S. et al. (2010). Fractional bit encoded spatial modulation (FBE–SM). *IEEE Communications Letters* **14** (5): 429–431.

59 Renzo, M.D. and Haas, H. (2010). Performance analysis of Spatial Modulation. International ICST Conference on Communications and Networking in China (CHINACOM), pp. 1–7.

60 Basar, E. (2016). Index modulation techniques for 5G wireless networks. *IEEE Communications Magazine* **54** (7): 168–175. doi: 10.1109/MCOM.2016.7509396.

61 Di Renzo, M., Haas, H., and Grant, P.M. (2011). Spatial modulation for multiple-antenna wireless systems: a survey. *IEEE Communications Magazine* **49** (11): 182–191.

62 Di Renzo, M., Haas, H., Ghrayeb, A. et al. (2014). Spatial modulation for generalized MIMO: challenges, opportunities, and implementation. *Proceedings of the IEEE* **102** (1): 56–103. doi: 10.1109/JPROC.2013.2287851.

63 Bian, Y., Cheng, X., Wen, M. et al. (2015). Differential spatial modulation. *IEEE Transactions on Vehicular Technology* **64** (7): 3262–3268. doi: 10.1109/TVT.2014.2348791.

64 Datta, T., Eshwaraiah, H.S., and Chockalingam, A. (2016). Generalized space-and-frequency index modulation. *IEEE Transactions on Vehicular Technology* **65** (7): 4911–4924. doi: 10.1109/TVT.2015.2451095.

65 Mesleh, R., Ikki, S., and Aggoune, H. (2015). Quadrature spatial modulation. *IEEE Transactions on Vehicular Technology* **64** (6): 2738–2742. doi: 10.1109/TVT.2014.2344036.

66 Mesleh, R., Elgala, H., and Haas, H. (2011). Optical spatial modulation. *IEEE/OSA Journal of Optical Communications and Networking* **3** (3): 234–244. doi: 10.1364/JOCN.3.000234.

67 Younis, A., Serafimovski, N., Mesleh, R., and Haas, H. (2010). Generalised spatial modulation. Asilomar Conference on Signals, Systems, and Computers, Pacific Grove, CA.

68 Jeganathan, J., Ghrayeb, A., and Szczecinski, L. (2008). Spatial modulation: optimal detection and performance analysis. *IEEE Communications Letters* **12** (8): 545–547. doi: 10.1109/LCOMM.2008.080739.

69 Jeganathan, J., Ghrayeb, A., and Szczecinski, L. (2008). Generalized space shift keying modulation for MIMO channels. Proceedings of the IEEE 19th International Symposium on Personal, Indoor and

Mobile Radio Communications PIMRC 2008, Cannes, France, pp. 1–5. doi: 10.1109/PIMRC.2008.4699782.

70 Mesleh, R., Ikki, S.S., and Aggoune, H.M. (2017). Quadrature spatial modulation-performance analysis and impact of imperfect channel knowledge. *Transactions on Emerging Telecommunications Technologies* **28** (1): e2905. doi: 10.1002/ett.2905.

71 Mesleh, R. (2007). Spatial modulation: a spatial multiplexing technique for efficient wireless data transmission. PhD thesis. Bremen, Germany: Jacobs University.

72 Younis, A. (2014). Spatial modulation: theory to practice. PhD thesis. University of Edinburgh.

73 Badarneh, O.S. and Mesleh, R. (2016). A comprehensive framework for quadrature spatial modulation in generalized fading scenarios. *IEEE Transactions on Communications* **64** (7): 2961–2970. doi: 10.1109/TCOMM.2016.2571285.

74 Mesleh, R., Badarneh, O.S., Younis, A., and Almehmadi, F.S. (2016). How significant is the assumption of the uniform channel phase distribution on the performance of spatial multiplexing MIMO system? *Wireless Networks* **23** (7): 2281–2288. doi: 10.1007/s11276-016-1286-z.

75 Mesleh, R., Badarneh, O.S., Younis, A., and Haas, H. (2015). Performance analysis of spatial modulation and space-shift keying with imperfect channel estimation over generalized $\eta - \mu$ fading channels. *IEEE Transactions on Vehicular Technology* **64** (1): 88–96. doi: 10.1109/TVT.2014.2321059.

76 Maleki, M., Bahrami, H., Beygi, S. et al. (2013). Space modulation with CSI: constellation design and performance evaluation. *IEEE Transactions on Vehicular Technology* **62** (4): 1623–1634. doi: 10.1109/TVT.2012.2232686.

77 Mesleh, R., Ikki, S.S., and Almehmadi, F.S. (2016). Impact of IQ imbalance on the performance of QSM multiple-input-multiple-output system. *IET Communications* **10** (17): 2391–2395. http://digital-library.theiet.org/content/journals/10.1049/iet-com.2016.0631, early access.

78 Sugiura, S., Chen, S., and Hanzo, L. (2011). Generalized space-time shift keying designed for flexible diversity-, multiplexing- and complexity-tradeoffs. *IEEE Transactions on Wireless Communications* **10** (4): 1144–1153. doi: 10.1109/TWC.2011.012411.100065.

79 Sugiura, S., Chen, S., and Hanzo, L. (2010). A unified MIMO architecture subsuming space shift keying, OSTBC, BLAST and LDC. IEEE Vehicular Technology Conference Fall (VTC 2010-Fall), pp. 1–5. doi: 10.1109/VETECF.2010.5594145.

80 Sugiura, S., Chen, S., and Hanzo, L. (2010). Coherent and differential space-time shift keying: a dispersion matrix approach. *IEEE Transactions on Communications* **58** (11): 3219–3230. doi: 10.1109/TCOMM.2010.093010.090730.

81 Wang, J., Jia, S., and Song, J. (2012). Generalised spatial modulation system with multiple active transmit antennas and low complexity detection scheme. *IEEE Transactions on Wireless Communications* **11** (4): 1605–1615. doi: 10.1109/TWC.2012.030512.111635.
82 Yang, P., Xiao, Y., Li, L. et al. (2012). An improved matched-filter based detection algorithm for space-time shift keying systems. *IEEE Signal Processing Letters* **19** (5): 271–274. doi: 10.1109/LSP.2012.2190059.
83 Yang, Y. and Aissa, S. (2011). Information-guided transmission in decode-and-forward relaying systems: spatial exploitation and throughput enhancement. *IEEE Transactions on Wireless Communications* **10** (7): 2341–2351. doi: 10.1109/TWC.2011.050511.101094.
84 Zhang, J., Wang, Y., Ding, L., and Zhang, N. (2014). Bit error probability of spatial modulation over measured indoor channels. *IEEE Transactions on Wireless Communications* **13** (3): 1380–1387. doi: 10.1109/TWC.2014.012814.130562.
85 Mesleh, R., Elgala, H., Mehmood, R., and Haas, H. (2011). Performance of optical spatial modulation with transmitters-receivers alignment. *IEEE Communications Letters* **15** (1): 79–81. doi: 10.1109/LCOMM.2010.01.101208.
86 Younis, A., Mesleh, R., Haas, H., and Grant, P.M. (2010). Reduced complexity sphere decoder for spatial modulation detection receivers. 2010 IEEE Global Telecommunications Conference (GLOBECOM), Miami, FL, pp. 1–5. doi: 10.1109/GLOCOM.2010.5683993.
87 Younis, A., Di Renzo, M., Mesleh, R., and Haas, H. (2011). Sphere decoding for spatial modulation. Proceedings of IEEE International Conference on Communications (ICC), Kyoto, Japan, pp. 1–6. doi: 10.1109/icc.2011.5963484.
88 Di Renzo, M. and Haas, H. (2013). On transmit diversity for spatial modulation MIMO: impact of spatial constellation diagram and shaping filters at the transmitter. *IEEE Transactions on Vehicular Technology* **62** (6): 2507–2531. doi: 10.1109/TVT.2013.2244927.
89 Jeganathan, J., Ghrayeb, A., Szczecinski, L., and Ceron, A. (2009). Space shift keying modulation for MIMO channels. *IEEE Transactions on Wireless Communications* **8** (7): 3692–3703. doi: 10.1109/TWC.2009.080910.
90 Sugiura, S., Chen, S., and Hanzo, L. (2010). Space-time shift keying: a unified MIMO architecture. Global Telecommunications Conference (GLOBECOM 2010), 2010 IEEE, pp. 1–5. doi: 10.1109/GLOCOM.2010.5684112.
91 Sugiura, S., Chen, S., and Hanzo, L. (2012). A universal space-time architecture for multiple-antenna aided systems. *IEEE Communications Surveys Tutorials* **14** (2): 401–420. doi: 10.1109/SURV.2011.041911.00105.
92 Sugiura, S. (2011). Dispersion matrix optimization for space-time shift keying. *IEEE Communications Letters* **15** (11): 1152–1155. doi: 10.1109/LCOMM.2011.100611.111770.

93 Rajashekar, R., Hari, K., and Hanzo, L. (2013). Structured dispersion matrices from division algebra codes for space-time shift keying. *IEEE Signal Processing Letters* **20** (4): 371–374. doi: 10.1109/LSP.2013.2247997.

94 Başar, E. (2015). Multiple-input multiple-output OFDM with index modulation. *IEEE Signal Processing Letters* **22** (12): 2259–2263. doi: 10.1109/LSP.2015.2475361.

95 Basar, E., Wen, M., Mesleh, R. et al. (2017). Index modulation techniques for next-generation wireless networks. *IEEE Access* **5** (99): 1–52. doi: 10.1109/ACCESS.2017.2737528.

96 Wen, M., Cheng, X., and Yang, L. (2017). *Index Modulation for 5G Wireless Communications*. Springer.

97 Mesleh, R., Althunibat, S., and Younis, A. (2017). Differential quadrature spatial modulation. *IEEE Transactions Communications* **65** (9): 3810–3817. doi: 10.1109/TCOMM.2017.2712720.

98 Zhang, M., Wen, M., Cheng, X., and Yang, L.Q. (2015). Differential spatial modulation for dual-hop amplify-and-forward relaying. 2015 IEEE International Conference on Communications (ICC), pp. 1518–1523. doi: 10.1109/ICC.2015.7248539.

99 Elgala, H., Mesleh, R., Haas, H., and Pricope, B. (2007). OFDM visible light wireless communication based on white LEDs. Proceedings of the 64th IEEE Vehicular Technology Conference (VTC), Dublin, Ireland.

100 Mesleh, R., Mehmood, R., Elgala, H., and Haas, H. (2010). An overview of indoor OFDM/DMT optical wireless communication systems. IEEE International Symposium on Communication Systems, Networks and Digital Signal Processing (CSNDSP), Newcastle, UK, pp. 1–5, to appear.

101 Elgala, H., Mesleh, R., and Haas, H. (2011). Indoor optical wireless communication: potential and state-of-the-art. *IEEE Communications Magazine* **49** (9): 56–62. doi: 10.1109/MCOM.2011.6011734.

102 Mesleh, R., Helgala, H., and Haas, H. (2012). Performance analysis of indoor OFDM optical wireless communication systems. Proceedings of the Wireless Communications and Networking Conference (WCNC), IEEE, IEEE, Paris, France, to appear.

103 Mesleh, R., Mehmood, R., Elgala, H., and Haas, H. (2010). Indoor MIMO optical wireless communication using spatial modulation. IEEE International Conference on Communications (ICC), Cape Town, South Africa, pp. 1–5.

104 Fakidis, J., Tsonev, D., and Haas, H. (2013). A comparison between DCO-OFDMA and synchronous one-dimensional OCDMA for optical wireless communications. Proceedings of the IEEE 24th International Symposium on Personal Indoor and Mobile Radio Communications (PIMRC), London, UK, pp. 3605–3609.

105 Ijaz, M., Tsonev, D., Stavridis, A. et al. (2014). Optical spatial modulation OFDM using micro LEDs. 48th Asilomar Conference on Signals, Systems and Computers, pp. 1734–1738. doi: 10.1109/ACSSC.2014.7094764.

106 Abaza, M., Mesleh, R., Mansour, A., and el Hadi Aggoune (2015). Performance analysis of MISO multi-hop FSO links over log-normal channels with fog and beam divergence attenuations. *Optics Communications* **334**: 247–252. doi: http://dx.doi.org/10.1016/j.optcom.2014.08.050.

107 Dimitrov, S. and Haas, H. (2015). *Principles of LED Light Communications: Towards Networked Li-Fi.* Cambridge University Press.

108 Bouchet, O., Sizun, H., Boisrobert, C. et al. (2010). *Free-Space Optics: Propagation and Communication.* Wiley-ISTE.

109 Uysal, M., Capsoni, C., Ghassemlooy, Z. et al. (2016). *Optical Wireless Communications: An Emerging Technology.* Springer.

110 Ghassemlooy, Z., Popoola, W., and Rajbhandari, S. (2017). *Optical Wireless Communications: System and Channel Modelling with MATLAB.* CRC Press.

111 Paulraj, A., Nabar, R., and Gore, D. (2003). *Introduction to Space-Time Wireless Communications.* Cambridge University Press, UK.

112 Wolniansky, P., Foschini, G., Golden, G., and Valenzuela, R. (1998). *V-BLAST: An Architecture for Realizing very High Data Rates over the Rich-Scattering Wireless Channel.* Unino Radio-Scientifique Internationale (URSI) International Symposium on Signals, Systems, and Electronics (ISSSE), pp. 295–300.

113 Foschini, G., Chizhik, D., Gans, M. et al. (2003). Analysis and performance of some basic space-time architectures. *IEEE Journal on Selected Areas in Communications [Invited Paper]* **21** (3): 303–320.

114 Viterbo, E. and Boutros, J. (1999). A universal lattice code decoder for fading channels. *IEEE Transactions on Information Theory* **45** (5): 1639–1642.

115 Damen, M., Abed-Meraim, K., and Belfiore, J.C. (2000). Generalised sphere decoder for asymmetrical space-time communication architecture. *Electronics Letters* **36** (2): 166–167. doi: 10.1049/el:20000168.

116 Kühn, V. (2006). *Wireless Communications over MIMO Channels.* Wiley.

117 Gesbert, D., Shafi, M., shan Shiu, D. et al. (2003). From theory to practice: an overview of MIMO space–time coded wireless systems. *IEEE Journal on Selected Areas in Communications* **21** (3): 281–302.

118 Rappaport, T.S. (2002). *Wireless Communications: Principles and Practice,* 2e. Prentice Hall PTR.

119 Rappaport, T.S. (2001). *Wireless Communications: Principles and Practice,* 2e. Prentice Hall. ISBN: 0130422320.

120 Proakis, J.G. (2000). *Digital Communications,* 4e. McGraw–Hill, New York.

121 Jafarkhani, H. (2005). *Space-Time Coding: Theory and Practice*. Cambridge University Press.
122 Rayleigh, L. (1880). On the resultant of a large number of vibrations of the same pitch and of arbitrary phase. *Philosophical Magazine* **10**: 73–78.
123 Simon, M.K. and Alouini, M. (2005). *Digital Communication Over Fading Channels*, Wiley Series in Telecommunications and Signal Processing, 2e. Wiley. ISBN: 978-0-471-64953-3.
124 Rice, S.O. (1944). Mathematical analysis of random noise. *Bell System Technical* **23**: 282–332. http://adsabs.harvard.edu/cgi-bin/nph-bib_query?bibcode=1944BSTJ...23..282R.
125 Godavarti, M., Hero, A., and Marzetta, T. (2001). Min-capacity of a multiple-antenna wireless channel in a static rician fading environment. Proceedings of the IEEE International Symposium on Information Theory ISIT' 2001, p. 57.
126 Nakagami, M. (1960). The m-distribution – a general formula of intensity distribution of rapid fading. In: *Statistical Methods in Radio Wave Propagation* (ed. W.C. Hoffmann), 3–6, 6a, 7–36. Elmsford, NY: Pergamon Press.
127 Yacoub, M., Fraidenraich, G., and Santos Filho, J. (2005). Nakagami-m phase-envelope joint distribution. *Electronics Letters* **41** (5): 259–261. doi: 10.1049/el:20057014.
128 Yacoub, M. (2010). Nakagami-m phase-envelope joint distribution: a new model. *IEEE Transactions on Vehicular Technology* **59** (3): 1552–1557. doi: 10.1109/TVT.2010.2040641.
129 da Costa, D.B. and Yacoub, M.D. (2007). The $\eta - \mu$ joint phase-envelope distribution. *IEEE Antennas and Wireless Propagation Letters* **6**: 195–198.
130 Dias, U.S., Yacoub, M.D., and da Costa, D.B. (2008). The $\kappa - \mu$ phase-envelope joint distribution. Proceedings of IEEE PIMRC, pp. 1–5.
131 Yacoub, M.D. (2007). The $\kappa - \mu$ distribution and the $\eta - \mu$ distribution. *IEEE Antennas and Propagation Magazine* **49** (1): 68–81.
132 Abramowitz, M. and Stegun, I.A. (1972). *Handbook of Mathematical Functions with Fomulas, Graphs, and Mathematical Tables*, 9e. Dover Publications.
133 Papazafeiropoulos, A.K. and Kotsopoulos, S.A. (2009). The $\alpha - \mu$ joint envelope-phase fading distribution. Proceedings ot the IEEE PIMRC, pp. 919–922.
134 Yacoub, M.D. (2002). The $\alpha - \mu$ distribution: a general fading distribution. Proceedings of the IEEE PIMRC, pp. 629.
135 Paulraj, A.J. and Papadias, C.B. (1997). Space-time processing for wireless communications. *IEEE Signal Processing Magazine* **14** (6): 49–83. doi: 10.1109/79.637317.
136 Hottinen, A., Tirkkonen, O., and Wichman, R. (2003). *Multi-Antenna Tansceiver Techniques for 3G and Beyond*. Wiley.

137 Forenza, A., Love, D., and Heath, R. Jr. (2004). A low complexity algorithm to simulate the spatial covariance matrix for clustered MIMO channel models. IEEE Vehicular Technology Conference (VTC Fall 2004), vol. 2, Los Angeles, CA, USA, pp. 889–893.

138 G. T. S. Group (2003). Spatial Channel Model, Spatial Channel Model AHG (Combined ad-hoc from 3GPP and 3GPP2).

139 Saleh, A. and Valenzuela, R. (1987). A statistical model for indoor multipath propagation. *IEEE Journal on Selected Areas in Communications* **5** (2): 128–137.

140 Spencer, Q., Jeffs, B., Jensen, M., and Swindlehurst, A. (2000). Modeling the statistical time and angle of arrival characteristics of an indoor multipath channel. *IEEE Journal on Selected Areas in Communications* **18** (3): 347–360.

141 Erceg, V. et al. (2004). Retrieved January 12, 2007 from, TGn Channel Models, IEEE P802.11 Wireless LANs http://www.nari.ee.ethz.ch/dsbaum/11-03-0940-04-000n-tgn-channel-models.pdf (accessed 20 December 2017).

142 Lee, W.Y. (1973). Effects on correlation between two mobile radio base-station antennas. *IEEE Transactions on Vehicular Technology* **22** (4): 130–140.

143 Adachi, F., Feeney, M.T., Parsons, J.D., and Williamson, A.G. (1986). Cross-correlation between the envelopes of 900 mhz signals received at a mobile radio base station site. *IEE Proceedings F Communications, Radar and Signal Processing* **133** (6): 506–512. doi: 10.1049/ip-f-1.1986.0083.

144 Salz, J. and Winters, J. (1994). Effect of fading correlation on adaptive arrays in digital mobile radio. *IEEE Transactions on Vehicular Technology* **43** (4): 1049–1057.

145 Pedersen, K., Mogensen, P., and Fleury, B. (1998). Spatial channel characteristics in outdoor environments and their impact on BS antenna system performance. Proceedings of the 1998 48th IEEE Vehicular Technology Conference, vol. 2, pp. 719–723.

146 Hsu, H. (1995). *Signals and Systems, Schaum's Outline Series.* McGraw-Hill.

147 Shin, H. and Lee, J.H. (2003). Capacity of multiple-antenna fading channels: spatial fading correlation, double scattering, and keyhole. *IEEE Transactions on Information Theory* **49** (10): 2636–2647.

148 Svantesson, T. and Ranheim, A. (2001). Mutual coupling effects on the capacity of multielement antenna systems. Proceedings of the IEEE International Conference on Acoustics, Speech, and Signal Processing (ICASSP '01), vol. 4, Salt Lake City, UT, USA, pp. 2485–2488.

149 Gupta, I. and Ksienski, A. (1983). Effect of mutual coupling on the performance of adaptive arrays. *IEEE Transactions on Antennas and Propagation* **31** (5): 785–791.

150 Janaswamy, R. (2002). Effect of element mutual coupling on the capacity of fixed length linear arrays. *IEEE Antennas and Wireless Propagation Letters* **1**: 157–160.

151 Cho, Y.S., Kim, J., Yang, W.Y., and Kang, C.G. (2010). Channel estimation. In: *MIMO-OFDM Wireless Communications with MATLAB*. Wiley-IEEE Press, pp. 187–207. doi: 10.1002/9780470825631.ch6.

152 Yoo, T. and Goldsmith, A. (2004). Capacity of fading MIMO channels with channel estimation error. 2004 IEEE International Conference on Communications, vol. 2, pp. 808–813. doi: 10.1109/ICC.2004.1312613.

153 Mesleh, R., Althunibat, S., and Younis, A. (2017). Differential quadrature spatial modulation. *IEEE Transactions on Communications* **65** (9): 3810–3817. doi: 10.1109/TCOMM.2017.2712720.

154 Mesleh, R., Di Renzo, M., Haas, H., and Grant, P.M. (2010). Trellis coded spatial modulation. *IEEE Transactions on Wireless Communications* **9** (7): 2349–2361. doi: 10.1109/TWC.2010.07.091526.

155 Fague, D.E. (2016). New RF DAC Broadens Software-Defined Radio Horizon. Tech. Rep. 50. Analog Dialouge. http://www.analog.com/en/analog-dialogue/articles/new-rf-dac-broadens-sdr-horizon.html (accessed 20 December 2017).

156 Soujeri, E. and Kaddoum, G. (2016). The impact of antenna switching time on spatial modulation. *IEEE Wireless Communications Letters* **5** (3): 256–259. doi: 10.1109/LWC.2016.2535318.

157 Agilent Technologies Understanding RF/Microwave Solid State Switches and their Applications. Tech. Rep. http://cp.literature.agilent.com/litweb/pdf/5989-7618EN.pdf (retreived online on 25 October 2016).

158 Skyworks Solutions, Inc. General purpose RF switches. http://www.skyworksinc.com/uploads/documents/PB-RFSwitches-PB121-15B.pdf (retrived online on 25 October 2016).

159 Chau, Y.A. and Yu, S.H. (2001). Space modulation on wireless fading channels. IEEE Vehicular Technology Conference (VTC Fall 2001), vol. 3, pp. 1668–1671. doi: 10.1109/VTC.2001.956483.

160 Xu, C., Sugiura, S., Ng, S.X., and Hanzo, L. (2013). Spatial modulation and space-time shift keying: optimal performance at a reduced detection complexity. *IEEE Transactions on Communications* **61** (1): 206–216. doi: 10.1109/TCOMM.2012.100312.120251.

161 Basşar, E., Aygölü, U., Panayirci, E., and Poor, H. (2010). Space-time block coding for spatial modulation. IEEE 21st International Symposium on Personal Indoor and Mobile Radio Communications (PIMRC), pp. 803–808. doi: 10.1109/PIMRC.2010.5671984.

162 Basar, E., Aygolu, U., Panayirci, E., and Poor, H.V. (2012). Super-orthogonal trellis-coded spatial modulation. *IET Communications* **6** (17): 2922–2932. doi: 10.1049/iet-com.2012.0355.

163 Hanzo, L., Liew, T., and Yeap, B. (2002). *Turbo Coding, Turbo Equalisation and Space-Time Coding for Transmission over Fading Channels*. Wiley.

164 Hassibi, B. and Hochwald, B. (2002). High-rate codes that are linear in space and time. *IEEE Transactions on Information Theory* **48** (7): 1804–1824. doi: 10.1109/TIT.2002.1013127.

165 Le, M.T., Ngo, V.D., Mai, H.A. et al. (2014). Spatially modulated orthogonal space-time block codes with non-vanishing determinants. *IEEE Transactions on Communications* **62** (1): 85–99. doi: 10.1109/TCOMM.2013.112913.130219.

166 Kohno, R. (1998). Spatial and temporal communication theory using adaptive antenna array. *IEEE Personal Communications [see also IEEE Wireless Communications]* **5** (1): 28–35. doi: 10.1109/98.656157.

167 Jafarkhani, H. (2001). A quasi-orthogonal space-time block code. *IEEE Transactions on Communications* **49** (1): 1–4. doi: 10.1109/26.898239.

168 Sugiura, S. and Hanzo, L. (2013). On the joint optimization of dispersion matrices and constellations for near-capacity irregular precoded space-time shift keying. *IEEE Transactions on Wireless Communications* **12** (1): 380–387. doi: 10.1109/TWC.2012.120412.120718.

169 Pless, V. (1997). *Introduction to the Theory of Error-Correcting Codes*, 3e. Wiley. ISBN: 978-0-471-19047-9.

170 Di Renzo, M., Mesleh, R., Haas, H., and Grant, P. (2010). Upper bounds for the analysis of trellis coded spatial modulation over correlated fading channels. IEEE 71st Vehicular Technology Conference (VTC 2010-Spring), pp. 1–5. doi: 10.1109/VETECS.2010.5493766.

171 Basar, E., Aygolu, U., Panayirci, E., and Poor, H.V. (2011). New trellis code design for spatial modulation. *IEEE Transactions on Wireless Communications* **10**: 2670–2680. doi: 10.1109/TWC.2011.061511.101745.

172 Vladeanu, C. (2012). Turbo trellis-coded spatial modulation. 2012 IEEE on Global Communications Conference (GLOBECOM), pp. 4024–4029. doi: 10.1109/GLOCOM.2012.6503746.

173 Ungerboeck, G. (1982). Channel coding with multilevel/phase signals. *IEEE Journal on Information Technology* **28** (1): 55–67.

174 Forney, G.D. and Ungerboeck, G. (1998). Modulation and coding for linear Gaussian channels. *IEEE Transactions on Information Theory* **44** (6): 2384–2415.

175 Golub, G.H. and van Loan, C.F. (1996). *Matrix Computations*. The John Hopkins University Press.

176 Kailath, T., Vikalo, H., and Hassibi, B. (2006). *Space-Time Wireless Systems: From Array Processing to MIMO Communications*. Cambridge University Press.

177 Hassibi, B. and Vikalo, H. (2005). On the sphere-decoding algorithm I. Expected complexity. *IEEE Transactions on Signal Processing* **53** (8): 2806–2818. doi: 10.1109/TSP.2005.850352.

178 Cui, T. and Tellambura, C. (2005). An efficient generalized sphere decoder for rank-deficient MIMO systems. *IEEE Communications Letters* **9** (5): 423–425. doi: 10.1109/LCOMM.2005.1431159.

179 Wang, P. and Le-Ngoc, T. (2009). A low-complexity generalized sphere decoding approach for underdetermined linear communication systems: performance and complexity evaluation. *IEEE Transactions on Communications* **57** (11): 3376–3388. doi: 10.1109/TCOMM.2009.11.060557.

180 Jalden, J., Barbero, L., Ottersten, B., and Thompson, J. (2009). The error probability of the fixed-complexity sphere decoder. *IEEE Transactions on Signal Processing* **57** (7): 2711–2720. doi: 10.1109/TSP.2009.2017574.

181 Xia, X., Hu, H., and Wang, H. (2007). Reduced initial searching radius for sphere decoder. Proceedings of the IEEE 18th International Symposium on Personal, Indoor and Mobile Radio Communications (PIMRC), Athens, Greece, pp. 1–4. doi: 10.1109/PIMRC.2007.4394469.

182 Auer, G., Giannini, V., Desset, C. et al. (2011). How much energy is needed to run a wireless network? *IEEE Wireless Communications* **18** (5): 40–49. doi: 10.1109/MWC.2011.6056691.

183 Ge, X., Cheng, H., Guizani, M., and Han, T. (2014). 5g wireless backhaul networks: challenges and research advances. *IEEE Network* **28** (6): 6–11. doi: 10.1109/MNET.2014.6963798.

184 Minicircuits (2017). SW SPDT. https://www.minicircuits.com/WebStore/dashboard.html?model=HSWA2-30DR%2B (accessed 20 December 2017).

185 Digikey (2017). RF transciver ICS. http://www.digikey.com/product-detail/en/analog-devices-inc/AD9364BBCZ/AD9364BBCZ-ND/4747823 (accessed 20 December 2017).

186 Microship (2017). 16-bit microcontrollers. http://eu.mouser.com/ProductDetail/Microchip-Technology/PIC24FJ64GB406-I-PT/?qs=w3MdF6xSSP5O%2Fs8hl8VR2A%3D%3D (accessed 20 December 2017).

187 Mouser (2017). Serial to parallel logic converters. http://eu.mouser.com/ProductDetail/Texas-Instruments/SN74LV8153PWR (accessed 20 December 2017).

188 Mesleh, R. and Ikki, S. (2012). On the effect of Gaussian imperfect channel estimations on the performance of space modulation techniques. IEEE 75th Vehicular Technology Conference (VTC Spring), pp. 1–5. doi: 10.1109/VETECS.2012.6239909.

189 Badarneh, O.S. and Mesleh, R. (2015). Performance analysis of space modulation techniques over $\alpha - \mu$ and $\kappa - \mu$ fading channels with imperfect channel estimation. *Transactions on Emerging Telecommunications Technologies* **28** (2): e2940. doi: 10.1002/ett.2940.

190 Alouini, M.S. and Goldsmith, A. (1999). A unified approach for calculating error rates of linearly modulated signals over generalized fading channels. *IEEE Transactions on Communications* **47** (9): 1324–1334.

191 Gradshteyn, I.S. and Ryzhik, I.M. (2007). *Table of Integrals, Series, and Products*, 7e. Academic Press. ISBN-10: 0123736374.

192 Wu, J. and Xiao, C. (2008). Optimal diversity combining based on linear estimation of rician fading channels. *IEEE Transactions on Communications* **56** (10): 1612–1615. doi: 10.1109/TCOMM.2008.060598.

193 Koca, M. and Sari, H. (2011). A general framework for performance analysis of spatial modulation over correlated fading channels. CoRR, abs/1109.5589.

194 Alshamali, A. and Quza, B. (2009). Performance of spatial modulation in correlated and uncorrelated Nakagami fading channel. *Journal of Communications* **4** (3): 170–174.

195 Di Renzo, M. and Haas, H. (2011). Bit error probability of space modulation over Nakagami-m fading: asymptotic analysis. *IEEE Communications Letters* **15** (10): 1026–1028. doi: 10.1109/LCOMM.2011.080811.110873.

196 Di Renzo, M. and Haas, H. (2012). Bit error probability of spatial modulation (SM) MIMO over generalized fading channels. *IEEE Transactions on Vehicular Technology* **61** (3): 1124–1144. doi: 10.1109/TVT.2012.2186158.

197 Di Renzo, M. and Haas, H. (2011). Bit error probability of space-shift keying MIMO over multiple-access independent fading channels. *IEEE Transactions on Vehicular Technology* **60** (8): 3694–3711. doi: 10.1109/TVT.2011.2167636.

198 Simon, M.K. and Alouini, M.S. (2000). *Digital Communication over Fading Channels: A Unified Approach to Performance Analysis*, 1e. Wiley.

199 Hedayat, A., Shah, H., and Nosratinia, A. (2005). Analysis of space-time coding in correlated fading channels. *IEEE Transactions on Wireless Communications* **4** (6): 2882–2891. doi: 10.1109/TWC.2005.858338.

200 Turin, G.L. (1960). The characteristic function of Hermitian quadratic forms in complex normal variables. *Biometrika* **47** (1/2): 199–201.

201 Alhassi, A., Abdelgader, A., Elsahli, H., and Mesleh, R. (2017). Performance of spatial multiplexing in the presence of channel estimation errors. *Almadar Journal for Communications, Information Technology, and Applicatoins* **4** (1): 15–19.

202 Yonghong, H., Pichao, W., Xiang, W. et al. (2013). Ergodic capacity analysis of spatially modulated systems. *China Communications* **10** (7): 118–125. doi: 10.1109/CC.2013.6571295.

203 Rajashekar, R., Hari, K., and Hanzo, L. (2014). Reduced-complexity ML detection and capacity-optimized training for spatial modulation systems. *IEEE Transactions on Communications* **62** (1): 112–125. doi: 10.1109/TCOMM.2013.120213.120850.

204 An, Z., Wang, J., Wang, J. et al. (2015). Mutual information analysis on spatial modulation multiple antenna system. *IEEE Transactions on Communications* **63** (3): 826–843. doi: 10.1109/TCOMM.2014.2387171.

205 Yang, Y. and Jiao, B. (2008). Information-guided channel-hopping for high data rate wireless communication. *IEEE Communications Letters* **12** (4): 225–227. doi: 10.1109/LCOMM.2008.071986.

206 Yang, P., Renzo, M.D., Xiao, Y. et al. (2015). Design guidelines for spatial modulation. *IEEE Communications Surveys Tutorials* **17** (1): 6–26. doi: 10.1109/COMST.2014.2327066.

207 Assaad, M. and Zeghlache, D. (2003). On the capacity of HSDPA. Global Telecommunications Conference, 2003. GLOBECOM '03. IEEE, vol. 1, pp. 60–64. doi: 10.1109/GLOCOM.2003.1258203.

208 Chung, S.T., Lozano, A., and Huang, H. (2001). Approaching eigenmode BLAST channel capacity using V-BLAST with rate and power feedback. Proceedings of the 54th Vehicular Technology Conference (VTC 01), vol. 2, Atlantic City, NJ, USA, pp. 915–919. doi: 10.1109/VTC.2001.956906.

209 Jayaweera, S. (2007). V-BLAST-based virtual MIMO for distributed wireless sensor networks. *IEEE Transactions on Communications* **55** (10): 1867–1872. doi: 10.1109/TCOMM.2007.906389.

210 Liu, P., Renzo, M.D., and Springer, A. (2016). Line-of-sight spatial modulation for indoor mmWave communication at 60 GHz. *IEEE Transactions on Wireless Communications* **15** (11): 7373–7389. doi: 10.1109/TWC.2016.2601616.

211 Fano, R.M. (1961). *Transmission of Information: A statistical Theory of Communications*. New York: Wiley.

212 Yang, Y. and Jiao, B. (2008). On the capacity of information-guided channel-hopping in multi-antenna system. IEEE INFOCOM Workshops 2008, pp. 1–5. doi: 10.1109/INFOCOM.2008.4544653.

213 Basnayaka, D.A., Renzo, M.D., and Haas, H. (2016). Massive but few active MIMO. *IEEE Transactions on Communications* **65** (9): 6861–6877.

214 Grimmett, G.R. and Stirzaker, D.R. (2001). *Probability and Random Processes*, 3e. Oxford University Press.

215 Gifford, W.M., Win, M.Z., and Chiani, M. (2005). Diversity with practical channel estimation. *IEEE Transactions on Wireless Communications* **4** (4): 1935–1947. doi: 10.1109/TWC.2005.852127.

216 Adinoyi, A. and Yanikomeroglu, H. (2007). Cooperative relaying in multi-antenna fixed relay networks. *IEEE Transactions on Wireless Communications* **6** (2): 533–544. doi: 10.1109/TWC.2007.05227.

217 Renk, T., Kloeck, C., Burgkhardt, D., and Jondral, F.K. (2007). Cooperative communications in wireless networks - a requested relaying protocol. 16th IST Mobile and Wireless Communications Summit, pp. 1–5. doi: 10.1109/ISTMWC.2007.4299037.

218 Pabst, R., Walke, B., Schultz, D. et al. (2004). Relay-based deployment concepts for wireless and mobile broadband radio. *IEEE Communications Magazine* **42** (9): 80–89. doi: 10.1109/MCOM.2004.1336724.

219 He, X., Luo, T., and Yue, G. (2010). Optimized distributed MIMO for cooperative relay networks. *IEEE Communications Letters* **14** (1): 9–11. doi: 10.1109/LCOMM.2010.01.091457.

220 Chen, D. and Laneman, J.N. (2006). Modulation and demodulation for cooperative diversity in wireless systems. *IEEE Transactions on Wireless Communications* **5**: 1785–1794.

221 Ng, C.T.K. and Huang, H. (2010). Linear precoding in cooperative MIMO cellular networks with limited coordination clusters. *IEEE Journal on Selected Areas in Communications* **28** (9): 1146–1454. doi: 10.1109/JSAC.2010.101206.

222 Chan, S. and Zukerman, M. (2002). Is max-min fairness achievable in the presence of insubordinate users? *IEEE Communications Letters* **6** (3): 120–122. doi: 10.1109/4234.991152.

223 Saraydar, C., Mandayam, N., and Goodman, D. (2002). Efficient power control via pricing in wireless data networks. *IEEE Transactions on Communications* **50** (2): 291–303. doi: 10.1109/26.983324.

224 Pischella, M. and Belfiore, J.C. (2008). Power control in distributed cooperative OFDMA cellular networks. *IEEE Transactions on Wireless Communications* **7** (5): 1900–1906. doi: 10.1109/TWC.2008.061039.

225 Hanzo, L., El-Hajjar, M., and Alamri, O. (2011). Near-capacity wireless transceivers and cooperative communications in the MIMO era: evolution of standards, waveform design, and future perspectives. *Proceedings of the IEEE* **99** (8): 1343–1385. doi: 10.1109/JPROC.2011.2148150.

226 Han, Z., Ji, Z., and Liu, K.J.R. (2007). Non-cooperative resource competition game by virtual referee in multi-cell OFDMA networks. *IEEE Journal on Selected Areas in Communications* **25** (6): 1079–1090. doi: 10.1109/JSAC.2007.070803.

227 Mietzner, J., Schober, R., Lampe, L. et al. (2009). Multiple-antenna techniques for wireless communications - a comprehensive literature survey. *IEEE Communication Surveys and Tutorials* **11** (2): 87–105. doi: 10.1109/SURV.2009.090207.

228 del Coso, A., Spagnolini, U., and Ibars, C. (2007). Cooperative distributed MIMO channels in wireless sensor networks. *IEEE Journal on Selected Areas in Communications* **25** (2): 402–414. doi: 10.1109/JSAC.2007.070215.

229 Kramer, G., Gastpar, M., and Gupta, P. (2005). Cooperative strategies and capacity theorems for relay networks. *IEEE Transactions on Information Theory* **51** (9): 3037–3063. doi: 10.1109/TIT.2005.853304.

230 Laneman, J.N., Tse, D.N.C., and Wornell, G.W. (2004). Cooperative diversity in wireless networks: efficient protocols and outage behavior. *IEEE Transactions on Information Theory* **50** (12): 3062–3080. doi: 10.1109/TIT.2004.838089.

231 Genc, V., Murphy, S., Yu, Y., and Murphy, J. (2008). IEEE 802.16J relay-based wireless access networks: an overview. *IEEE Transactions on Wireless Communications* **15** (5): 56–63. doi: PDF.

232 Nokia (2005). E-utra link adaption: consideration on MIMO.

233 Mesleh, R., Ikki, S., and Alwakeel, M. (2011). Performance analysis of space shift keying with amplify and forward relaying. *IEEE Communciations Letters* **15** (12): 1350–1352. doi: 10.1109/LCOMM.2011.100611.111690.

234 Mesleh, R., Ikki, S.S., Tumar, I., and Alouneh, S. (2017). Decode-and-forward with quadrature spatial modulation in the presence of imperfect channel estimation. *Physical Communication* **24**: 103–111. doi: https://10.1016/j.phycom.2017.06.005.

235 Mesleh, R., Ikki, S.S., Aggoune, E.H.M., and Mansour, A. (2012). Performance analysis of space shift keying (SSK) modulation with multiple cooperative relays. *EURASIP Journal on Advances in Signal Processing* **2012** (1): 201. doi: 10.1186/1687-6180-2012-201.

236 Wen, M., Cheng, X., Poor, H.V., and Jiao, B. (2014). Use of SSK modulation in two-way amplify-and-forward relaying. *IEEE Transactions on Vehicular Technology* **63** (3): 1498–1504. doi: 10.1109/TVT.2013.2277553.

237 Mesleh, R. and Ikki, S.S. (2013). Performance analysis of spatial modulation with multiple decode and forward relays. *IEEE Wireless Communications Letters* **2** (4): 423–426. doi: 10.1109/WCL.2013.051513.130256.

238 Stavridis, A., Basnayaka, D., Sinanovic, S. et al. (2014). A virtual MIMO dual-hop architecture based on hybrid spatial modulation. *IEEE Transactions on Communications* **62** (9): 3161–3179. doi: 10.1109/TCOMM.2014.2343999.

239 Narayanan, S., Renzo, M.D., Graziosi, F., and Haas, H. (2016). Distributed spatial modulation: a cooperative diversity protocol for half-duplex relay-aided wireless networks. *IEEE Transactions on Vehicular Technology* **65** (5): 2947–2964. doi: 10.1109/TVT.2015.2442754.

240 Som, P. and Chockalingam, A. (2015). Performance analysis of space-shift keying in decode-and-forward multihop MIMO networks. *IEEE Transactions on Vehicular Technology* **64** (1): 132–146. doi: 10.1109/TVT.2014.2318437.

241 Afana, A., Mesleh, R., Ikki, S., and Atawi, I.E. (2016). Performance of quadrature spatial modulation in amplify-and-forward cooperative relaying. *IEEE Communications Letters* **20** (2): 240–243. doi: 10.1109/LCOMM.2015.2509975.

242 Zhang, J., Li, Q., Kim, K.J. et al. (2016). On the performance of full-duplex two-way relay channels with spatial modulation. *IEEE Transactions on Communications* **64** (12): 4966–4982. doi: 10.1109/TCOMM.2016.2600661.

243 Mesleh, R. and Ikki, S.S. (2015). Space shift keying with amplify-and-forward MIMO? relaying. *Transactions on Emerging Telecommunications Technologies* **26** (4): 520–531. doi: 10.1002/ett.2611.

244 Altın, G., Aygölü, Ü., Basar, E., and Çelebi, M. (2017). Multiple-input–multiple-output cooperative spatial modulation systems. *IET Communications* **11** (15): 2289–2296.

245 Hasna, M.O. and Alouini, M.S. (2003). End-to-end performance of transmission systems with relays over Rayleigh-fading channels. *IEEE Transactions on Wireless Communications* **2** (6): 1126–1131. doi: 10.1109/TWC.2003.819030.

246 Hasna, M.O. and Alouini, M.S. (2004). A performance study of dual-hop transmissions with fixed gain relays. *IEEE Transactions on Wireless Communications* **3** (6): 1963–1968. doi: 10.1109/TWC.2004.837470.

247 Raed, M., Salama, I., Hadi, A., and Mansour, A. (2012). Performance analysis of space shift keying (SSK) modulation with multiple cooperative relays. *EURASIP Journal on Advances in Signal Processing* **2012** (1). doi: 10.1186/1687-6180-2012-201.

248 Ikki, S.S. and Ahmed, M.H. (2010). Performance analysis of adaptive decode-and-forward cooperative diversity networks with best-relay selection. *IEEE Transactions on Communications* **58** (1): 68–72. doi: 10.1109/TCOMM.2010.01.080080.

249 Beaulieu, N.C. and Hu, J. (2006). A closed-form expression for the outage probability of decode-and-forward relaying in dissimilar Rayleigh fading channels. *IEEE Communications Letters* **10** (12): 813–815. doi: 10.1109/LCOMM.2006.061048.

250 Masnick, B. and Wolf, J. (1967). On linear unequal error protection codes. *IEEE Transactions on Information Theory* **13** (4): 600–607. doi: 10.1109/TIT.1967.1054054.

251 Rajashekar, R., Hari, K.V.S., and Hanzo, L. (2015). Quantifying the transmit diversity order of Euclidean distance based antenna selection in spatial modulation. *IEEE Signal Processing Letters* **22** (9): 1434–1437. doi: 10.1109/LSP.2015.2408574.

252 Lee, I.H. and Kim, D. (2007). BER analysis for decode-and-forward relaying in dissimilar Rayleigh fading channels. *IEEE Communications Letters* **11** (1): 52–54. doi: 10.1109/LCOMM.2007.061375.

253 Chen, H., Liu, J., Zheng, L. et al. (2010). An improved selection cooperation scheme for decode-and-forward relaying. *IEEE Communciations Letters* **14** (12): 1143–1145. doi: 10.1109/LCOMM.2010.102610.101115.

254 Thompson, J.S., Grant, P.M., and Mulgrew, B. (1996). Smart antenna arrays for CDMA systems. *IEEE [see also IEEE Wireless Communications] Personal Communications* **3** (5): 16–25. doi: 10.1109/98.542234.

255 Papavassiliou, S. and Tassiulus, L. (1998). Improving the capacity in wireless networks through integrated channel base station and power assignment. *IEEE Transactions on Vehicular Technology* **47** (2): 417–427. doi: 10.1109/25.669080.

256 Ju, H., Oh, E., and Hong, D. (2009). Catching resource-devouring worms in next-generation wireless relay systems: two-way relay and full-duplex relay. *IEEE Communications Magazine* **47** (9): 58–65. doi: 10.1109/MCOM.2009.5277456.

257 Pi, Z. and Khan, F. (2011). An introduction to millimeter-wave mobile broadband systems. *IEEE Communications Magazine* **49** (6): 101–107. doi: 10.1109/MCOM.2011.5783993.

258 Rappaport, T., Sun, S., Mayzus, R. et al. (2013). Millimeter wave mobile communications for 5G cellular: it will work! *IEEE Access* **1**: 335–349. doi: 10.1109/ACCESS.2013.2260813.

259 Samimi, M.K. and Rappaport, T.S. (2016). 3-D millimeter-wave statistical channel model for 5G wireless system design. *IEEE Transactions on Microwave Theory and Techniques* **64** (7): 2207–2225. doi: 10.1109/TMTT.2016.2574851.

260 Wells, J. (2009). Faster than fiber: the future of multi-G/S wireless. *IEEE Microwave Magazine* **10** (3): 104–112. doi: 10.1109/MMM.2009.932081.

261 Wells, J. (2006). *Multigigabit wireless technology at 70 GHz, 80 GHz and 90 GHz*. Defense Electronics Magazine.

262 Nie, S., MacCartney, G.R., Sun, S., and Rappaport, T.S. (2013). 72 GHz millimeter wave indoor measurements for wireless and backhaul communications. IEEE 24th Annual International Symposium on Personal, Indoor and Mobile Radio Communications (PIMRC), pp. 2429–2433. doi: 10.1109/PIMRC.2013.6666553.

263 MacCartney, G.R. and Rappaport, T.S. (2014). 73 GHz millimeter wave propagation measurements for outdoor urban mobile and Backhaul communications in New York City. IEEE International Conference on Communication (ICC), pp. 4862–4867. doi: 10.1109/ICC.2014.6884090.

264 ITU-R (2017). Characteristics of Precipitation for Propagation Modelling. Recommendation ITU-R P.837–7.

265 IEEE 802.15.3c-2009 (2009). mmWave WPAN. Amendment to IEEE Std 802.15.3-2003.

266 IEEE Standard 802.11ad (2012). Wigig.

267 WirelessHD (2010). http://www.wirelesshd.org/ (accessed 12 December 2017).

268 Ahmadi-Shokouh, J., Rafi, R., Taeb, A., and Safavi-Naeini, S. (2015). Empirical MIMO beamforming and channel measurements at 57–64?GHz frequencies. *Transactions on Emerginig Telecommunications Technologies* **26** (6): 1003–1009. doi: 10.1002/ett.2794.

269 Maltsev, A., Sadri, A., Cordeiro, C., and Pudeyev, A. (2015). Practical LOS MIMO technique for short-range millimeter-wave systems. IEEE International Conference on Ubiquitous Wireless Broadband (ICUWB), pp. 1–6. doi: 10.1109/ICUWB.2015.7324501.

270 Torkildson, E., Zhang, H., and Madhow, U. (2010). Channel modeling for millimeter wave MIMO. Information Theory and Applications Workshop (ITA), pp. 1–8. doi: 10.1109/ITA.2010.5454109.

271 Shah, S.T., Kim, J.S., Bae, E.S. et al. (201). Radio resource management for 5G mobile communication systems with massive antenna structure. *Transactions on Emerging Telecommunications Technologies* **27** (4): 504–518. doi: 10.1002/ett.2986.

272 Torkildson, E., Madhow, U., and Rodwell, M. (2011). Indoor millimeter wave MIMO: feasibility and performance. *IEEE Transactions on Wireless Communications* **10** (12): 4150–4160. doi: 10.1109/TWC.2011.092911.101843.

273 Zhou, L. and Ohashi, Y. (2015). Performance analysis of mmWave LOS-MIMO systems with uniform circular arrays. IEEE 81st Vehicular Technology Conference (VTC Spring), pp. 1–5. doi: 10.1109/VTC-Spring.2015.7146001.

274 Liu, P. and Springer, A. (2015). Space shift keying for LOS communication at mmWave frequencies. *IEEE Wireless Communications Letters* **4** (2): 121–124.

275 Ishikawa, N., Rajashekar, R., Sugiura, S., and Hanzo, L. (2016). Generalized spatial modulation based reduced-RF-chain millimeter-wave communications. *IEEE Transactions on Vehicular Technology* **99**: 1.

276 Liu, P., Renzo, M.D., and Springer, A. (2017). Variable- N_u generalized spatial modulation for indoor LOS mmWave communication: performance optimization and novel switching structure. *IEEE Transactions on Communications* **65** (6): 2625–2640. doi: 10.1109/TCOMM.2017.2676818.

277 Cui, Y., Fang, X., and Yan, L. (2016). Hybrid spatial modulation beamforming for mmWave railway communication systems. *IEEE Transactions on Vehicular Technology* **65** (12): 9597–9606. doi: 10.1109/TVT.2016.2614005.

278 Thomas, T.A., Nguyen, H.C., MacCartney, G.R., and Rappaport, T.S. (2014). 3D mmWave channel model proposal. IEEE 80th Vehicular Technology Conference (VTC Fall), pp. 1–6. doi: 10.1109/VTC-Fall.2014.6965800.

279 Rappaport, M.K.S.S.S.T.S. (2016). MIMO Channel Modeling and Capacity Analysis for 5G Millimeter-Wave Wireless Systems. 10th European Conference on Antennas and Propagation (EuCAP'2016).

280 Samimi, M.K. and Rappaport, T.S. (2015). 3-D statistical channel model for millimeter-wave outdoor mobile broadband communications. Proceeding of the IEEE International Conference on Communciations. doi: 10.1109/ICC.2008.976.

281 Jammalamadaka, S.R. and Sengupta, A. (2001). *Multivariate Analysis*, Topics in Circular Statistics, vol. **5**. World Scientific Pub Co Inc.

282 Gesbert, D., Bolcskei, H., Gore, D., and Paulraj, A. (2002). Outdoor MIMO wireless channels: models and performance prediction. *IEEE Transactions on Communications* **50** (12): 1926–1934.

283 Steinbauer, M., Molisch, A.F., and Bonek, E. (2001). The double-directional radio channel. *IEEE Antennas and Propagation Magazine* **43** (4): 51–63. doi: 10.1109/74.951559.

284 Forenza, A., Love, D.J., and Heath, R.W. (2007). Simplified spatial correlation models for clustered MIMO channels with different array configurations. *IEEE Transactions on Vehicular Technology* **56** (4): 1924–1934. doi: 10.1109/TVT.2007.897212.

285 Molisch, A.F., Steinbauer, M., Toeltsch, M. et al. (2002). Capacity of MIMO systems based on measured wireless channels. *IEEE Journal on Selected Areas in Communications* **20** (3): 561–569. doi: 10.1109/49.995515.

286 Karttunen, P., Kalliola, K., Laakso, T., and Vainikainen, P. (1998). Measurement analysis of spatial and temporal correlation in wideband radio channels with adaptive antenna array. IEEE 1998 International Conference on Universal Personal Communications (ICUPC '98), vol. 1, pp. 671–675. doi: 10.1109/ICUPC.1998.733053.

287 Christodoulou, C.G., Tawk, Y., Lane, S.A., and Erwin, S.R. (2012). Reconfigurable antennas for wireless and space applications. *Proceedings of the IEEE* **100** (7): 2250–2261.

288 Ourir, A., Rachedi, K., Phan-Huy, D.T. et al. (2017). Compact reconfigurable antenna with radiation pattern diversity for spatial modulation. 11th European Conference on Antennas and Propagation, Paris, France.

289 Phan-Huy, D.T., Kokar, Y., Rioult, J. et al. (2017). First visual demonstration of transmit and receive spatial modulations using the radio wave display. 21st International ITG Workshop on Smart Antennas, Berlin, Germany.

290 Bouida, Z., El-Sallabi, H., Abdallah, M. et al. (2016). Reconfigurable antenna-based space-shift keying for spectrum sharing systems under rician fading. *IEEE Transactions on Communications* **64** (9): 3970–3980.

291 Ishibashi, K. and Sugiura, S. (2014). Effects of antenna switching on band-limited spatial modulation. *IEEE Wireless Communications Letters* **3** (4): 345–348. doi: 10.1109/LWC.2014.2315819.

Index

Space Modulation Techniques, First Edition. Raed Mesleh and Abdelhamid Alhassi.